IET TELECOMMUNICATIONS SERIES 109

Digital Twins for 6G

Other volumes in this series:

Volume 9	**Phase Noise in Signal Sources** W.P. Robins	
Volume 12	**Spread Spectrum in Communications** R. Skaug and J.F. Hjelmstad	
Volume 13	**Advanced Signal Processing** D.J. Creasey (Editor)	
Volume 19	**Telecommunications Traffic, Tariffs and Costs** R.E. Farr	
Volume 20	**An Introduction to Satellite Communications** D.I. Dalgleish	
Volume 26	**Common-Channel Signalling** R.J. Manterfield	
Volume 28	**Very Small Aperture Terminals (VSATs)** J.L. Everett (Editor)	
Volume 29	**ATM: The broadband telecommunications solution** L.G. Cuthbert and J.C. Sapanel	
Volume 31	**Data Communications and Networks, 3rd Edition** R.L. Brewster (Editor)	
Volume 32	**Analogue Optical Fibre Communications** B. Wilson, Z. Ghassemlooy and I.Z. Darwazeh (Editors)	
Volume 33	**Modern Personal Radio Systems** R.C.V. Macario (Editor)	
Volume 34	**Digital Broadcasting** P. Dambacher	
Volume 35	**Principles of Performance Engineering for Telecommunication and Information Systems** M. Ghanbari, C.J. Hughes, M.C. Sinclair and J.P. Eade	
Volume 36	**Telecommunication Networks, 2nd Edition** J.E. Flood (Editor)	
Volume 37	**Optical Communication Receiver Design** S.B. Alexander	
Volume 38	**Satellite Communication Systems, 3rd Edition** B.G. Evans (Editor)	
Volume 40	**Spread Spectrum in Mobile Communication** O. Berg, T. Berg, J.F. Hjelmstad, S. Haavik and R. Skaug	
Volume 41	**World Telecommunications Economics** J.J. Wheatley	
Volume 43	**Telecommunications Signalling** R.J. Manterfield	
Volume 44	**Digital Signal Filtering, Analysis and Restoration** J. Jan	
Volume 45	**Radio Spectrum Management, 2nd Edition** D.J. Withers	
Volume 46	**Intelligent Networks: Principles and applications** J.R. Anderson	
Volume 47	**Local Access Network Technologies** P. France	
Volume 48	**Telecommunications Quality of Service Management** A.P. Oodan (Editor)	
Volume 49	**Standard Codecs: Image compression to advanced video coding** M. Ghanbari	
Volume 50	**Telecommunications Regulation** J. Buckley	
Volume 51	**Security for Mobility** C. Mitchell (Editor)	
Volume 52	**Understanding Telecommunications Networks** A. Valdar	
Volume 53	**Video Compression Systems: From first principles to concatenated codecs** A. Bock	
Volume 54	**Standard Codecs: Image compression to advanced video coding, 3rd Edition** M. Ghanbari	
Volume 59	**Dynamic Ad Hoc Networks** H. Rashvand and H. Chao (Editors)	
Volume 60	**Understanding Telecommunications Business** A. Valdar and I. Morfett	
Volume 65	**Advances in Body-Centric Wireless Communication: Applications and state-of-the-art** Q.H. Abbasi, M.U. Rehman, K. Qaraqe and A. Alomainy (Editors)	
Volume 67	**Managing the Internet of Things: Architectures, theories and applications** J. Huang and K. Hua (Editors)	
Volume 68	**Advanced Relay Technologies in Next Generation Wireless Communications** I. Krikidis and G. Zheng	
Volume 69	**5G Wireless Technologies** A. Alexiou (Editor)	
Volume 70	**Cloud and Fog Computing in 5G Mobile Networks** E. Markakis, G. Mastorakis, C.X. Mavromoustakis and E. Pallis (Editors)	
Volume 71	**Understanding Telecommunications Networks, 2nd Edition** A. Valdar	
Volume 72	**Introduction to Digital Wireless Communications** Hong-Chuan Yang	
Volume 73	**Network as a Service for Next Generation Internet** Q. Duan and S. Wang (Editors)	
Volume 74	**Access, Fronthaul and Backhaul Networks for 5G & Beyond** M.A. Imran, S.A.R. Zaidi and M.Z. Shakir (Editors)	
Volume 75	**Digital Television Fundamentals** S. Mozar and K. Glasman (Editors)	
Volume 76	**Trusted Communications with Physical Layer Security for 5G and Beyond** T.Q. Duong, X. Zhou and H.V. Poor (Editors)	

Volume 77	**Network Design, Modelling and Performance Evaluation** Q. Vien	
Volume 78	**Principles and Applications of Free Space Optical Communications** A.K. Majumdar, Z. Ghassemlooy, A.A.B. Raj (Editors)	
Volume 79	**Satellite Communications in the 5G Era** S.K. Sharma, S. Chatzinotas and D. Arapoglou	
Volume 80	**Transceiver and System Design for Digital Communications, 5th Edition** Scott R. Bullock	
Volume 81	**Applications of Machine Learning in Wireless Communications** R. He and Z. Ding (Editors)	
Volume 83	**Microstrip and Printed Antenna Design, 3rd Edition** R. Bancroft	
Volume 84	**Low Electromagnetic Emission Wireless Network Technologies: 5G and beyond** M.A. Imran, F. Héliot and Y.A. Sambo (Editors)	
Volume 86	**Advances in Communications Satellite Systems Proceedings of the 36th International Communications Satellite Systems Conference (ICSSC-2018)** I. Otung, T. Butash and P. Garland (Editors)	
Volume 87	**Real Time Convex Optimisation for 5G Networks and Beyond** T.Q. Duong, L.D. Nguyen and H.D. Tuan	
Volume 89	**Information and Communication Technologies for Humanitarian Services** M.N. Islam (Editor)	
Volume 90	**Communication Technologies for Networked Smart Cities** S.K. Sharma, N. Jayakody, S. Chatzinotas and A. Anpalagan (Editors)	
Volume 91	**Green Communications for Energy-Efficient Wireless Systems and Networks** Himal Asanga Suraweera, Jing Yang, Alessio Zappone and John S. Thompson (Editors)	
Volume 92	**Flexible and Cognitive Radio Access Technologies for 5G and Beyond** H. Arslan and E. Başar (Editors)	
Volume 93	**Antennas and Propagation for 5G and Beyond** Q. Abbasi, S.F. Jilani, A. Alomainy and M.A. Imran (Editors)	
Volume 94	**Intelligent Wireless Communications** G. Mastorakis, C.X. Mavromoustakis, J.M. Batalla and E. Pallis (Editors)	
Volume 95	**ISDN Applications in Education and Training** R. Mason and P.D. Bacsich	
Volume 96	**Edge Caching for Mobile Networks** H. Vincent Poor and Wei Chen (Editors)	
Volume 98	**Artificial Intelligence Applied to Satellite-based Remote Sensing Data for Earth Observation** M.P. Del Rosso, A. Sebastianelli and S.L. Ullo (Editors)	
Volume 99	**Metrology for 5G and Emerging Wireless Technologies** T.H. Loh (Editor)	
Volume 101	**Wireless Mesh Networks for IoT and Smart Cities: Technologies and applications** L. Davoli and Gi. Ferrari (Editors)	
Volume 104	**Enabling Technologies for Social Distancing: Fundamentals, concepts and solutions** D.N. Nguyen, D.T. Hoang, T.X. Vu, E. Dutkiewicz, S. Chatzinotas and B. Ottersten	
Volume 105	**Non-Geostationary Satellite Communications Systems** E. Lagunas, S. Chatzinotas, K. An and B.F. Beidas	
Volume 107	**Physical Layer Security for 6G Networks** T.Q. Duong, J. Zhang, N. Yang, X. Zhou and V. Sharma	
Volume 108	**The Role of 6G and Beyond on the Road to Net-Zero Carbon** M. Ali Imran, A. Taha, S. Ansari, M. Usman and Q.H. Abbasi (Editors)	

Digital Twins for 6G

Fundamental theory, technology and applications

Edited by
Hamed Ahmadi, Trung Q. Duong, Avishek Nag,
Vishal Sharma, Berk Canberk and Octavia A. Dobre

The Institution of Engineering and Technology

Published by The Institution of Engineering and Technology, London, United Kingdom

The Institution of Engineering and Technology is registered as a Charity in England & Wales (no. 211014) and Scotland (no. SC038698).

© The Institution of Engineering and Technology 2024

First published 2024

This publication is copyright under the Berne Convention and the Universal Copyright Convention. All rights reserved. Apart from any fair dealing for the purposes of research or private study, or criticism or review, as permitted under the Copyright, Designs and Patents Act 1988, this publication may be reproduced, stored or transmitted, in any form or by any means, only with the prior permission in writing of the publishers, or in the case of reprographic reproduction in accordance with the terms of licences issued by the Copyright Licensing Agency. Enquiries concerning reproduction outside those terms should be sent to the publisher at the undermentioned address:

The Institution of Engineering and Technology
Futures Place
Kings Way, Stevenage
Hertfordshire, SG1 2UA, United Kingdom

www.theiet.org

While the authors and publisher believe that the information and guidance given in this work are correct, all parties must rely upon their own skill and judgement when making use of them. Neither the authors nor publisher assumes any liability to anyone for any loss or damage caused by any error or omission in the work, whether such an error or omission is the result of negligence or any other cause. Any and all such liability is disclaimed.

The moral rights of the authors to be identified as authors of this work have been asserted by them in accordance with the Copyright, Designs and Patents Act 1988.

British Library Cataloguing in Publication Data
A catalogue record for this product is available from the British Library

ISBN 978-1-83953-745-5 (hardback)
ISBN 978-1-83953-746-2 (PDF)

Typeset in India by MPS Limited

Cover Image: Yuichiro Chino/Moment via Getty Images

Contents

About the editors	xv
Preface	xix

1 Digital twins for resilient and reliable 6G networks   1
Fahad Alaklabi, Ahmed Al-Tahmeesschi, Avishek Nag and Hamed Ahmadi

1.1	Introduction	1
	1.1.1 6G KPIs	3
	1.1.2 6G technologies	4
	1.1.3 6G applications	8
1.2	DTs and 6G	10
	1.2.1 DTs and higher frequency technologies	11
	1.2.2 DTs of non-terrestrial networks	12
	1.2.3 New physical layer and multi-antenna techniques	14
	1.2.4 DT and new network technologies	15
	1.2.5 DT and intelligent networks (AI-based networking)	17
1.3	DT and Internet of Things (IoT)	17
1.4	Low latency DT	18
	1.4.1 Low latency communications	18
1.5	DT deployment challenges	19
1.6	Conclusion	20
	References	20

2 Digital twin-enabled aerial edge networks with ultra-reliable low-latency communications   27
Dang Van Huynh, Yijiu Li, Tan Do-Duy, Emi Garcia-Palacios and Trung Q. Duong

2.1	Introduction	27
	2.1.1 Literature review	27
	2.1.2 Motivations and main contributions	29
2.2	System model and problem formulation	30
	2.2.1 DT-empowering URLLC-based edge networks model	30
	2.2.2 Transmission model	30
	2.2.3 DT empowered task offloading model	32
	2.2.4 Energy and power consumption model	34
	2.2.5 UAV deployment	34

		2.2.6 Problem formulation	34
	2.3	Proposed solutions	35
		2.3.1 Transmit power and computation resource optimisation	36
		2.3.2 Task offloading optimisation	38
		2.3.3 Proposed algorithm	38
	2.4	Numerical simulations	38
		2.4.1 Simulations setup	38
		2.4.2 Results and discussion	39
	2.5	Conclusion	42
	References		44

3 AI-enabled data management for digital twin networks — 49
Elif Ak, Gökhan Yurdakul, Ahmed Al-Dubai and Berk Canberk

	3.1	Introduction	49
		3.1.1 Importance of data management in DTNs	49
		3.1.2 Explanation of AI's role in data management for DTNs	51
		3.1.3 Challenges in data management	52
		3.1.4 Three states of DTD	53
	3.2	The twinning process: AI-driven data acquisition, preprocessing, modeling, and data storage for DTNs	55
		3.2.1 Understanding the data acquisition	55
		3.2.2 Techniques for collecting data from various sources in DTNs	59
		3.2.3 Data preprocessing methods for cleaning, filtering, and transforming raw data	65
		3.2.4 Digital twin ontology and data modeling	67
		3.2.5 Storage architectures for managing large-scale data in DTNs	70
	3.3	AI-enabled data analysis and interpretation	72
		3.3.1 Overview of AI algorithms for data analytics in DTNs	72
	3.4	Ethical considerations and security aspects	74
		3.4.1 Ethical considerations of using AI in data management	74
		3.4.2 Security aspects, including data privacy and protection	75
		3.4.3 Privacy-preserving techniques for protecting sensitive data in DTNs	75
	3.5	Conclusion and future directions	76
		3.5.1 Summary of the chapter	76
		3.5.2 Remaining challenges and open research questions in AI-enabled data management for DTNs	76
	References		77

4 AI-based traffic analysis in digital twin networks — 83
Sarah Al-Shareeda, Khayal Huseynov, Lal Verda Cakir, Craig Thomson, Mehmet Ozdem and Berk Canberk

	4.1	DTNs ecosystem	83

	4.2	DTNs development efforts: literature review	86
		4.2.1 Networks in general	86
		4.2.2 Cellular networks: 5G and beyond	87
		4.2.3 Wireless networks	87
		4.2.4 Optical networks	88
		4.2.5 Satellite and aeronautic networks	88
		4.2.6 Vehicular networks	89
		4.2.7 Industrial IoT networks	89
	4.3	Key tasks in DTNs analysis	89
		4.3.1 AI-based network performance enhancement	91
		4.3.2 AI-based network management	94
		4.3.3 AI-based communication enhancement	98
		4.3.4 AI-based prediction analysis	100
		4.3.5 AI-based fault and anomaly detection	102
		4.3.6 AI-based security and privacy preservation	102
	4.4	Main AI models and tools harnessed by DTNs	105
		4.4.1 ML tools and models	106
		4.4.2 DL models and techniques	108
		4.4.3 RL and optimization techniques	110
		4.4.4 FL and collaborative learning	111
		4.4.5 Graph and network analysis techniques	112
	4.5	Main challenges in AI-based DTNs	113
		4.5.1 Key challenges	113
		4.5.2 Responsible AI considerations	114
	4.6	Conclusion and key points	115
	References		117
5	**Digital twin empowered Open RAN of 6G networks**		**133**
	Antonino Masaracchia, Vishal Sharma, Muhammad Fahim, Octavia A. Dobre and Trung Q. Duong		
	5.1	Introduction	133
		5.1.1 Motivation and contribution	135
	5.2	Background on O-RAN	136
		5.2.1 RAN functionalities, building blocks, and disaggregation	136
		5.2.2 Toward the concept of Open RAN	139
		5.2.3 Definition of O-RAN architecture	141
		5.2.4 O-RAN as an enabler for 6G deployment	143
	5.3	The concept of digital twin	144
		5.3.1 Definition of digital twin	144
		5.3.2 General architecture of a DT system	144
		5.3.3 DT as an accelerator toward digitalization	145
	5.4	DT on O-RAN architecture: use cases	146
		5.4.1 Channel modeling for RAN optimization	147
		5.4.2 Network traffic forecasting and mobility management	147

		5.4.3	Security and threat detection	148
		5.4.4	Network fault detection	149
	5.5	Challenges and future directions		149
		5.5.1	Real-time synchronization	150
		5.5.2	Data flow security and privacy	150
		5.5.3	Data annotation	151
		5.5.4	Compliance	151
	5.6	Conclusions		152
	Acknowledgments			152
	References			152

6 Potentials of the digital twin in 6G communication systems — 155
Bin Han, Mohammad Asif Habibi, Nandish Kuruvatti, Sanket Partani, Amina Fellan and Hans D. Schotten

6.1	Introduction	155
6.2	Optimized planning, service testing, and rapid development of the 6G network	156
6.3	Simplifying and accelerating the site deployment configuration	157
6.4	Testing the impact of configuration and function changes	159
6.5	Building platforms to train AI models for the 6G system	159
6.6	Tackling the security and resiliency issues in 6G	160
6.7	Efficient network slice management and orchestration	161
6.8	Enabling 6G RAN optimization and effective traffic management	162
6.9	Optimizing the 6G radio resource management	163
6.10	Terahertz wave analysis in support of reconfigurable intelligent surfaces for enhanced 6G performance	164
6.11	Enhancing the operation of mobile edge clouds in 6G	166
6.12	Enabling 6G-based IIoT and industrial 6G use cases	167
6.13	Conclusion	167
Acknowledgments		167
References		167

7 Digital twins for optical networks — 171
Agastya Raj, Dan Kilper and Marco Ruffini

7.1	Introduction		171
7.2	Current issues in optical networks		173
	7.2.1	Issue of suboptimal network operation	174
	7.2.2	Issue of limited automation	176
7.3	DT development for optical networks		177
	7.3.1	Data collection layer	178
	7.3.2	Data fusion layer	179
	7.3.3	Modeling layer	179
	7.3.4	Simulation and virtualization	186
	7.3.5	Application layer	187
	7.3.6	Recent work and case studies on optical DTs	189

7.4	Open testbeds, open emulation tools, and open data		190
	7.4.1	Large-scale testbeds	191
	7.4.2	Open source emulation tools	194
	7.4.3	Open software and open data	194
7.5	Conclusions		195
Appendix A			196
References			198

8 **Dynamic decomposition of service function chain using a deep reinforcement learning approach**   203
Swarna B. Chetty, Hamed Ahmadi, Massimo Tornatore and Avishek Nag

8.1	Introduction		204
	8.1.1	NFV as we know	204
	8.1.2	Microservices decomposed NFVs	205
	8.1.3	Digital twin	206
8.2	Literature review		207
8.3	Problem statement		209
	8.3.1	Objective	209
	8.3.2	Constraints	209
8.4	Deep RL solution for microservice decomposition		211
	8.4.1	Reinforcement learning	211
	8.4.2	Environment	213
	8.4.3	State space	214
	8.4.4	Action space	215
	8.4.5	Reward function	215
	8.4.6	Decomposition Identifier	217
	8.4.7	Granularity criteria	218
	8.4.8	Re-architecture of VNF-FG	219
	8.4.9	Overview of the proposed model	220
8.5	DNN architecture		222
8.6	Simulation results		223
	8.6.1	Heuristic model	225
	8.6.2	Time complexity	225
	8.6.3	Netrail Topology	226
	8.6.4	BtEurope topology	229
	8.6.5	Nodal capacity	230
8.7	Conclusions		231
References			233

9 **An Optimization-as-a-Service platform for 6G exploiting network digital twins**   237
Oriol Sallent, José-Manuel Martínez-Caro, Javier Baliosian, Luis Diez, Luis M. Contreras, Jordi Pérez-Romero, Juan Luis Gorricho, Matías Richart, Ramón Agüero, Joan Serrat, Pablo Pavón-Mariño and Irene Vilà

9.1	OaaS platform: architectural overview		239
	9.1.1	Functional architecture	240

		9.1.2 OaaS system APIs	243
		9.1.3 A workflow example	246
	9.2	OaaS platform: network model	248
		9.2.1 Overview of standardized network models	248
		9.2.2 A transport and computing infrastructure model	253
	9.3	OaaS platform: use cases	255
		9.3.1 Dimensioning problems	255
		9.3.2 Operational problems	261
	Acknowledgments		265
	References		265

10 Robotics digital twin for 6G 269
Milan Groshev, Carlos Guimarães and Antonio de la Oliva

10.1	Introduction	269
10.2	The Shift from Industry 4.0 to Industry 5.0	269
10.3	Digital Twin as the Pillar of Industry 5.0	270
10.4	ICT technologies and adaptation for Industry 5.0	271
10.5	Unified role of Industrial E2E digital twin systems in Industry 5.0	273
10.6	Fundamentals and challenges of digital twins for robotic systems	274
10.7	Digital twins in real industrial environments	274
10.8	From digital twins in Industry 4.0 to Industrial E2E digital twin systems in Industry 5.0	275
10.9	The infrastructure behind industrial digital twins	276
	10.9.1 Computing and storage	277
	10.9.2 Connectivity	277
10.10	Enablers for industrial digital twins	278
	10.10.1 Cloud-to-robot continuum for digital twins	278
	10.10.2 Computation offloading for digital twin	279
	10.10.3 Digital twin as a service	279
	10.10.4 Robot operating system framework	280
	10.10.5 Resource and service federation	281
10.11	Open challenges to achieve Industrial E2E digital twin systems	282
10.12	6G enablers and their applicability to E2E digital twins systems	284
10.13	Context awareness	284
10.14	Joint communication and sensing	285
10.15	Semantic orchestration	286
10.16	Distributed ledger technology federation	287
10.17	Artificial intelligence	288
10.18	Industrial E2E digital twin systems in collaborative robotic applications	290
10.19	Manufacturing: localization and material inspection	290
10.20	Warehouse: material handling and logistics	291

10.21	Construction: safety takeover	292
10.22	Healthcare: patient rehabilitation	292
10.23	Conclusions	293
Acknowledgments		295
References		295

Index   **305**

About the editors

Hamed Ahmadi is a reader in digital engineering at the School of Physics, Engineering and Technology, University of York, UK. He is also an adjunct academic at the School of Electrical and Electronic Engineering, University College Dublin, Ireland. He received his PhD from the National University of Singapore in 2012 where he was a SINGA PhD scholar at the Institute for Infocomm Research, A-STAR. Since then, he worked in different academic and industrial positions in the Republic of Ireland and the UK. Dr Ahmadi has published more than 100 peer-reviewed book chapters, journal, and conference papers. He is the associate editor in chief of *IEEE Communication Standards* magazine, a senior member of IEEE, and a Fellow of UK Higher Education Academy. He has been the Networks working group chair of COST Actions CA15104 (IRACON) and CA20120 (INTERACT). He had chairing roles in organizing and technical program committees of several IEEE major conferences including IEEE ICC 2024, EUCNC 2019, and PIMRC 2024 and 2019. He is also the treasurer of the IEEE UK and Ireland Diversity, Equity, and Inclusion Committee. His current research interests include the design, analysis, and optimization of wireless communications networks, the application of machine learning in wireless networks, open radio access and networking, green networks, airborne networks, digital twins of networks, and Internet of Things.

Trung Q. Duong is a Canada Excellence Research Chair (CERC) and a full professor at Memorial University of Newfoundland, Canada. He is also the adjunct chair professor in telecommunications at Queen's University Belfast, UK. His current research interests include quantum communications, wireless communications, machine learning, and real-time optimization. He has received two prestigious awards from the Royal Academy of Engineering (RAEng): RAEng Research Chair (2021–2025) and the RAEng Research Fellow (2015–2020). He is the recipient of the prestigious Newton Prize 2017. He is a Fellow of IEEE and a Fellow of AAIA.

Avishek Nag received the BE (Honors) degree from Jadavpur University, Kolkata, India, in 2005; the MTech degree from the Indian Institute of Technology, Kharagpur, India, in 2007; and the PhD degree from the University of California, Davis, in 2012. Dr Nag was the recipient of the Best Paper Award at the 2nd IEEE Advanced Networks and Telecommunication Symposium in 2008 and has published over 100 publications including journals, conference proceedings, and book chapters. His research interests include, but are not limited to, cross-layer optimization in wired and wireless networks, network reliability, mathematics of networks (optimization, graph theory), network virtualization, software-defined networks, machine learning, data analytics, blockchain, and the Internet of Things. Dr Nag is a senior member of the Institute of Electronics and Electrical Engineers (IEEE) and also the outreach lead for Ireland for the IEEE UK and Ireland Blockchain Group.

Vishal Sharma is a senior lecturer in the School of Electronics, Electrical Engineering and Computer Science (EEECS) at Queen's University Belfast (QUB), Northern Ireland, UK. At QUB, he focuses on cyber defense and security and leads research on drone security, digital twins (DT), and blockchain systems. Since 2022, he has been affiliated with the Global Innovation Institute as a Fellow and works with the Centre for Secure Information Technologies (CSIT) and Centre for Data Science and Scalable Computing (DSSC). He is the director of the Innovation-by-design lab at QUB. He leads the British Computer Society (BCS) Accreditation for QUB and is the chair of the Computer Science Programme Review Working Group. He is the co-investigator for the Northern Ireland Advanced Research and Engineering Centre (ARC). He has authored/co-authored more than 130 journal/conference articles and book chapters, co-edited three books, and won seven best paper awards. He has served on the editorial board of *IEEE Communications Magazine* and as section editor-in-chief of *Drones* journal. Currently, he serves as associate editor of *CAAI Transactions on Intelligence Technology*, *IET Networks*, and *ICT Express*. He is also the co-chair of the IEEE UK and Ireland Diversity, Equity, and Inclusion Committee. He is a senior member of IEEE and a professional member of ACM. He received his PhD in Computer Science and Engineering from Thapar University, India.

Berk Canberk is a professor in the School of Computing, Engineering and The Built Environment at Edinburgh Napier University, UK. He is also an adjunct professor in Artificial Intelligence and Data Engineering Department at Istanbul Technical University, Turkey. He is also an adjunct professor in the Electrical and Computer Engineering Department at Northeastern University, Boston, USA. His research focus includes digital twin networks, AI-enabled autonomous 5G+ systems, data-driven networks, and unmanned aerial networks. He is a senior member of the IEEE, a member of the ACM, and a member of the IET. He serves as an editor for *IEEE Transactions in Vehicular Technology*, *Elsevier Computer Networks*, and *Elsevier Computer Communications Journal*. He is a board member and observer of the IEEE 1900.5 Cognitive Radio for Dynamic Spectrum Access and IEEE Tactile Internet Standardization Working Groups. He is a group member of the Internet Research Task Force's Network Management Research Group (IRTF NMRG). He has been involved in several international conferences as technical program co-chair, symposium chair, demo/poster chair, regional chair, publicity chair, tutorial chair, and TPC member. He holds a PhD in Computer Science from Istanbul Technical University, Turkey.

Octavia A. Dobre is a professor and Canada Research Chair Tier 1 in the Faculty of Engineering and Applied Science at Memorial University, Canada. Her research focus is on wireless communication technologies. She serves as the vice president of publications of the IEEE Communications Society. She has served as general chair, technical program co-chair, and tutorial co-chair for numerous conferences. She was a Fulbright scholar, Royal Society scholar, and distinguished lecturer of the IEEE

Communications Society. She is an elected member of the European Academy of Sciences and Arts, a Fellow of the Engineering Institute of Canada, a Fellow of the Canadian Academy of Engineering, and a Fellow of the IEEE. She holds a PhD from the Polytechnic Institute of Bucharest, Romania.

Preface

The evolution of wireless communication technologies has been marked by significant leaps, each generation bringing forth transformative advancements. As we stand on the cusp of the sixth generation, or 6G, the convergence of cutting-edge technologies promises to redefine the way we connect, communicate, and interact. At the heart of this transformative journey lies the concept of digital twins (DT), a multidimensional paradigm that is set to play a pivotal role in shaping the foundation, technology, and applications of 6G networks.

The concept of DTs is not new, having gained prominence in various industries like manufacturing and healthcare. It involves creating a digital replica or model that mirrors a physical object, system, or process. In the context of 6G, DTs take on a new dimension, encapsulating the virtual representation of the entire communication network. This goes beyond mere simulation; it is a real-time, dynamic representation that includes both the network infrastructure and the connected devices.

The fundamental theory of DTs for 6G is grounded in the principles of real-time data synchronization, machine learning, and advanced analytics. The digital replica of the network allows for continuous monitoring, analysis, and optimization. Through a feedback loop, insights from the physical network are fed into the digital model, facilitating predictive maintenance, resource allocation, and the adaptation of the network to changing conditions.

The integration of artificial intelligence (AI) and machine learning (ML) algorithms forms the crux of the fundamental theory. These technologies empower DTs to learn from historical data, predict future network behavior, and proactively address challenges such as congestion, latency, and energy efficiency. The result is an autonomous and self-optimizing network capable of delivering unprecedented performance.

DTs for 6G leverage a spectrum of technologies to bring the virtual network to life. The core components include: (a) sensor networks and IoT devices: these gather real-time data from the physical network, providing a continuous stream of information to update the DT. (b) 5G infrastructure: the foundation laid by 5G networks forms the base upon which DTs for 6G build. The high-speed, low-latency connectivity of 5G is essential for the seamless integration of digital and physical networks. (c) Edge computing: the decentralization of computing power at the network edge facilitates rapid data processing, enabling DTs to operate in real-time and respond swiftly to changing conditions. (d) Blockchain: ensuring the security and integrity of the DT, blockchain technology plays a crucial role in protecting sensitive network information and maintaining trust in the system. (e) Advanced analytics platforms: robust analytics platforms process the vast amounts of data generated by the DT, extracting actionable insights and facilitating data-driven decision-making.

The applications of DTs in the realm of 6G are diverse and far-reaching. For example, DTs can aid in network optimization by continuously analyzing network performance, predicting potential issues, and optimizing parameters such as bandwidth allocation and routing in real-time. DTs can also help in dynamic resource allocation. Through AI-driven insights, DTs enable the dynamic allocation of network resources based on demand, ensuring efficient use of available bandwidth. DTs can also enhance ultra-reliable low latency communication (URLLC) by minimizing latency through predictive analytics, ensuring mission-critical applications operate with the utmost reliability. Another obvious applications of DTs can be in smart cities and infrastructure, whereby extending the DT concept to urban environments, 6G networks can support the development of smart cities, optimizing energy consumption, traffic management, and public services. Finally, DTs contribute to the creation of immersive augmented reality (AR) and virtual reality (VR) experiences, pushing the boundaries of what is possible in terms of connectivity and user interaction.

DTs for 6G represent a paradigm shift in the way we conceive and manage wireless communication networks. The fusion of real-time data, advanced analytics, and autonomous decision-making ushers in an era of unparalleled connectivity, where networks adapt and evolve in tandem with the physical world. As we delve deeper into the realm of 6G, the integration of DTs not only promises enhanced network performance but also opens the door to innovative applications that have the potential to reshape the way we experience and interact with the digital landscape.

This book captures the essence of how DTs can be leveraged for better understanding of the evolution of 6G networks and how both the technologies, i.e., 6G and DTs, can transform each other. The book is divided into ten chapters that broadly talk about how resilient and reliable 6G networks can be empowered by digital twins (Chapters 1, 5, and 6), how DTs can help in the complex real-time optimization of 6G networks (Chapters 2, 8, and 9), how specific enablers (e.g., open radio access networks (O-RAN) and optical networks) of 6G networks can be aided using DTs technology (Chapters 5 and 7), how AI can specifically play a key role in DT-assisted 6G networks (Chapters 3 and 4), and, finally, how 6G can benefit the hosting of DTs for industrial systems (Chapter 10).

Chapter 1 reviews 6G and its relevant technologies and applications, and then focuses on evaluating the role that DT technology can play in each of the specified 6G technologies and application scenarios as well as the main challenges in DT deployment. In Chapter 2, an unmanned aerial vehicle (UAV)-based edge network architecture is proposed and a DT of such a network is also envisaged. The authors formulate the latency minimization problem of UAV-based edge networks with URLLC in the DT regime and establish a strong case of DTs can play an empowering role in minimizing the task-offloading latency leading to overall latency minimization. Chapter 3 nicely establishes the power of AI and how this can play a pivotal role in managing (i.e., collecting, storing, integrating, analyzing, and utilizing) the voluminous data generated by DTs and network of DTs. Highlighting the role of AI further in the context of DTs and 6G networks, Chapter 4 describes the key AI-based tasks in networks of DTs, more suitably termed as DT networks (DTNs). Chapter 4 also describes in detail the typical AI models applied for the AI-based tasks in DTNs and pin-points the challenges of such models quite elaboratively.

In Chapter 5, a comprehensive perspective on how DT and O-RAN constitute two synergistic concepts is presented. In particular, it illustrates how their mutual integration holds the potential to facilitate the deployment of a smart and resilient 6G RAN. Notably, the DT concept will play a pivotal role in enhancing the core principles of intelligence, autonomy, and openness that underline O-RAN. The chapter begins with a concise overview of both O-RAN and DT concepts. It then proceeds to illustrate and discuss potential use cases and services achievable through a DT-based O-RAN architecture. Finally, the chapter concludes by outlining current challenges and discussing future research direction toward the implementation of such innovative network architecture. In Chapter 6, the authors provide a general but detailed overview of the applications of DT technology to 6G network design, planning, and deployment. In particular, the authors discuss how the DT technology can be utilized across diverse domains and solution domains within the proposed 6G communication system to improve system performance and streamline operational processes. Chapter 7 presents the possibilities that DTs can bring in terms of operations and management of optical networks. Optical networks form the backbone of the network infrastructure and an important enabler of 6G networks. Accordingly, it is imperative to investigate the role of DTs in optical networks and include them in a book that deals with the synergy between DTs and 6G networks.

Chapter 8 focuses on the deployment of virtual network functions (VNFs) and service function chains onto substrate networks. Optimal deployment of VNF service function chains considering the latency of accessing the services, substrate network physical resources, etc., is a complex optimization problem and the authors propose a deep Q-learning framework to solve such problems. This is important in the context of evaluating the significance of technologies like DT in solving dynamic optimization problems in complex connected systems e.g., the 6G networks. In Chapter 9, the concept of DT-aided optimization in networks is further bolstered by proposing an operational network DT integrated with ML and AI capabilities allowing the execution of advanced optimization algorithms in a flexible manner. Finally, in Chapter 10, a more explicit use case, e.g., industrial end-to-end DT systems in collaborative robotics applications, is discussed. The role of 6G networks to enable such practical DT systems is analyzed in detail and several challenges are enumerated.

Overall, this book with its ten contributed chapters, strongly establishes the synergy between 6G networks and DTs. These chapters are contributed by research teams, both from industry and academia, who are experts and early thought leaders in this domain. We believe the knowledge presented in this book will be greatly beneficial to all stakeholders working in the collaborative domain of 6G, DT, AI, and network optimization.

This book could not have been possible without the support of many people. The editors would like to thank the chapter authors for their contributions. They also acknowledge the many reviewers who contributed to the accuracy of each chapter, and last but not least, the valuable support of the IET staff during the preparation of this book.

Chapter 1
Digital twins for resilient and reliable 6G networks

Fahad Alaklabi[1], Ahmed Al-Tahmeesschi[1], Avishek Nag[2] and Hamed Ahmadi[1]

Our communication systems continue to fall short of our somewhat existing aspirations, prompting us to reconsider our future communication systems, given that other technologies and applications are also under development. In this work, we describe resilient and reliable sixth generation of mobile communications and networks (6G), starting with the drawbacks of fifth generation's (5G) key performance indicators (KPIs) and comparing them with 6G KPIs, exploring potential applications and scenarios for 6G. To motivate the potential role of digital twin (DT) in this new domain, we introduce key technologies for 6G, such as terahertz communication (Terahertz (THz)) and optical wireless communication (OWC) as new spectrums. We review the enhanced air interface through the design of new waveforms, modulations, and information theory methods, non-orthogonal multiple access (NOMA), and ultra-massive multiple input multiple output (MIMO), along with their compatibility with other technologies. We delve into more capable technologies, including integrated sensing and communication (ISAC), semantic communication (SC), and artificial intelligence (AI). Having gained a general overview of 6G and its relevant technologies and applications, we turn our focus on evaluating the role that DT technology can play in each of the specified 6G technologies and application scenarios as well as the main challenges in DT deployment.

1.1 Introduction

In today's world, there are many clear examples that have made commoners and entrepreneurs wonder about the limitations of our current formal mobile communication systems performance (i.e., 5G KPIs), ranging from slow Internet in crowded places, to lose Internet connection during high-speed train trips. One can explain that by attributing to the limitations of the current wireless systems, which have three times spectrum efficiency and 10–100× times network energy efficiency compared

---
[1] School of Physics, Engineering and Technology, University of York, UK
[2] School of Computer Science, University College Dublin, Ireland

to 4G systems [1]. While the 5G-oriented system offers many advantages for urban rail transit, it also presents numerous obstacles in wireless coverage, channel modelling, train control, sensor accuracy, and high-speed situations [2]. Meanwhile, the future applications of internet of everything (IoE) (e.g., full-sensory extended reality (XR) services and remote diagnosis and treatment of patients) will undermine the initial 5G purpose of delivering short-packet, sensing-based ultra-reliable low latency communication (URLLC) capabilities [3]. Table 1.1 enumerates URLLC employed for the evaluation and compensation of performance in both 5G and 6G [3]. This is expected to achieve a high level of performance and feasibility as the 6G standard matures, driven by the 6G KPIs. In 2030 and beyond, 6G will be required to enable disruptive use cases and applications, demanding more capabilities and increased network performance beyond 5G [1]. For instance, to meet the complete proliferation of IoE applications (e.g., autonomous driving and remote medical operations) that operate at frequencies beyond sub-6 GHz as well as demanding self-sustaining and smart network systems [4]. Therefore, the 6G whose architecture is fundamentally geared to the performance needs of IoE applications and their associated technical advancements can help solve these criteria [4]. However, resilience in network systems is an important aim that cannot be met without proactive preparation and planning [5]. Organisations may guarantee that their networks remain operational and dependable in the face of possible interruptions by investing in resilience. DT modularity gives

Table 1.1 KPIs for 5G and 6G compensation

KPIs	5G	6G
Peak data rate	Downlink: 20 Gbps & Uplink: 10 Gbps [6]	1 Tbps [3]
User experienced data rate	Downlink: 100 Mbps & Uplink: 50 Mbps [6]	10 Gbps [3]
Latency	4 ms for eMBB & 1 ms for URLLC [6]	0.1 ms [3]
Delay Jitter	1 ms [3]	1 μs [3]
Area traffic capacity	10 Mbps/km^2 [3]	10 Gbps/km^2 [3]
Connection density	10^6 devices/km^2 [3]	10^8 devices/km^2 [3]
Coverage	10% [3]	99% [3]
Spectrum efficiency	Downlink: 30 bps/Hz & Uplink: 15bps/Hz [6]	≥90 bps/Hz [3]
Network energy efficiency	10^7 bit/J [3]	10^9 bit/J [3]
Cost efficiency	10 Gb/$ [3]	500 Gb/$ [3]
Mobility	500 km/h [3]	1000 km/h [3]
Battery life	10 years [3]	20 years [3]
Reliability	99.999% [3]	99.99999% [3]
Positioning	1 m & 10 m [3]	10 cm & 1 m [3]
Sensing/imaging resolution	1 m [3]	1 mm [3]
Security capacity	Low [3]	High [3]
Intelligence level	Low [3]	High [3]

a strategic benefit in terms of optimising resource allocation, increasing resilience, and ensuring that systems stay flexible and responsive to changing conditions and requirements [5]. This 'on-demand' flexibility and redundancy can considerably contribute to the overall stability and dependability of complex systems [5]. A DT can be described as a digital portrayal of an asset that offers a record of the asset's past conditions and simultaneously provides up-to-date information about its present state [5]. The aim of this chapter is to comprehensively examine several key aspects related to 6G and DT, these aspects include:

- **6G key performance indicators (KPIs)**: Analysing and discussing the essential performance metrics and benchmarks that will define the success of 6G. This involves assessing factors such as data rates, latency, reliability, and energy efficiency.
- **Review of major potential 6G technologies**: Conducting an in-depth review of the significant technological innovations and advancements expected to underpin 6G. This could encompass higher frequency bands, intelligent surfaces, new modulation schemes, and more.
- **6G applications**: Exploring the various real-world applications that 6G technology is expected to support. This may include remote robotic surgery, autonomous transportation, XR, rescue communications, and DT.
- **Evaluation of 6G key performance indicators and application scenarios**: Assessing how the identified 6G KPIs align with and support the anticipated application scenarios. Evaluating how these KPIs impact the performance and feasibility of 6G-enabled applications.
- **The role of DTs**: Exploring how DT technology can play a pivotal role in each of the identified 6G technologies and application scenarios. This involves examining how DT can be utilised for simulation, optimisation, and enhancing the performance of 6G and applications.

In summary, this study seeks to provide a comprehensive understanding of 6G, covering performance metrics, technological innovations, real-world applications, and DT with minimal latency to enhance the resilience and dependability of 6G.

1.1.1 6G KPIs

Based on research and analysis, the study [3] presented 17 quantitative and qualitative KPIs for 6G and identifies their reference numbers. The suggested KPIs are divided into four categories: data rate and delay, capacity and coverage, service efficiency, and diversified service evaluation [3]. The data rate and delay category contain KPIs such as peak data rate, user-experienced data rate, latency, and delay jitter. The capacity and coverage category contain KPIs such as area traffic capacity, connection density, and coverage. The service efficiency category contains KPIs such as spectrum efficiency, network energy efficiency, and cost efficiency. The diversified service evaluation category contains KPIs such as mobility, battery life, reliability, positioning, sensing and imaging resolution, security capacity, and intelligence level.

However, it is important to note that the utilisation of certain KPIs from Table 1.1 in 6G is specific to a single-usage scenario [6].

Since the development of 6G technologies is still ongoing, the authors of [3] believe that the existing literature on KPIs do not comprehensively cover all areas of 6G, particularly in areas like security and intelligence. The authors expect that the 6G would address many security and trust issues using new technologies (e.g., quantum communication and blockchain) allowing the system to be more secure [3]. In addition, the intelligence level of the 6G is predicted to be considerably enhanced with the more rapid permeation of AI [3].

1.1.2 6G technologies

To pave the way for 6G, a diverse array of innovative technologies has been proposed, these technologies encompass:

- **Higher frequency bands and intelligent surfaces**: The utilisation of higher frequency bands, particularly in the **terahertz!** (**terahertz!**) range, along with the deployment of intelligent surfaces (i.e., intelligent reflective surfaces (IRS)), is pivotal components of 6G [5].
- **Integrated terrestrial and non-terrestrial networks**: 6G involves the integration of terrestrial and satellite networks, expanding the coverage and capabilities of the network infrastructure [5].
- **Machine learning (ML) and artificial intelligence (AI)**: machine learning (ML) and AI are extensively integrated into 6G, enhancing network management, optimisation, and intelligence [5].
- **Optical wireless communication (OWC)**: To meet the escalating demand for data services, OWC plays a crucial role in 6G [3].
- **Advanced waveforms, modulation, coding, and information theory**: Innovative approaches to waveforms, modulation, coding [3], and electromagnetic information theory (EIT) [7] are being developed to support 6G.
- **NOMA and MIMO**: Two significant air interface technologies, NOMA and MIMO, are anticipated to enhance communication system capacity and efficiency in 6G [3].
- **Integrated sensing and communication (ISAC)**: ISAC technology is poised to enable simultaneous sensing and communication capabilities in 6G [3].
- **Semantic communication**: SC is another emerging technology that promises to play a role in 6G, allowing for more context-aware and intelligent communication [3].

This diverse set of technologies is instrumental in shaping the landscape of 6G, with further details about each technology provided in the subsequent sections.

1.1.2.1 THz communication

THz (i.e., 100 GHz and 10 THz) have promised to support high data rate communication from indoor-to-outdoor scenarios (e.g., vehicle-to-vehicle communication

(V2V)) and space applications like inter-satellite connection as well as accommodating uses beyond communication (e.g., explosive detection and gas sensing) [3]. To overcome 0.3 THz (i.e., the initial range of THz), there is a great requirement to improve the suitability of hardware devices at 5G [3]. For instance, complementary metal-oxide semiconductor (CMOS) technologies battle with 0.3 THz frequency needs [3]. In addition, 6G must enable new services such as immersive remote presence, holographic teleportation, networked robots, XR, and DT, which would necessitate a 1000× increase in bandwidth capacity over 5G [8]. Thus, technical requirements indicate a growing demand for the development of THz and 6G. However, the absence of propagation losses and power restrictions of short communication distances in the THz band as well as health and safety issues have to be investigated and considered [3].

1.1.2.2 Optical wireless communication

OWC systems offer high-density broadband services with low latency, innate physical layer security, free electromagnetic interference, unlicensed spectrum, low costs, and complementing RF-based systems [3]. OWC (i.e., visible light communications (VLC), light fidelity (LiFi), optical camera communications (OCC), free-space optical (FSO), and light detection and ranging (LiDAR) are regarded as promising for ultra-short to ultra-long distance communications [3]. For instance, wireless fidelity (WiFi) with LiFi may deliver illuminating and multiuser communication services at the same time. FSOs are utilised for high-data-rate transmission in data centres, deep space communication, and underwater devices. VLC offers enormous potential for use in indoor, underwater, vehicle communications, and localisation systems. LiDAR is an intriguing optical distant sensing technology that has enormous promise for usage in transportation, aerial communication, and autonomous vehicular communication. OCC specialises in vehicle-to-everything (V2x) communications, indoor positioning, digital signs, and virtual reality (VR) [3]. However, despite the fact that a tiny fraction of the optical spectrum (i.e., comprising middle ultraviolet (UV), a part of VL, and near-infrared) is now used, further study is needed to optimise the efficiency of these subcategories [9]. In addition, UV radiation may have a detrimental influence on health and safety, which should be thoroughly investigated before practical implementation [3].

1.1.2.3 Design of fresh waveform, modulation, coding, and information theory methods

The interplay of information theory with electromagnetism has resulted in the birth of a new area known as EIT [7]. Present information-theoretic methods are insufficient to properly characterise the performance limits of 6G technologies (e.g., IRSs and orbital angular momentum (OAM)) due to assumptions such as scalar-quantity, far-field, planar wavefront, and physically incorrect assumptions [7]. EIT is predicted to play a critical role in characterising these basic constraints and perhaps developing better physically consistent communication system models [7]. The mathematical tools provided by EIT will offer guidance to designers, enabling them to efficiently create antennas for extensive intelligent surfaces (e.g., IRSs) and surface wave communication (e.g., OAM) [7]. In addition to new information theory, novel waveform design techniques can be employed to achieve greater spectrum efficiency such as spectrally efficient frequency domain multiplexing (SEFDM) and overlapping x domain multiplexing (OVXDM) [3].

In high mobility applications, traditional orthogonal frequency division multiplexing (OFDM) encodes the signal information in the time and frequency domain, but this approach struggles with managing substantial delays and Doppler spread [10]. On the other hand, orthogonal time–frequency space (OTFS) encodes the signal information across delay and Doppler domains, which makes it an appealing solution for mitigating Doppler spread in high mobility communication systems [10].

Along with new information theory and waveforms, modulation techniques such as index modulation (IM) could improve spectral efficiency without requiring more complex hardware or additional bandwidth [11]. The advantage of IM is to use indices of available resources (e.g., subcarriers, antennas, or time slots) to deliver more information in a more energy-efficient manner [11].

In light of the previous discussion, it becomes necessary to develop new codes, particularly tailored for applications that involve transmitting small data packets (i.e., internet of things (IoT)) [12]. This is due to the extended non-sleep mode, which can pose challenges for low-energy devices in handling automatic repeat request (ARQ) mechanisms, potentially leading to increased energy consumption [12]. Therefore, the new coding techniques should include both forward error correction and innovative iterative retransmission/feedback mechanisms as well as ML-based approaches [12].

1.1.2.4 Non-orthogonal multiple access (NOMA)

NOMA promotes numerous terminals to share the same radio resources by presenting interference information and demodulating with superposition coding (SC), rate splitting (RS), successive interference cancellation (SIC), and message passing (MP) among others [13]. However, while NOMA provides benefits such as enhanced spectral efficiency and fairness, it does so at the expense of increased latency and computing complexity owing to the requirement for serial detection and user ordering, particularly in scenarios with a large number of users [13].

1.1.2.5 Ultra-massive MIMO

6G employs ultra-massive MIMO, which utilises hundreds or thousands of antennas, to produce greater spectral and energy efficiency, larger network coverage, and improved positioning accuracy over a wide range of frequency bands [3]. In addition, ultra-massive MIMO is intended to use higher frequency bands, such as mmWave and THz, to make use of abundant spectrum resources [3,14]. Effective deployment of ultra-massive MIMO technology in higher frequency bands for 6G relies on addressing key areas such as integrated circuit design, channel acquisition [15], and modulation techniques [3].

1.1.2.6 Integrated sensing and communication (ISAC)

ISAC has significance for a 6G, since it improves perception and communication systems while also improving hardware, spectrum, time, and energy economy [3]. Both (i.e., sensing and communication) employ consistent high-frequency and large-aperture antennas and are intended to use similar signals and data processing technologies [3]. The future direction of this technology includes development of high-precision ISAC measurement equipment, design of realistic measurement scenarios, selection of efficient transmission frequency bands, and creating correct ISAC channel models [3].

1.1.2.7 Semantic communication (SC)

SC has the promise of improving communication efficiency, quality of human-centred services, and realising full seamless intelligent connectivity of everything [3]. SC entails extracting and encoding semantic information for transmission over a noisy channel, with the goal of constructing a widespread and understandable semantic knowledge base and break past transmission bottlenecks and free communication networks from existing data protocols and formats [3]. The future direction of this technology is focused on the critical correctness of its core semantic components in order to assure dependability in actual applications [3]. However, there is still difficulty in identifying techniques for developing robust error–tolerant processes and quickly applying adaptable solutions inside resource-limited systems [3].

1.1.2.8 Intelligent reflective surfaces (IRSs)

IRS is a platform consisting of programmable 2D meta-materials that dynamically alter signals to generate electromagnetic fields with adjustable amplitudes and phases [3]. IRS is simpler to implement compared to traditional relay structures and it can greatly boost network transmission rate, improve signal coverage, and improve frequency, energy, and cost efficiency [3]. In addition, IRSs are effective when the signal is unable to reach the end user with sufficient strength (i.e., non-line-of-sight) since they can be installed on buildings, advertisement panels, and automobiles [16]. Hence, the future direction of 6G networks is to incorporate numerous IRSs given their benefits [16].

1.1.2.9 Terrestrial and non-terrestrial networks collaboration

Linking terrestrial, airborne and satellite networks into one integrated wireless system will be critical for 6G, since both airborne and terrestrial networks may require communication with low Earth orbit (LEO) satellites to offer backhaul support and broad coverage area [4]. However, these networks necessitate consistent and reliable communication among numerous nodes functioning in various environments and domains, making interoperability crucial due to variations in communication protocols, hardware, software architectures, and network structures [17].

1.1.2.10 Artificial intelligence and machine learning

Machine learning and AI will enhance 6G by improving design, deployment, and operating phases, enabling extreme connection needs [18]. For instance, AI and machine learning-based network automation will be essential for improving network administration and optimisation as the network (i.e., 6G) shifts to programmable and flexible cloud native implementation [18]. Furthermore, AI can be utilised for network demands prediction and caching, which can assist in minimising latency and operating fees [3]. Due to the high computational overhead of traditional channel estimation approaches, considerable attempts are being made to conduct channel estimation and signal identification using AI [3]. However, the challenges of using AI in the context of 6G include inefficient data management schemes and the significant overhead associated with exchanging information among communication participants [19,20].

1.1.3   6G applications
1.1.3.1   Tele-medicine and remote robotic surgery
As the number of patients infected with the coronavirus grows, it is becoming clear that there is a shortage of competent medical care due to a shortage of doctors, and transportation (i.e., critical with major injuries). Tele-medicine and remote consultancy can help to alleviate these issues. Nevertheless, current network technologies (i.e., 5G) are still inadequate for such applications due to their inherited limitations in throughput and delay [16]. Tele-medicine can take full advantage of 6G network since it can provide high data rates above 1 Tbps and a low end-to-end latency of less than 1 ms which are essential for gathering large data from sensors in real-time manner [16].

In addition, 6G KPIs may also enable remote robotic surgery through using tactile Internet [16]. Remote robots will operate as multi-modal avatars of people through the tactile Internet, allowing them to transmit "skills" with perception and synchronisation activities [3].

1.1.3.2   Autonomous transportation
With a 7 trillion dollar market, fully autonomous transportation systems provide safer travel, enhanced traffic management and infotainment [21]. Autonomous driving enables vehicles to autonomously navigate and operate, in the absence of human involvement [22]. Before achieving autonomous driving, various levels of automation need to be reached, supported not only by communication capabilities but also by additional factors like localisation and sensing [22]. For instance, autonomous cars demand a high level of reliability and low latency of greater than 99.99999% and less than 1 ms to ensure passenger safety [21]. 6G enables a variety of key services to enable autonomous transportation, such as V2x communications, information collection for vehicle surroundings, and traffic forecasting, all of which contribute to the development of a new generation of smart transportation [23].

1.1.3.3   Extended reality (XR)
XR is an umbrella word that encompasses augmented reality (AR), VR, and mixed reality (MR) [3]. 5G has limited practical uses for XR while 6G will enable immersive XR with continued development. 6G will provide fast data rate, big bandwidth,

low latency, dependability, and high image resolutions to enable immersive XR applications in entertainment, telemedicine, and remote industrial control [3].

1.1.3.4 Rescue communications

Considering the right rescue time into account, the 3D complete space coverage in 6G will be introduced on demand (e.g., unmanned aerial vehicles (UAVs)) and satellite communication networks), enabling emergency communications to help in fast search and rescue (e.g., earthquakes and floods) [3]. For example, the authors in [24] stated that each aircraft could functions as a router and delivers packets to others, utilising DT technology to control AANET complexity and avoid AI-based methodological problems.

1.1.3.5 DT

6G are required to enable potentially high-demanding and rapidly increasing DT applications since 6G can provide reliable and ubiquitous communication at 100 Gb/s [5]. For instance, the future large-scale industrial IoT applications assisted by DT [5]. DT are projected to be used in a variety of disciplines, including urban infrastructure, transportation networks, and urban ecosystems [3]. In addition, DT body area network may mimic a virtual human body utilising 6G and information and communications technology (ICT), enabling real-time tracking, illness prediction, and modelling of procedures and medicines [3]. However, this study [25] shows that the reliability of the communication between the physical-to-physical and physical-to-virtual digital twin network (DTN) is low as well as the latency is high in today's communication systems. Therefore, 6G can be an enabler of widespread DT deployment [5,25].

1.1.3.6 Evaluate 6G KPIs and applications

Analysis and evaluation of critical 6G KPIs for applications priory discussed in terms of scalability, coverage, data rate, latency, and cost requirements are shown in Table 1.2.

For remote robotic surgery, the transmission of tactile information has extremely high requirements on the network latency and reliability for remote robotic surgery utilising tactile Internet [3]. Given that the robotic would be in an urban location, coverage and scalability are less significant, while the cost may be covered by the government for civil people. For autonomous transportation, millions of self-driving cars and drones will be on the road by 2030, providing safe, efficient, and environmentally friendly transportation for both individuals and goods [26]. To ensure the safety of passengers and pedestrians, these vehicles must have high reliability and low latency [26]. With today's quick progress in this industry, the cost may be the same or less than what we have presently, but the coverage may be crucial outside of the metropolitan region in case. For XR, to deliver an immersive experience, high-quality video with the resolution, frame rate, colour depth, and dynamic range is required, needing more than 1.6 Gbps per device [26]. In addition, interactive XR applications such as gaming, surgery, and industrial control need low latency and

Table 1.2 6G KPIs and applications

Application	Scalability	Coverage	Data rate	Low latency	Reliability	Affordable
Remote robotic surgery	–	–	✓✓✓	✓✓✓	✓✓✓	–
Autonomous transportation	✓✓✓	✓✓	✓✓	✓✓✓	✓✓✓	✓
XR	✓✓	–	✓✓✓	✓✓✓	✓✓	✓✓
Rescue communications	✓	✓✓✓	✓	–	✓	✓
DT	✓✓✓	✓	✓✓	✓✓✓	✓✓	✓

high reliability [26]. However, with today's rapid advancement in this business, the cost might be less than what we have now, while the scalability may be expanded to include more future deployment applications. For rescue communications, the world population is expected to reach 8.5 billion by 2030, 9.7 billion by 2050, and 10.4 billion by 2100 [27] while lacks network coverage in oceans, deserts, and natural disasters [3]. A high-bandwidth network with enough coverage is required to ensure golden rescue time (i.e., the right rescue time (can you explain more on the golden rescue time definition, few more words)) while satellite networks have high battery needs in order to achieve low power consumption and longer emergency communication system support [3]. For DT, 6G are necessary for meeting potentially high-demanding and fast-expanding DT applications with high-throughput, reliable, pervasive communication at 100 Gb/s [5], and low latency compared to today's communication [25].

1.2 DTs and 6G

The concept of DT originated during the 1970 Apollo 13 mission when NASA employed mirrored systems for simulations to guide emergency response and return the crew safely to Earth, subsequently leading to the development of various informal DT definitions [28]. Recently, new DT taxonomies have been produced, which are grouped into three groups with diverse viewpoints and priorities and provide the DTN definition [25]. The first one emphasises a physical object's mirror model, implying that no automatic data flows between the actual thing and the virtual model and ignores their co-evolution. The second one, a computational model, simulation, or software used to comprehend, predict, and optimise physical objects is referred to as DT that regards the data flow between the real thing and its digital model as unidirectional, involving changes to the physical object impacting the virtual model but not inversely [25]. The third one is a complete system that includes a real physical entity, its virtual twin, data, services, and connections. Based on gathered online data and information, the virtual twin continually adjusts to operational changes and anticipates the status of the actual thing. Note that the physical entity can be a gadget, a machine, or a robot, as well as an industrial process or a complicated physical system [25].

DTN enables one-to-one mapping connections of DTs to be expanded using modern data processing, computing, and communications technologies to construct a complex large-scale system with greater reliability and efficiency [25]. Therefore, information exchange and exact status sensing, real-time analysis, fast decision-making, and precise execution on physical objects are made possible [25]. Consequently, the introduction of DTN opens up the potential for overcoming security, spectrum efficiency, intelligence, energy efficiency, and customisation in the 6G, all while altering and speeding its growth [25]. DTN includes communications (i.e. P2P, P2V, and V2V), physical data processing (e.g., data confusion and data visualisation), DT modelling (e.g. unique model, multiple dimensions model, and general model), cloud computing (i.e., storage, resources, and sharing of computing), and edge computing (i.e., user privacy, power consumption, latency, cost, and the availability of wireless links) [25]. However, the authors in [25] indicate that the reliability of the communication between the physical-to-physical and physical-to-virtual is low as well as the latency is high in today's communication systems. On the other hand, scalability refers to achieving latency requirements for a large number of 6G devices and adding new nodes without affecting system latency performance [29]. As a result, the roles of DT as an enabler for 6G technologies such as high-frequency communications, non-terrestrial networks, MIMO, RIS, AI, new network technologies, and IoT are required to be considered and analysed. The debate explores the functions of DT as a facilitator for expanding 6G, with these technologies elaborated on in the following sections.

1.2.1 DTs and higher frequency technologies

High-frequency communication paradigms have advantages in latency and data rates, but have limitations (e.g., limited communication distances, limited spatial multiplexing gains, and signal blockage) due to high path-loss, strong channel sparsity, and line-of-sight dominance respectively [30]. The potentials of DT as an enabler for high-frequency communications, such as THz and VLC, are discussed below.

1.2.1.1 THz

Although THz are prone to deep fading and blocking, DT can improve the performance of THz even in networks that are extremely dynamic (e.g., in indoor environments communication) by increasing context awareness and adaptive beam steering [30]. DT provides a rich environment for testing and evaluating THz channels with heuristic techniques and theoretically developed models for increased accuracy, accomplished by modelling, forecasting and manipulating THz channel parameters, with the goal of maximising received signal-to-noise ratio (SNR) at the receiver end and determine the best signal path [30,31].

High-fidelity ray tracing models are required for developing DTN, capturing accurate physical characteristics (e.g., geometries, materials, properties, lighting, behaviours, and rules), and simulating radio wave propagation in true-to-reality environments, particularly for 6G with ultra-wide signal bandwidth and IRSs [32]. On the other hand, THz enable the evolution of DTs by facilitating the transfer of vast amounts of information with low delays and high reliability, crucial for representing not only assets but also the environment and interactions between DTs [33,34]. In addition, THz provide accurate positioning and mapping between the real and digital worlds, improving human–machine interaction and assuring high reliability and efficiency [33]. This might enhance DT decision-making or be beneficial for future DT applications.

1.2.1.2 Visible light communications (VLC)

VLC using light emitting diodes (LEDs) is a green technology known for its high security and the availability of license-free bandwidth. To design, evaluate performance and test VLC systems, a reliable channel model is crucial, whereas two main types of channel modelling schemes: one based on static scene data configuration, relying on expert knowledge, and the other based on neural network static modelling [35]. Existing schemes have limitations, primarily due to their reliance on control standards and historical experience rather than real-time data [35]. A DT generates a dynamic digital representation of physical equipment inside a digital sphere by continually monitoring real-time equipment data, enabling efficient and cost-effective dynamic modelling and control of physical equipment, resulting in improved performance and maintenance management [35]. In addition, DT obtains relevant data from user equipment (UE) measurements (e.g., LiDAR data, ray-tracing simulations), estimate the Channel State Information (CSI), and then minimises the overhead [36].

On the other hand, DT uses VLC for 3D-geometry model visualisation since typical 2D projections on computers do not account for user perspective or human stereoscopic vision in geocentric visualisation of DT [37].

1.2.2 DTs of non-terrestrial networks

The upcoming 6G aim to offer widespread coverage with enhanced reliability and lower latency by integrating both terrestrial and space networks (e.g., low Earth orbit (LEO) satellites (LEO) and UAVs). LEO are seen as a crucial technology for various applications, especially in smart cities, where they can efficiently connect numerous users across different locations while maintaining sustainability [30]. On the other hand, UAVs have garnered considerable interest and have been increasingly utilised in wireless networks for tasks like surveillance, tracking objects, and extending network coverage [30]. DT can analyse coverage gaps, connection budgets, and factors like satellite paths, antenna patterns, and atmospheric conditions, enhancing beamforming and satellite constellations [30]. In the following parts, more details given about how DT are connected to networks beyond Earth, such as LEO and UAVs. The upcoming 6G aim to offer widespread coverage with enhanced reliability and lower latency by integrating both terrestrial and space networks (e.g., LEO and UAVs). LEO are seen as a crucial technology for various applications, especially in smart cities, where they can efficiently connect numerous users across different locations while maintaining sustainability [30]. On the other hand, UAVs have garnered considerable interest and have been increasingly utilised in wireless networks for tasks like surveillance, tracking objects, and extending network coverage [30]. DT can analyse coverage gaps, connection budgets, and factors like satellite paths, antenna patterns, and atmospheric conditions, enhancing beamforming and satellite constellations [30]. In the following parts, more details given about how DT are connected to networks beyond Earth, such as LEO and UAVs.

1.2.2.1 Satellite communications

The rapid mobility of LEO satellites restricts their wireless communications, resulting in frequent handovers and service disruptions in satellite wireless communications. The current satellite handover techniques depend on optimising a single measure for frequent handover, which is unscalable and inadequate, whereas many characteristics are required for optimised handover and enhanced QoS for ground users [30]. DT allows for improved handover strategy design by taking into account system aspects such as elevation angle, service time, closest position, and satellite speed [30]. The utilisation of DT in satellite communication handover can minimise transmission latency, enhance data delivery quality, and potentially improve the QoS for satellite-terrestrial communications [30]. In [38], DTN used to map satellite networks in the virtual to reduce the number of satellite handovers. Subsequently, genetic algorithms are utilised to enhance the satellite's data delivery quality (i.e., intersatellite connection latency).

Aside from the function of DT in satellite communication, field trials are difficult when the device under test is a radio placed in aerial platforms or, more specifically, in a satellite for non-terrestrial networks (NTN) [39]. DT approach enables controlled, repeatable radio performance testing in difficult environments, and simulating field trials [39]. In addition, researchers have integrated a machine learning framework with classical dynamic models to address orbit prediction errors and enhance accuracy. However, this strategy has drawbacks such as computational cost, update issues, and model overfitting [40]. To solve these challenges, a DT system for satellite orbit is presented, which employs a temporal convolution network (TCN) to construct a unique orbital error prediction model [40].

On the other hand, satellite air–ground networks may benefit DT by collecting real-time data from mobile nodes even in remote places and delivering that to DTs for DTN services [41]. Stable connections are provided by high-frequency wireless spectrum, while satellites provide worldwide coverage for DTN and 6G [41].

1.2.2.2 UAV communications and networks

A swarm of networked UAVs provides improved coverage and services, but covering a large geographical area necessitates a large number of UAVs, increasing management and coordination difficulty, whereas optimised cross-layer scheduling considers multiple KPIs to meet reliability and latency requirements, including UAV status, instantaneous resources between UAVs and network [30,42]. DT provides cross-layer network resource optimisation, with an emphasis on spectrum and power allocation, data routing, and UAV trajectory design, resulting in efficient resource management and effective routing [30,43]. DT is crucial for achieving real-time control and synchronisation between UAVs and the network through the cloud twin (CT) (i.e., a virtual part of twin) [30].

UAVs, on the other hand, can be employed in the air to offer low-cost rural area coverage, gathering data and transferring that data to DTN through space–air–ground integrated networks (SAGIN) [41]. Within SAGIN, UAVs can serve as aerial mobile base stations, offering additional connectivity to users to ensure the continuity of reliable data delivery [42]. In disaster rescue scenarios for instance, UAVs equipped with DTs can assess the connectivity of rescue equipment, optimising communication resources for critical devices to enhance long-term Quality of Service (QoS) and outperform stationary terrestrial infrastructures by offering flexible deployment and extensive coverage [42].

1.2.3 New physical layer and multi-antenna techniques

Massive MIMO technology improves wireless communication and network capacity by expanding antenna deployment at base stations, but its maximum deployment bottleneck limits 6G expansion [30]. Beyond 5G, smart buildings aim to be intelligent both internally and externally, supporting various indoor and outdoor applications (e.g., reflective intelligent surfaces (RISs)) [30]. In the following parts, more details presented about how DT are connected to emerging physical layer technologies such as massive MIMO, cell-free massive MIMO, and RISs.

1.2.3.1 Massive MIMO and cell-free massive MIMO

Physical layer processes rely on wireless channels, with tasks such as MIMO precoding and link adaptation requiring partial or complete knowledge of these channels, whereas channel acquisition is often linked with substantial overhead in large-dimensional systems, which reduces total system efficiency [44]. Real-time DTs present new possibilities for transforming the channel acquisition procedure [44]. When the accuracy of real-time 3D maps and ray tracing computations reaches a certain level (i.e., sufficiently accurate), DT can be directly employed to identify the channels, minimising or completely removing the need for channel acquisition overhead [44]. For instance, a massive MIMO base station (BS) employing frequency division duplex (FDD) can leverage the real-time DT to forecast the downlink channel, or, at the very least, its primary subspace, leading to significant reductions in channel training and feedback overhead [44]. These DTs can also serve to assess the signal-to-noise ratio (SNR) of the communication links and enhance the choice of modulation and coding schemes [44]. He et al. [45] introduced a conditional generative adversarial network (C-GAN)-based DTN to model the relationship between beamforming and system performance, while also aligning the distribution of system performance across various UE positions for a specific beamforming configuration. The suggested DTN can perform data augmentation and enhance pre-validation accuracy by supplementing generated system performance samples, all without modifying the neural network (NN) structure or making adjustments to network hyperparameters [45].

The user-centric cell-free massive MIMO network structure serves fewer users in dense deployments and eliminates cell borders to ensure consistent performance and coverage [31]. High-mobility users, on the other hand, experience a problem known as "channel ageing," in which the estimated channel quality does not match the actual data transmission time [31]. DT of the distributed antenna system is proposed to overcome this, allowing dynamic clustering of distributed units (DUs) and resource allocation, consider mobility features, user history, and channel ageing effects via time-varying channel modelling, accounting for user velocity, frequency, propagation geometry, and antenna parameters [31].

Both massive MIMO and cell-free MIMO are two novel radio access technologies that provide high-throughput, low-latency data pipelines between sensors and DTs, which are critical for keeping DTs up to date [46]. Using new spectrum from mmWave to sub-THz bands gives a huge bandwidth for enormous twinning with enough data flow and connection density, as well as shortens the minimal transmission time gap, allowing for more precise synchronisation among vast DTs [46].

1.2.3.2 RISs

To properly utilise RISs in 6G, adaptive information of RISs (i.e., the ideal amplitude and phase of each element) must be retained throughout the network lifespan, allowing for smart network vision [30]. The operational twin can manage computational overhead, complicated optimisation, and continual reconfiguration of RIS properties, giving the adaptability and flexibility required for intelligent network [30]. In addition, a planning DT can develop ideal RIS placement and orientation schemes for numerous users, evaluating locations and orientations over a Cyber-Twin for coverage, interference management, and security, which is especially beneficial for big surfaces that need an extensive amount of money and time to deploy at buildings [30]. In this study [47], the research addresses the optimisation of total end-to-end latency in a RIS-assisted mobile edge computing (MEC) system, taking into account constraints related to URLLC, and focusses on jointly optimising beamforming design, power allocation, bandwidth allocation, processing rates, and task offloading parameters using a DT architecture [47].

On the other hand, the air interface for DT serves as the communication link between DT and physical objects like sensors and wireless users, with a focus on two key design aspects (i.e., selecting the frequency band for wireless access and choosing the access scheme) [48]. In this context, the wireless access can leverage IRSs comprised of metasurface elements capable of modifying signal properties (polarisation, frequency, amplitude, or phase) to facilitate effective communication between non-line-of-sight transmitter–receiver pairs [48].

1.2.4 DT and new network technologies

The complexity of overseeing wireless networks arises from the presence of numerous similar interactive devices, including both stationary and mobile smart objects. Thanks to recent advancements in virtual services, the physical and digital worlds are increasingly integrated. The primary objective of this digital platform is to facilitate precise network asset planning and the implementation of required actions through distributed control capabilities [30]. By combining historical data from the CT (i.e., a virtual part of twin) with real-time data from the PT (i.e., a physical part of twin), then merge them with simulated scenarios [30]. This integration facilitates aiding in the identification of potential bottlenecks and the prediction of events that could affect network operations [30]. These DT are intended to create data relevant to predicted future occurrences, allowing for the construction of more adaptable AI models capable of managing a wide range of network scenarios [30]. In the upcoming sections, additional details will be provided regarding the relationship between DT and emerging network technologies like cloud radio access networks (C-RAN), open radio access network (O-RAN), network slicing, and AI-driven networking.

1.2.4.1 New radio access network (RAN) technologies

Physical-level DTs provide fine-grained information about communication connections, boosting RAN modelling accuracy and increasing efficiency and reliability in network-level operations [44]. C-RAN is an architectural approach in wireless communication systems, specifically 4G and 5G networks, offering benefits such as cost reduction, improved system performance, enhanced mobility support, extended coverage, and increased energy efficiency [49]. However, C-RAN implementation faces challenges including the need for high-capacity fronthaul networks, virtualisation for efficient resource management, and meeting stringent timing and latency requirements for accurate synchronisation in a virtualised environment [49].

Ren *et al.* [50] proposed a DTN-based SLA quality closed-loop management scheme that focusses on intelligent and low-cost end-to-end (E2E) network service deployment and service level agreement (SLA) quality assurance in the two-level C-RAN. This method enhances network automation as well as operational and maintenance (O&M) capabilities [50].

O-RAN architecture employs E2 interface for KPIs telemetry reporting and offline rApp/xApp models training at non-RT/near-RT RIC levels [51]. However, since these models are trained on certain network KPIs snapshots, they might not be reliable in a variety of channel conditions [51]. A DT-based O-RAN architecture is appropriate to training and testing scenarios based on real-time channel conditions [51]. In a more elaborate explanation, create digital replicas of real-time channel conditions and testing scenarios to allow rApps/xApps choose the optimal actions instead of relying on pre-recorded training data [51].

Open-RAN, on the other hand, enhances DT communication capability and decision-making [44]. For instance, communication devices (e.g., sensors) can exchange data and access DT by communicating with an O-RAN application on the edge or in the cloud [44].

1.2.4.2 Network slicing

Slice-enabled communication networks (SeCNs) employ advanced technologies like software-defined networking (SDN) and network function virtualisation (NFV) to establish multiple virtual networks, with each slice being under the ownership of a logical operator on a shared physical infrastructure. Due to the presence of multiple slices operating on a shared physical infrastructure based to a pre-determined service level agreement (SLA), managing the software, virtual infrastructure, and varying QoS levels in slice-enabled communication networks (SeCN) becomes noticeably complicated and posing significant challenges for network orchestration [52]. DT can play an important role in SeCNs by successfully mimicking the operation and performance of real-world SeCNs, providing optimal resource allocation in addition to the ability to assess the impact of changes to the optimised network setup by offering important insights and allowing informed decision-making processes [52].

On the other hand, a 6G built upon DT technology will leverage network slicing along with other technologies such as data decoupling, interfacing, blockchain, proactive analytics, and optimisation, to enhance control and ensure efficient operations [48].

1.2.5 DT and intelligent networks (AI-based networking)

In AI-native 6G, accessing real-world data is essential for optimising online training and minimising latency, but obtaining a sufficiently diverse dataset that encompasses various network operating conditions and operational parameters can be challenging [32]. DTN will play a pivotal role by offering faster-than-real-time simulation capabilities for real-time AI-based network optimisation and control [32]. For instance, DTN are able to make sandboxed decisions (i.e., decisions made in a confined area under strict supervision and safety), by comparing real operational system data and synthetic data (i.e., annotated information generated by simulations within DTN), highlighting the probable deviations between the two models, and providing a valuable dataset to address limitations in data diversity [32,53]. In wireless communication scenarios, a cohesive system is established in which AI-powered channel estimation and signal detection leverage machine learning capabilities for adjustment in response to changing conditions [54], while DT-based sandboxed decisions may ensure that AI-based decisions remain within predefined constraints to maintain system reliability and stability.

On the other hand, AI models can be employed to conduct continuous predictive "what-if" simulations in the DTN, enabling the pre-assessment of potential outcomes of hypothetical actions, thus facilitating impact analysis of changes, outage prevention, and the identification of optimal configurations for future operations [32].

1.3 DT and Internet of Things (IoT)

The integration of IoT within the 5G is revolutionising various aspects of human life, such as smart healthcare, education, industry, homes, cities, and autonomous transportation, which are characterised by significant data, computational demands, and latency sensitivity [55]. In traditional methods, IoT's objective is to accumulate data from the physical world, transfer it to remote computing systems for analysis, and subsequently use feedback to make decisions [56]. In this manner, data is stored in backlogs of IoT devices for later evaluation and improvements to be made afterwards, which not only lacks immediate feedback but also could lead to significant malfunctions [56]. DT is a promising paradigm for IoT services due to its ability to create high-fidelity replicas of real-world objects, enabling prediction, monitoring, control, and decision-making [56]. For instance, DT provides a comprehensive perspective that integrates both the time-based (covering the entire lifecycle) and data-focused (including real-time and past data) facets of an asset and makes that a valuable asset to enhance their operational efficiency [57]. In addition, DT serves as an abstraction layer that facilitates continuous communication between IoT applications and devices [57]. For instance, DT enables complex modelling by massive data transport [56].

On the other hand, IoT sensors (e.g., cameras, radars, LiDARs, and location) complement one another, allowing for real-time recording of high-fidelity data [58], which is a key enabler for DT [44]. This might enhance DT decision-making or be valuable for future DT functions. For instance, DT integrate domain knowledge and real-time IoT data collection provide cost-efficient remote testing, making them widely adopted in various applications [56].

1.4 Low latency DT

DT and DTN require a reliable and resilient infrastructure for real-time interactions between the real and virtual worlds. Minimal latency and ultra-high transmission reliability are crucial for interactions between physical and virtual objects which is a key enabler for dynamic high-fidelity modelling in DT [59]. Recently, multi-tier computing complements cloud and edge computing by offloading and distributing computational (i.e., communication and caching) tasks and multi-dimensional resources across the cloud-to-things continuum. Multi-tier computing efficiently divides the processing of DT service components to delegate delay-sensitive elements to the network edge for low-latency handling, while delay–tolerant and computation-intensive parts are directed to the cloud to leverage its powerful computational resources [59]. This technique guarantees that each DT service component is processed where it works best, balancing low-latency requirements with the need for powerful computing capabilities [59].

1.4.1 Low latency communications

The 3GPP Release 16 [60] specifications intend to increase the reliability and reduce latency of URLLC services by offering 0.5–1 ms latency and an error rate of 10^{-6}. These guidelines enable new use cases such as smart factories and immersive AR/VR in applications such as DT and metaverse [59]. In the realm of 6G URLLC and DT-assisted edge intelligence, addressing the optimisation of both communication and computation resources involves considering a broad array of factors. On the communication side, this encompasses elements like the allocation of bandwidth, the management of power, and the design of beamforming strategies [59]. From a computational standpoint, recent studies have delved into issues such as making decisions about task offloading, selecting the appropriate edge nodes, determining the ideal processing rates for UEs and edge servers (ESs), accounting for queuing-related latency, and adapting to user mobility [59]. However, solving these optimisation problems, which encompass diverse communication and computation variables, is an intricate task due to their highly complex and non-linear nature [59]. Machine learning-based optimisation approaches that are fast and efficient can be an effective tool for dealing with reasonable computational time for real-time operation and low-latency needs [59]. To mirror and optimise physical things, the digital representation has to capture real-time data from the physical object and utilise machine learning for dynamic analysis, estimate, and prediction [41].

1.5 DT deployment challenges

As the adaptation of DTs accelerates given their numerous applications in industries such as construction, manufacturing, urban planning, and healthcare. A new set of challenges is accompanied by DTs creation and maintenance, including technical, financial, data ownership, access rights, privacy, and security. These challenges are mainly due to differences in ownership of the physical entity and the DT platform [5]. For instance, a fitness tracker is owned by an individual, whereas the generated data are stored in cloud servers owned by the application provider. The individual can solely access the data through a dedicated application interface, without the option for data extraction. Nonetheless, the individual can disconnect the fitness tracker at any desired time. Given that DT requires a complex data structure with continuous information to be collected from the physical objects, several challenges need to be addressed for a successful data deployment:

- **Technical**: The backbone of a DTN is the closed-loop communication between the digital and physical entities. Which requires a continuous stream of data to be communicated among them. The DT relies on the data for simulation analysis and decision-making. But DTs lack support/counter-measures for sudden disturbances and abnormalities in the collected data [61].
- **Financial**: The DT financial aspect encompasses not only the initial deployment costs but also maintenance, human resources, data storage, acquisition, and processing. Nevertheless, the potential outcomes can outweigh these expenses.
- **Data ownership**: Is defined as the ability to create, access, modify, and gain profit from selling data [62]. One of the foundation layers of a DT is the creation of a large database which includes collecting data from public and private providers. As a result, the matter of data ownership becomes an issue, given that data can be owned by multiple entities [63].
- **Access rights**: By default, the owners of DT have access to it and they might grant access to research, collaboration, or even commercial purposes. Therefore, licensing agreements should clearly define access and usage rights or it can lead to legal disputes.
- **Privacy and security**: The DT requires a continuous communication with the physical entity. Therefore, the DT and the required storage space should be kept in a secure network and encryption. In addition, designing and implementing a DT needs to comply with regional and international regulations for users'

data privacy and security. In Europe, general data protection rules (GDPR) has defined strict rules [57] to secure the communication of DTs which must be taken into consideration during the design phase of DT.

1.6 Conclusion

This chapter focused on the role of DT in the context of resilient and reliable 6G networks. DT offers a versatile testing environment for various 6G technologies, including THz, VLC, massive MIMO, cell-free massive MIMO, and satellite–terrestrial communications. DT can significantly reduce handover, transmission latency, enhance data delivery quality, and potentially enhance the QoS provided. Furthermore, DT supports cross-layer network resource optimisation, emphasising aspects like spectrum and power allocation, data routing, and UAV trajectory design, leading to efficient resource management and effective routing. DT also manages complex tasks like computational overhead, optimisation, and reconfiguration of IRSs properties, providing the adaptability required for intelligent networks. DT plays a pivotal role in identifying potential bottlenecks and predicting events that could impact new radio access network (RAN) technologies like C-RAN and O-RAN. It optimises resource allocation and can assess the impact of changes to the network setup, such as network slicing. Additionally, DT offers real-time simulation capabilities for AI-based network optimisation. DT is a promising paradigm for IoT services, creating high-fidelity replicas of real-world objects, enabling prediction, monitoring, control, and decision-making. In the context of 6G URLLC and DT-assisted edge intelligence, optimisation involves considering various communication and computation factors. This includes bandwidth allocation, power management, and beamforming strategies on the communication side, as well as decisions about task offloading, edge node selection, processing rates for UEs and ESs, queuing-related latency, and adapting to user mobility on the computational side. These optimisation challenges are intricate due to their complex and non-linear nature. ML-based optimisation approaches offer a fast and efficient solution, addressing real-time operation and low-latency requirements while dealing with the complexity of the problem. Nevertheless, the deployment and maintenance of DTN is hindered by several challenges including technical, financial, data ownership, access rights, privacy, and security which needs to be addressed to enable a successful deployment.

References

[1] Z. Zhang, Y. Xiao, Z. Ma, *et al.*, "6G wireless networks: Vision, requirements, architecture, and key technologies," *IEEE Vehicular Technology Magazine*, vol. 14, no. 3, pp. 28–41, 2019.

[2] J. Zhao, J. Liu, L. Yang, B. Ai, and S. Ni, "Future 5G-oriented system for urban rail transit: Opportunities and challenges," *China Communications*, vol. 18, no. 2, pp. 1–12, 2021.

[3] C.-X. Wang, X. You, *et al.*, "On the road to 6G: Visions, requirements, key technologies, and testbeds," *IEEE Communications Surveys & Tutorials*, vol. 25, no. 2, pp. 905–974, 2023.

[4] W. Saad, M. Bennis, and M. Chen, "A vision of 6G wireless systems: Applications, trends, technologies, and open research problems," *IEEE Network*, vol. 34, no. 3, pp. 134–142, 2020.

[5] H. Ahmadi, A. Nag, Z. Khar, K. Sayrafian, and S. Rahardja, "Networked twins and twins of networks: An overview on the relationship between digital twins and 6G," *IEEE Communications Standards Magazine*, vol. 5, no. 4, pp. 154–160, 2021.

[6] G. Liu, Y. Huang, N. Li, J. Dong, J. Jin, Q. Wang, and N. Li, "Vision, requirements and network architecture of 6G mobile network beyond 2030," *China Communications*, vol. 17, no. 9, pp. 92–104, 2020.

[7] M. Chafii, L. Bariah, S. Muhaidat, and M. Debbah, "Twelve scientific challenges for 6G: Rethinking the foundations of communications theory," *IEEE Communications Surveys & Tutorials*, vol. 25, no. 2, pp. 868–904, 2023.

[8] C. Chaccour, M. N. Soorki, W. Saad, M. Bennis, P. Popovski, and M. Debbah, "Seven defining features of terahertz (THz) wireless systems: A fellowship of communication and sensing," *IEEE Communications Surveys & Tutorials*, vol. 24, no. 2, pp. 967–993, 2022.

[9] K. K. Vaigandla, "Communication technologies and challenges on 6G networks for the Internet: Internet of things (IoT) based analysis," in *2022 2nd International Conference on Innovative Practices in Technology and Management (ICIPTM)*, vol. 2, 2022, pp. 27–31.

[10] Z. Wei, W. Yuan, S. Li, J. Yuan, G. Bharatula, R. Hadani, and L. Hanzo, "Orthogonal time-frequency space modulation: A promising next-generation waveform," *IEEE Wireless Communications*, vol. 28, no. 4, pp. 136–144, 2021.

[11] E. Basar, M. Wen, R. Mesleh, M. Di Renzo, Y. Xiao, and H. Haas, "Index modulation techniques for next-generation wireless networks," *IEEE Access*, vol. 5, pp. 16693–16746, 2017.

[12] H. Tataria, M. Shafi, A. F. Molisch, M. Dohler, H. Sjöland, and F. Tufvesson, "6G wireless systems: Vision, requirements, challenges, insights, and opportunities," *Proceedings of the IEEE*, vol. 109, no. 7, pp. 1166–1199, 2021.

[13] Y. Liu, S. Zhang, X. Mu, Z. Ding, R. Schober, N. Al-Dhahir, E. Hossain, and X. Shen, "Evolution of NOMA toward next generation multiple access (NGMA) for 6G," *IEEE Journal on Selected Areas in Communications*, vol. 40, no. 4, pp. 1037–1071, 2022.

[14] W. K. Alsaedi, H. Ahmadi, Z. Khan, and D. Grace, "Spectrum options and allocations for 6G: A regulatory and standardization review," *IEEE Open Journal of the Communications Society*, vol. 4, pp. 1787–1812, 2023.

[15] J. Huang, C.-X. Wang, H. Chang, J. Sun, and X. Gao, "Multi-frequency multi-scenario millimeter wave MIMO channel measurements and modeling for

B5G wireless communication systems," *IEEE Journal on Selected Areas in Communications*, vol. 38, no. 9, pp. 2010–2025, 2020.

[16] M. Banafaa, I. Shayea, J. bin Din, M. H. Azmi, A. Alashbi, Y. I. Daradkeh, and A. Alhammadi, "6G mobile communication technology: Requirements, targets, applications, challenges, advantages, and opportunities," *Alexandria Engineering Journal*, vol. 64, pp. 245–274, 2022.

[17] S. Sharif, S. Zeadally, and W. Ejaz, "Space-aerial-ground-sea integrated networks: Resource optimization and challenges in 6G," *Journal of Network and Computer Applications*, vol. 215, p. 103647, 2023.

[18] J. Hoydis, F. A. Aoudia, A. Valcarce, and H. Viswanathan, "Toward a 6G AI-Native air interface," *IEEE Communications Magazine*, vol. 59, no. 5, pp. 76–81, 2021.

[19] Y. Zuo, J. Guo, N. Gao, Y. Zhu, S. Jin, and X. Li, "A survey of blockchain and artificial intelligence for 6G wireless communications," *IEEE Communications Surveys & Tutorials*, vol. 30, pp. 1–1, 2023.

[20] N. Afraz, F. Wilhelmi, H. Ahmadi, and M. Ruffini, "Blockchain and smart contracts for telecommunications: Requirements vs. cost analysis," *IEEE Access*, vol. 11, pp. 95653–95666, 2023.

[21] M. Giordani, M. Polese, M. Mezzavilla, S. Rangan, and M. Zorzi, "Toward 6G networks: Use cases and technologies," *IEEE Communications Magazine*, vol. 58, no. 3, pp. 55–61, 2020.

[22] R. Liu, M. Hua, K. Guan, *et al.*, "6G enabled advanced transportation systems," ArXiv. https://doi.org/10.1109/TITS.2024.3362515 2023.

[23] X. Xu, B. Li, and Y. Wang, "Exploration of the principle of 6G communication technology and its development prospect," in *2022 International Conference on Electronics and Devices, Computational Science (ICEDCS)*, 2022, pp. 100–103.

[24] T. Bilen, B. Canberk, and T. Q. Duong, "Digital twin evolution for hard-to-follow aeronautical Ad-Hoc networks in beyond 5G," *IEEE Communications Standards Magazine*, vol. 7, no. 1, pp. 4–12, 2023.

[25] Y. Wu, K. Zhang, and Y. Zhang, "Digital twin networks: A survey," *IEEE Internet of Things Journal*, vol. 8, no. 18, pp. 13789–13804, 2021.

[26] W. Jiang, B. Han, M. A. Habibi, and H. D. Schotten, "The road towards 6G: A comprehensive survey," *IEEE Open Journal of the Communications Society*, vol. 2, pp. 334–366, 2021.

[27] Global issues population (2022a) un.org, https://www.un.org/en/global-issues/population#:ïtext=the%20world%20population%20is%20projected,and%2010.4%20billion%20by%202100. Accessed: 10 October 2023.

[28] A. Masaracchia, V. Sharma, B. Canberk, O. A. Dobre, and T. Q. Duong, "Digital twin for 6G: Taxonomy, research challenges, and the road ahead," *IEEE Open Journal of the Communications Society*, vol. 3, pp. 2137–2150, 2022.

[29] L. U. Khan, W. Saad, D. Niyato, Z. Han, and C. S. Hong, "Digital-twin-enabled 6G: Vision, architectural trends, and future directions," *IEEE Communications Magazine*, vol. 60, no. 1, pp. 74–80, 2022.

[30] L. Bariah, H. Sari, and M. Debbah, "Digital twin-empowered communications: A new frontier of wireless networks," *ArXiv*, vol. abs/2307.00973, 2023. Available: https://api.semanticscholar.org/CorpusID:259316908

[31] N. P. Kuruvatti, M. A. Habibi, S. Partani, B. Han, A. Fellan, and H. D. Schotten, "Empowering 6G communication systems with digital twin technology: A comprehensive survey," *IEEE Access*, vol. 10, pp. 112158–112186, 2022.

[32] X. Lin, L. Kundu, C. Dick, E. Obiodu, T. Mostak, and M. Flaxman, "6G digital twin networks: From theory to practice," *IEEE Communications Magazine*, vol. 61, pp. 1–7, 2023.

[33] Future technology trends of terrestrial international mobile telecommunications systems towards 2030 and beyond (2022a) itu.int, https://www.itu.int/dms_pub/itu-r/opb/rep/r-rep-m.2516-2022-pdf-e.pdf. Accessed: 10 October 2023.

[34] L. Zhang, H. Wan, X. Zheng, M. Tian, Y. Liu, and L. Su, "Digital-twin prediction of metamorphic object transportation by multi-robots with THz communication framework," *IEEE Transactions on Intelligent Transportation Systems*, vol. 24, no. 7, pp. 7757–7765, 2023.

[35] Y. Huang, M. Zhang, Y. Zhu, L. Wang, Q. Wang, and C. Yang, "Channel modeling for indoor visible light communication enabled by digital twin," in *2022 Asia Communications and Photonics Conference (ACP)*, 2022, pp. 516–520.

[36] H. B. Eldeeb, S. Naser, L. Bariah, S. Muhaidat and M. Uysal, "Digital twin-assisted OWC: towards smart and autonomous 6G networks," in *IEEE Network*, doi:10.1109/MNET.2024.3374370.

[37] M. Schnierle and S. Röck, "Latency and sampling compensation in mixed-reality-in-the-loop simulations of production systems," *Production Engineering*, vol. 17, no. 3-4, pp. 341–353, 2023.

[38] L. Zhao, C. Wang, K. Zhao, D. Tarchi, S. Wan, and N. Kumar, "INTERLINK: A digital twin-assisted storage strategy for satellite-terrestrial networks," *IEEE Transactions on Aerospace and Electronic Systems*, vol. 58, no. 5, pp. 3746–3759, 2022.

[39] H. Gao, P. Kyosti, X. Zhang, and W. Fan, "Digital twin enabled 6G radio testing: Concepts, challenges and solutions," *IEEE Communications Magazine*, vol. 61, pp. 1–7, 2023.

[40] X. Xu, H. Wen, H. Song, and Y. Zhao, "A DT machine learning-based satellite orbit prediction for IoT applications," *IEEE Internet of Things Magazine*, vol. 6, no. 2, pp. 96–100, 2023.

[41] Q. Guo, F. Tang, T. K. Rodrigues, and N. Kato, "Five disruptive technologies in 6G to support digital twin networks," *IEEE Wireless Communications*, vol. 31, pp. 1–8, 2023.

[42] W. Sun, S. Lian, H. Zhang, and Y. Zhang, "Lightweight digital twin and federated learning with distributed incentive in air-ground 6G networks," *IEEE Transactions on Network Science and Engineering*, vol. 10, no. 3, pp. 1214–1227, 2023.

[43] G. Shen, L. Lei, Z. Li, et al., "Deep reinforcement learning for flocking motion of multi-UAV systems: Learn from a digital twin," *IEEE Internet of Things Journal*, vol. 9, no. 13, pp. 11141–11153, 2022.

[44] A. Alkhateeb, S. Jiang, and G. Charan, "Real-time digital twins: Vision and research directions for 6G and beyond," *ArXiv*, vol. abs/2301.11283, 2023. Available: https://api.semanticscholar.org/CorpusID:256274643

[45] W. He, C. Zhang, J. Deng, Q. Zheng, Y. Huang, and X. You, "Conditional generative adversarial network aided digital twin network modeling for Massive MIMO optimization," in *2023 IEEE Wireless Communications and Networking Conference (WCNC)*, 2023, pp. 1–5.

[46] B. Han, M. A. Habibi, B. Richerzhagen, et al., "Digital twins for industry 4.0 in the 6G era," *arXiv preprint arXiv:2210.08970*, 2022.

[47] S. Kurma, K. Singh, M. Katwe, S. Mumtaz, and C.-P. Li, "RIS-empowered MEC for URLLC systems with digital-twin-driven architecture," in *IEEE INFOCOM 2023 – IEEE Conference on Computer Communications Workshops (INFOCOM WKSHPS)*, 2023, pp. 1–6.

[48] L. U. Khan, Z. Han, W. Saad, E. Hossain, M. Guizani, and C. S. Hong, "Digital twin of wireless systems: Overview, taxonomy, challenges, and opportunities," *IEEE Communications Surveys & Tutorials*, vol. 24, no. 4, pp. 2230–2254, 2022.

[49] A. Checko, H. L. Christiansen, Y. Yan, L. Scolari, G. Kardaras, M. S. Berger, and L. Dittmann, "Cloud RAN for mobile networks—a technology overview," *IEEE Communications Surveys & Tutorials*, vol. 17, no. 1, pp. 405–426, 2015.

[50] Y. Ren, S. Guo, B. Cao, and X. Qiu, "End-to-end network SLA quality assurance for C-RAN: A closed-loop management method based on digital twin network," *IEEE Transactions on Mobile Computing*, pp. 1–18, 2023.

[51] A. Masaracchia, V. Sharma, M. Fahim, O. A. Dobre, and T. Q. Duong, "Digital twin for Open RAN: Towards intelligent and resilient 6G radio access networks," *IEEE Communications Magazine*, vol. 61, pp. 1–6, 2023.

[52] M. Abdel-Basset, H. Hawash, K. M. Sallam, I. Elgendi, and K. Munasinghe, "Digital twin for optimization of slicing-enabled communication networks: A federated graph learning approach," *IEEE Communications Magazine*, vol. 61, pp. 1–7, 2023.

[53] N. Apostolakis, L. E. Chatzieleftheriou, D. Bega, M. Gramaglia, and A. Banchs, "Digital twins for next-generation mobile networks: Applications and solutions," *IEEE Communications Magazine*, vol. 61, pp. 1–7, 2023.

[54] U. Challita, H. Ryden, and H. Tullberg, "When machine learning meets wireless cellular networks: Deployment, challenges, and applications," *IEEE Communications Magazine*, vol. 58, no. 6, pp. 12–18, 2020.

[55] F. Guo, F. R. Yu, H. Zhang, X. Li, H. Ji, and V. C. M. Leung, "Enabling massive IoT toward 6G: A comprehensive survey," *IEEE Internet of Things Journal*, vol. 8, no. 15, pp. 11891–11915, 2021.

[56] C. Wang, Z. Cai, and Y. Li, "Sustainable Blockchain-based digital twin management architecture for IoT devices," *IEEE Internet of Things Journal*, vol. 10, no. 8, pp. 6535–6548, 2023.

[57] S. Muralidharan, B. Yoo, and H. Ko, "Designing a semantic digital twin model for IoT," in *2020 IEEE International Conference on Consumer Electronics (ICCE)*, 2020, pp. 1–2.

[58] M. Aloqaily, O. Bouachir, F. Karray, I. A. Ridhawi, and A. E. Saddik, "Integrating digital twin and advanced intelligent technologies to realize the metaverse," *IEEE Consumer Electronics Magazine*, vol. 12, pp. 1–8, 2022.

[59] T. Q. Duong, D. Van Huynh, S. R. Khosravirad, V. Sharma, O. A. Dobre, and H. Shin, "From digital twin to metaverse: The role of 6G ultra-reliable and low-latency communications with multi-tier computing," *IEEE Wireless Communications*, vol. 30, no. 3, pp. 140–146, 2023.

[60] Release 16 (2020a) 3gpp.org, https://www.3gpp.org/specifications-technologies/releases/release-16. Accessed: 10 October 2023.

[61] R. Zhang, F. Wang, J. Cai, Y. Wang, H. Guo, and J. Zheng, "Digital twin and its applications: A survey," *The International Journal of Advanced Manufacturing Technology*, vol. 123, no. 11–12, p. 4123–4136, 2022.

[62] S. M. Bazaz, M. Lohtander, and J. Varis, "5-dimensional definition for a manufacturing digital twin," *Procedia Manufacturing*, vol. 38, pp. 1705–1712, 2019, *29th International Conference on Flexible Automation and Intelligent Manufacturing (FAIM 2019)*, June 24–28, 2019, Limerick, Ireland, Beyond Industry 4.0: Industrial Advances, Engineering Education and Intelligent Manufacturing. Available: https://www.sciencedirect.com/science/article/pii/S2351978920301086

[63] S. M. Bazaz, M. Lohtander, and J. Varis, "Availability of manufacturing data resources in digital twin," *Procedia Manufacturing*, vol. 51, pp. 1125–1131, 2020, *30th International Conference on Flexible Automation and Intelligent Manufacturing (FAIM2021)*. Available: https://www.sciencedirect.com/science/article/pii/S2351978920320151

Chapter 2
Digital twin-enabled aerial edge networks with ultra-reliable low-latency communications

Dang Van Huynh[1], Yijiu Li[1], Tan Do-Duy[2], Emi Garcia-Palacios[1] and Trung Q. Duong[1]

This chapter studies unmanned aerial vehicles (UAVs) edge networks assisted by digital twin (DT) with ultra-reliable low-latency communications (URLLC) to enable delay-sensitive and mission-critical applications. In particular, we formulate a min–max fairness latency minimisation and jointly optimise various communication and computation variables such as transmission power, offloading portions, and the processing rate of users and edge servers. To deal with the formulated problem, we propose an efficient alternating optimisation (AO) algorithm. The effectiveness of the proposed solution is clearly validated by extensive simulation results.[*]

2.1 Introduction

2.1.1 Literature review

Digital twin (DT) is an emerging technology that is able to create virtual twins of physical objects in order to facilitate the processing of control and manage cyber-physical systems. DT can be exploited in networking, and communications for many aspects such as system modelling, physical data processing, cloud computing, and edge computing [2]. Therefore, studies of DT are attracting much attention from active researchers [3–6]. More specifically, in [3], a DT-assisted task offloading in mobile edge computing (MEC) was investigated to address the problem of minimising power and time overhead. Another work in reducing offloading latency for DT edge network was introduced in [4]. In [5], DT was proposed for intelligent authorisation in the beyond 5G smart grid applications. DT was exploited in [6] to empower edge networks for the industrial Internet of Things (IoT) environment. These representative studies demonstrate the huge potential of DT in various domains, especially in networked systems.

---
[1]School of Electronics, Electrical Engineering and Computer Science, Queen's University Belfast, UK
[2]Department of Computer and Communications Engineering, HCMC University of Technology and Education, Vietnam
*This chapter has been published partly in [1].

In recent years, unmanned aerial vehicles (UAVs) have been under the spotlight due to their flexible configuration and mobile characteristics [7,8]. Numerous research has been conducted to enhance the control performance of UAVs. In [9], authors carried out the ground test of UAV, evaluating the performance of its entire flight control system through rotor speed, roll attitude, etc. The flight test [10] and collision avoidance [11] have also been studied to give a stronger control over this smart vehicle. UAVs achieve even better performance in many research areas by combining with other advanced technologies. Integrate with the intelligent reflecting surface (IRS) [12]; a well-performed UAV-assisted IRS symbiotic radio system has been formed. The system performs better as data information is transferred via UAV by optimising the UAV trajectory and the IRS phase shifts. Applying a deep reinforcement learning (RL) algorithm [13], the decision-making of UAVs can be autonomous rather than pre-planned. UAV's low consumption of energy also attracts public attention. Through optimisation algorithms, resource allocation can be optimised, and the total energy consumption in a multi-UAV network framework can be minimised [14,15]. Based on their outstanding advantages, UAVs are now being developed and used in military and civil applications [16]. These applications are in many fields such as environment monitoring, traffic control, public safety, damaged buildings detection and industrial automation [17,18]. Particularly, with the help of UAV, the authors of [19] develop a method for high-precision geolocation of distant targets, which is more effective than the conventional one-shot localisation way. Search and rescue operations with UAVs participation increase the speed of rescue and thus improve the survival rate of people [20,21]. In [22], UAVs are used as flying base stations to ensure the connectivity of communication networks in unexpected disasters. These aerial vehicles also play a role in smart cities [18].

With the rapid development of the 5G network, ultra-reliable and low-latency communication (URLLC) emerges as a promising paradigm to ensure a certain quality of service (QoS). With strict requirements of extremely low latency (from 1 ms to few milliseconds) and ultra-high reliability (over 99.999%) [23], URLLC plays an indispensable role in remote healthcare, autonomous driving, immersive virtual reality, cloud robotics, deterministic communication, and many other areas [24]. This novel communication service uses short packet transport, which allows optimising the transmission of control information [25].

Evolved from cloud computing, MEC has been widely considered as a key application in 5G communication. This promising technology extends the capabilities of cloud computing at the network edge [26] and performs excellently in smart manufacturing, industrial IoT, as well as many other areas [27,28]. In recent years, many efforts have been put into MEC. In paper [29], the author demonstrates a well-established MEC architecture and integrates an application deployment use-case, establishing a proof of concept which is very similar to the actual deployment of MEC system in a 5G environment. Combining with the optimising method, MEC is an appropriate solution to improve the quality of service. In [30], the author proposes an RL-based optimisation framework to minimise the cost of delay and energy consumptions for user equipment in a time-variant dynamic MEC system. The considered MEC system outperforms other baseline solutions according to the demonstration. Whereas the authors in [31] adopt an offloading strategy in MEC that considers delay and energy consumptions cost optimisation. Two schemes named optimised OMA and hybrid NOMA are proposed in [32] for solving the problem of joint power and time allocation for MEC offloading. Through these extensive studies of MEC, this evolving technology has been used in various use cases, e.g., an audience meter [33]. In this particular use case, MEC modules are used to improve the algorithm's performance, detecting the estimated number of participants in an event over the entire time period.

MEC and URLLC techniques are closely related to each other. Under the sufficiently powerful computing of MEC, applications can be processed in real-time. By reducing the transmission processing time and reception processing time, MEC can reduce latency in 5G systems significantly. Since finite blocklength (FBL) is adopted to satisfy the latency constraints of URLLC, the paper [34] proposes a MEC network where one MEC server is used to minimise the error probability between users under FBL and energy consumption constraints. This FBL scheme is also considered in a MEC-enabled vehicular network to support URLLC [35]. The placement of MEC server that promises URLLC requirements has also been studied. An algorithm called LowMEP has been proposed to find a minimum number of MEC servers satisfying the quality of service [36]. The combination of these two promising techniques is frequently used in various fields. For industrial applications, the paper [37] proposes a 5G MEC gateway system capable of supporting URLLC to enable communication in factories. For autonomous vehicles application, vehicle's latency and energy cost functions have been established to investigate URLLC resource scheduling for edge computing [38].

Additionally, UAV-assisted communications can be effectively combined with MEC to enable time-sensitive, and computation-intensive services for a wide range of IoT applications [39]. UAV-based edge networks not only be able to minimise the energy consumption of IoT devices with optimal offloading decisions [40] but also reduce the latency by providing edge caching solutions [41]. More importantly, ultra-reliable and low-latency communications (URLLC) recently emerged as a promising technology for mission-critical applications [42]. Combining UAV-enabled MEC with URLLC opens many opportunities, as well as challenges for the next generation of IoT applications [43].

2.1.2 Motivations and main contributions

Recently, combined merging technologies including MEC, DT, UAV, and URLLC are attracting many active research groups [3,4,39]. In particular, the DT-assisted task offloading based on edge collaboration has been investigated in [3] with a DRL-based approach. The edge selection and offloading optimisation have been addressed in this chapter; however, the communication resource optimisation has not taken into consideration. Similarly, in papers [4,39], the joint computation and communication resources have not been fully addressed to obtain the optimal latency of task offloading. Hence, the integration of DT and MEC with URLLC in industrial scenarios presents a promising avenue for research and could significantly contribute to this field of study.

Moving beyond above background, this chapter addresses the problem of combining these important technologies, namely DT, MEC, and URLLC, in UAV-assisted IoT systems. Both communication and computation factors, including transmit power, the processing rate of IoT devices or user equipment (UE), edge servers, and offloading policies, are carefully taken into consideration to reduce the end-to-end (e2e) latency. Main contributions of this chapter are summarised as follows:

- We first formulate the latency minimisation problem of UAV-based edge network with URLLC in the DT regime. The addressed problem fully considers both communication and computation factors in reducing the task offloading latency.
- In order to solve the problem, we propose the AO-IA algorithm with two sub-problems, namely, transmit power and computation resources optimisation and offloading portions optimisation.
- Finally, intensive simulations have been conducted to demonstrate the effectiveness of the proposed solution.

The rest of the chapter is structured as follows: Section 2.2 presents the system model and problem formulation for attaining minimum e2e latency. Section 2.3 resolves the problem formed in the previous section using an algorithmic solution. Numerical results are discussed in Section 2.4. Finally, Section 2.5 concludes the chapter.

2.2 System model and problem formulation

This section leads towards the problem to be solved for minimising the e2e latency by expressing the basic network model, transmission model, DT-empowered offloading, associated energy and power consumption model, and the UAV deployment.

2.2.1 DT-empowering URLLC-based edge networks model

Figure 2.1 presents a DT-enabled UAV-based edge network architecture with URLLC. The physical layer consists of IoT devices (IoT), a.k.a. UEs, and UAVs. These physical devices connect via URLLC links to ensure stringent reliability and low-latency communications in mission-critical applications.

Let $\mathcal{M} = \{1, 2, ..., M\}$ be the set of M IoT devices and $\mathcal{K} = \{1, 2, ..., K\}$ be the set of K UAVs. There are K UAV-IoT groups, in which the kth UAV serves M_k IoT devices in each group. Each UAV can act as an access point (AP) with the capability to perform as an edge server (ES). We assume that the UAV deployment and network planning are performed in advance.

2.2.2 Transmission model
2.2.2.1 Channel model
The air-to-ground (ATG) channels between UAVs and UEs are also dominant by light-of-sight (LoS) propagation but these are more complex due to the effects of

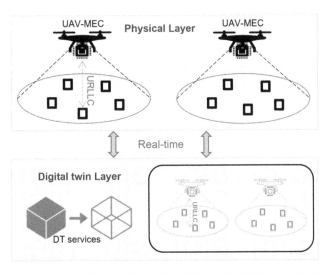

Figure 2.1 An exemplary illustration of the DT-enabled UAV-based edge networks with URLLC. ©IET 2022. Reprinted with permission from [1].

propagation attenuation by blockage geometry and shadowing [44]. As such, the path loss of the link between the kth UAV and the (m, k)th UE can be written as

$$g_{mk} = PL_{mk} + \eta^{LoS} P^{LoS}_{m,k} + \eta^{NLoS} P^{NLoS}_{mk}, \qquad (2.1)$$

where η^{LoS} and η^{NLoS} are the average additional losses for LoS and non-LoS (NLoS), respectively. The path loss with respect to the distance (PL_{mk}) is given by

$$PL_{mk} = 10 \log \left(\frac{4\pi f_c r_{mk}}{c} \right)^{\beta}, \qquad (2.2)$$

where f_c is the carrier frequency (Hz), c is the speed of light (m/s), $\beta \geq 2$ is the path loss exponent, $r_{mk} = \sqrt{d_{mk}^2 + Z_k^2}$; d_{mk} is the Euclidean distance between the mth UE and the kth UAV, and Z_k is the antenna height of the kth UAV. The probability of LoS and NLoS can be shown as

$$P^{LoS}_{mk} = \frac{1}{1 + a \exp\left[-b\left(\arctan\left(\frac{Z_k}{d_{mk}}\right) - a\right)\right]} \qquad (2.3)$$

$$P^{NLoS}_{mk} = 1 - P^{LoS}_{mk}, \qquad (2.4)$$

where the constants a and b depend on the specific arrangement of the environment.

Each UAV is equipped with L antennas to serve M_k single-antenna IoT devices. Let $\mathbf{h}_{mk} \in \mathbb{C}^{L \times 1}$ be the channel vector between the kth UAV and the m-th IoT, can be modelled as $\mathbf{h}_{mk} = \sqrt{g_{mk}} \bar{\mathbf{h}}_{mk}$. Here, g_{mk} denotes the large-scale channel coefficient defined in (2.1), and $\bar{\mathbf{h}}_{mk}$ is the small-scale fading following the distribution of $\mathcal{CN}(0, \mathbf{I})$. Let $\mathbf{H}_k \in \mathbb{C}^{L \times M_k}$ be the channel matrix from M_k devices to the kth UAV, with $\mathbf{H}_k = [\mathbf{h}_{k1}, \mathbf{h}_{k2}, .., \mathbf{h}_{kM_k}]$. Under the shared wireless medium, the $L \times 1$ received signal vector at the kth UAV is given by $\mathbf{y}_k = \sum_{m=1}^{M_k} \mathbf{h}_{mk} \sqrt{p_{mk}} s_{mk} + \mathbf{n}_k$, where p_{mk} is the payload power of the (m, k)th device, s_{mk} is the zero mean and unit variance Gaussian information message from the (m, k)th IoT, and $\mathbf{n}_k \sim \mathscr{CN}(\mathbf{0}, N_0 \mathbf{I}_L)$ is the additive white Gaussian noise (AWGN) during the data transmission with N_0 being the noise power.

In this chapter, we consider the uplink transmission from IoTs to UAVs to perform task offloading. We apply the maximum-ratio combining (MRC) at the UAV to improve the performance gain. Moreover, to guarantee fairness among all IoT devices, and further improve wireless transmission performance, we additionally adopt the matched filter and successive interference cancellation (MF-SIC) technique at the UAVs. In particular, by using MF-SIC, we assume that the decoding order follows IoTs' index by arranging the channel vector as $\|h_{1k}\|^2 \geq \|h_{2k}\|^2 ... \geq \|h_{M_k k}\|^2, \forall k$. Consequently, the signal-to-interference-plus-noise (SINR) at the kth UAV of the signal from the (m, k)th IoT device can be expressed as

$$\gamma_{mk}(\mathbf{p}) = \frac{p_{mk} \|\mathbf{h}_{mk}\|^2}{\mathscr{I}_{mk}(\mathbf{p}) + N_0}, \tag{2.5}$$

where $\mathscr{I}_{mk}(\mathbf{p}) = \sum_{n>m}^{M} p_{nk} \frac{|h_{mk}^H h_{nk}|^2}{\|h_{mk}\|^2}$ is the interference power caused by IoT devices $n > m$, and $\mathbf{p} = \{p_m\}_{\forall m}$.

2.2.2.2 URLLC-based uplink transmission rate

The approximation of achievable transmission rate (bit/s) in URLLC finite blocklength is [45,46]:

$$R_{mk}(\mathbf{p}) \approx B \log_2 [1 + \gamma_{mk}(\mathbf{p})] - B \sqrt{\frac{V_{mk}(\mathbf{p})}{N}} \frac{Q^{-1}(\varepsilon)}{\ln 2}, \tag{2.6}$$

where $\omega_k = M_k/N, \forall k$, N is the blocklength, which can be written as $N = \delta B$, with B as the bandwidth and δ as the transmission time interval; ε is the decoding error probability, $\gamma_{mk}(\mathbf{p})$ denotes the SINR, $Q^{-1}(\cdot)$ is the inverse function $Q(x) = \frac{1}{\sqrt{2\pi}} \int_x^\infty \exp\left(\frac{-t^2}{2}\right) dt$, and V is the channel dispersion given by $V_{mk}(\mathbf{p}) = 1 - [1 + \gamma_{mk}(\mathbf{p})]^{-2}$. When the blocklength N approaches to infinity, the data rate R_{mk} approaches $B \log_2(1 + \gamma_{mk}(\mathbf{p}))$, which is the classic Shannon's equation.

2.2.3 DT empowered task offloading model

A particular task from the (m, k)th IoT device is represented by a tuple $J_{mk} = \{D_{mk}, C_{mk}, T_{mk}\}$, where D_{mk} is the data size (bits), C_{mk} is the required computation resource (cycles), and T_{mk} (s) is the minimum required latency for task J_{mk}.

Let $\alpha = \{\alpha_{mk}\}_{\forall m,k}$ be the amount of the task processed locally, which satisfies $0 \leq \alpha_m \leq 1$.

2.2.3.1 Local processing

For the (m, k)th IoT device, its DT (DT_{mk}) can be expressed as

$$\text{DT}_{mk} = (f^{\text{loc}}_{mk}, \hat{f}^{\text{loc}}_{mk}), \tag{2.7}$$

where f^{loc}_{mk} is the estimated processing rate of the physical IoT device, and $\hat{f}^{\text{loc}}_{mk}$ is the deviation between the real device and its DT.

The DT layer has the estimated processing rate f^{loc}_{mk} to replicate the behaviours of IoT devices and trigger decisions on optimising physical devices configuration.

The (m, k)th IoT executes α_{mk} portion of task J_{mk} with the estimated processing rate f^{loc}_{mk}, and the estimated time required to execute the task locally is given by

$$\tilde{T}^{\text{loc}}_{mk}\left(\alpha_{mk}, f^{\text{loc}}_{mk}\right) = \frac{\alpha_{mk} C_{mk}}{f^{\text{loc}}_{mk}}. \tag{2.8}$$

Assuming that the deviation between the physical IoT ($\mathcal{M}$) and $\tilde{\mathcal{M}}$ in DT can be acquired in advance, the computing latency gap between real value and DT estimation is computed as

$$\Delta T^{\text{loc}}_{mk}\left(\alpha_{mk}, f^{\text{loc}}_{mk}\right) = \frac{\alpha_{mk} C_{mk} \hat{f}^{\text{loc}}_{mk}}{f^{\text{loc}}_{mk}\left(f^{\text{loc}}_{mk} - \hat{f}^{\text{loc}}_{mk}\right)} \tag{2.9}$$

The actual time for local computing is expressed as

$$T^{\text{loc}}_{mk} = \Delta T^{\text{loc}}_{mk} + \tilde{T}^{\text{loc}}_{mk}. \tag{2.10}$$

2.2.3.2 Edge processing

Given the estimated processing rate of the kth ES for executing the offloaded task from the (m, k)th IoT device is f^{es}_{mk}, the estimated latency of the kth ES to execute task J_m is given by

$$\tilde{T}^{\text{es}}_{mk}\left(\alpha_{mk}, f^{\text{es}}_{k}\right) = \frac{(1 - \alpha_{mk}) C_{mk}}{f^{\text{es}}_{mk}}. \tag{2.11}$$

Then, the latency gap $\Delta T^{\text{es}}_{mk}$ between the real value and DT estimation can be expressed as

$$\Delta T^{\text{es}}_{mk}\left(\alpha_{mk}, f^{\text{es}}_{k}\right) = \frac{(1 - \alpha_{mk}) C_m \hat{f}^{\text{es}}_{mk}}{f^{\text{es}}_{mk}\left(f^{\text{es}}_{mk} - \hat{f}^{\text{es}}_{mk}\right)}. \tag{2.12}$$

As a result, the actual latency for executing at edge DT can be expressed as

$$T^{\text{es}}_{mk} = \Delta T^{\text{es}}_{mk} + \tilde{T}^{\text{es}}_{mk}. \tag{2.13}$$

The total DT latency in the system can be expressed as follows:

$$T^{\text{tot}}_{mk} = T^{\text{loc}}_{mk} + T^{\text{comm}}_{mk} + T^{\text{es}}_{mk} = \frac{\alpha_{mk} C_m}{f^{\text{loc}}_{mk} - \hat{f}^{\text{loc}}_{mk}} + \frac{(1 - \alpha_{mk}) D_{mk}}{R_{mk}(\mathbf{p})} + \frac{(1 - \alpha_{mk}) C_{mk}}{f^{\text{es}}_{mk} - \hat{f}^{\text{es}}_{mk}}. \tag{2.14}$$

The latency comprises three main components, namely, local processing latency (T^{loc}_{mk}), uplink transmission latency (T^{comm}_{mk}), and edge processing latency (T^{es}_{mk}). Since the response messages from UAVs to IoTs are typically small (e.g. control packets), the downlink transmission latency is negligible [47,48].

2.2.4 Energy and power consumption model

Total energy consumption of the (m,k)th IoT includes energy for transmission and computation:

$$E_{mk}^{tot}(\alpha_{mk}, \boldsymbol{\beta}, \mathbf{p}) = E_{mk}^{comp} + E_{mk}^{comm} = \frac{\theta_m}{2}\alpha_{mk}C_{mk}(f_{mk}^{loc} - \hat{f}_{mk}^{loc})^2 + \frac{(1-\alpha_{mk})p_{mk}D_{mk}}{R_{mk}(\mathbf{p})}, \quad (2.15)$$

where $\theta_m/2$ represents the average switched capacitance, and the average activity factor of the mth IoT [49].

The power consumption of the kth UAV for processing the uploaded tasks is modelled as follows [41]:

$$P_k(\alpha_{mk}, f_{mk}^{es}) = \sum_{m \in \mathcal{M}_k}(1-\alpha_{mk})(f_{mk}^{es})^3\theta_k, \quad (2.16)$$

where θ_k represents the average switched capacitance and the average activity factor of the kth UAV.

2.2.5 UAV deployment

In this section, we present the clustering algorithm for UAV deployment by considering an efficient QoS-constrained K-means clustering approach [50–52]. In particular, the constrained clustering method considers whether the mth UE can be grouped in the kth cluster based on two types of pairwise constraints, namely must-link constraints and cannot-link constraints [51], which represent the satisfied QoS constraints, and the violated QoS constraints, respectively.

The constrained K-means clustering algorithm can be briefly described as follows. At the initial stage, the locations of UAVs are randomly set within the deployment area as the centroid location. First, based on the Euclidean distance between the UEs and the UAVs, the mth UE is assigned to an appropriate cluster with the smallest distance. Then, if QoS constraints are not satisfied for any UEs, the altitude of the corresponding UAV must be adjusted. Finally, the centroid location for each cluster is updated. Such procedure is repeated until the cluster members are stable or the number of iterations exceeds a predefined threshold.

As an illustrative case, we set the number of UEs randomly located in a critical area at $M = 6$ and the number of UAVs at $K = 2$ with the path loss threshold corresponding to the QoS requirement $\gamma_{QoS} = 110$ dB. Figure 2.2 shows the clustering result after implementing the QoS-constrained K-means clustering algorithm.

2.2.6 Problem formulation

Here, the worst-case of the total DT latency is minimised by optimising offloading policies, transmit power, and estimated processing rates of IoT and ESs. By defining the following notations $\mathscr{D} \triangleq \{\alpha_{mk}, \forall m, k | 0 \leq \alpha_{mk} \leq 1, \forall m, k\}$, $\mathscr{P} \triangleq \{p_{mk}, \forall m, k | 0 \leq p_{mk} \leq P_{mk}^{max}, \forall m\}$, and $\mathscr{F} \triangleq \{f_{mk}^{loc}, f_{mk}^{es}, \forall m, k | 0 \leq f_{mk}^{loc} \leq F_{max}^{loc}, \forall m; 0 \leq$

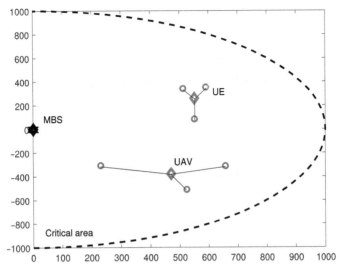

Figure 2.2 An illustrative system model with M = 6 UEs and K = 2 UAVs after the implementation of QoS-constrained K-means clustering algorithm. ©IET 2022. Reprinted with permission from [1].

$f_{mk}^{es} \leq F_{max}^{es}, \forall k\}$ as the set constraints of offloading decisions, uplink transmission power, processing rates, respectively, the problem is formulated as follows:

$$\min_{\alpha,\mathbf{p},\mathbf{f}} \max_{\forall m \in \mathcal{M}} \left\{ T_m^{tot}(\alpha_{mk}, \mathbf{f}, \mathbf{p}) \right\}, \tag{2.17a}$$

$$\text{s.t.} \quad T_{mk}^{tot}(\alpha_{mk}, \mathbf{f}, \mathbf{p}) \leq T_{mk}^{max}, \forall mk, \tag{2.17b}$$

$$R_{mk}(\mathbf{p}) \geq R_{min}, \forall m, k, \tag{2.17c}$$

$$E_{mk}^{tot}(\alpha_{mk}, \mathbf{p}) \leq E_{max}, \forall m, \tag{2.17d}$$

$$\sum_{m \in \mathcal{M}_k} (1 - \alpha_{mk}) f_{mk}^{es} \leq F_{max}^{es}, \forall k, \tag{2.17e}$$

$$P_k(\alpha_{mk}, f_k^{es}) \leq P_{max}^{es}, \forall k, \tag{2.17f}$$

$$\alpha, \beta \in \mathcal{D}, \mathbf{p} \in \mathcal{P}, \mathbf{f} \in \mathcal{F}, \tag{2.17g}$$

where constraint (2.17b) presents maximum latency constraint for every incoming task. Constraints (2.17c) and (2.17d) are the minimum transmission rate requirement for uplink transmission and the maximum energy consumption requirement of IoT, respectively. Finally, constraints (2.17e) and (2.17f) refer to maximum available computation resource and power budget of the UAVs.

2.3 Proposed solutions

As we can observe from (2.17), the objective function is non-concave and non-smooth while other constraints (2.17c), (2.17d), (2.17e), and (2.17f) are also highly complex non-convex constraints. This results in solving the problem directly is computationally challenging. Therefore, to solve the problem (2.17), we first replace the objective function with an upper bound function with introduced variables $\mathbf{t} \triangleq \{t_{lc}, t_{cm}, t_{ed}\}$ satisfying $\tau_m(t_{lc}, t_{cm}, t_{ed}) \triangleq t_{lc} + t_{cm} + t_{ed}$ to equivalently transform (2.17) to

$$\min_{\alpha,\mathbf{p},\mathbf{f},\mathbf{t}} \quad \max_{\forall m,k}\{\tau_{mk}(\mathbf{t})\}, \tag{2.18a}$$

$$\text{s.t.} \quad (2.17c) - (2.17g) \tag{2.18b}$$

$$\tau_m(\mathbf{t}) \leq T_m^{\max}, \forall m, \tag{2.18c}$$

$$t_{lc} \geq \frac{\alpha_{mk} C_{mk}}{f_{mk}^{loc} - \hat{f}_{mk}^{loc}}, \forall m, k \tag{2.18d}$$

$$t_{cm} \geq \frac{(1 - \alpha_{mk}) D_{mk}}{R_{mk}(\mathbf{p})}, \forall m, k \tag{2.18e}$$

$$t_{ed} \geq \frac{(1 - \alpha_{mk}) C_{mk}}{f_{mk}^{es} - \hat{f}_{mk}^{es}}, \forall m, k, \tag{2.18f}$$

Due to the complexity of the non-convex problem (2.18), we decompose (2.18) into two sub-problems and solve the problem in the fashion of alternating optimisation (AO) approach and inner approximation (AO-IA) framework [53,54]. The following subsections fully present the development of our proposed solution.

2.3.1 Transmit power and computation resource optimisation

In this subsection, we solve (2.18) for given $(\alpha^{(i)})$ to obtain the next optimal values of $(\mathbf{p}^{(i+1)}, \mathbf{f}^{(i+1)})$

$$\min_{\mathbf{p},\mathbf{f},\mathbf{t}|\alpha^{(i+1)}} \quad \max_{\forall m,k}\{\tau_{mk}(\mathbf{t})\}, \tag{2.19a}$$

$$\text{s.t.} \quad (2.17c) - (2.17g), (2.18c), (2.18d), (2.18e), (2.18f). \tag{2.19b}$$

As we can observe from the sub-problem (2.19), the constraints (2.17c), (2.17d), and (2.18e) are non-convex. We are now in the position to approximate these constraints.

Convexify of (2.17c): To address constraint (2.17c), we first rewrite that $\gamma_{mk}(\mathbf{p}) = \frac{p_{mk}}{q_{mk}(\mathbf{p})}$, where $q_{mk}(\mathbf{p})$ is defined as

$$q_{mk}(\mathbf{p}) \triangleq \frac{\mathscr{I}_{mk}(\mathbf{p}) + N_0}{\|\mathbf{h}_{mk}\|^2} \tag{2.20}$$

Following the appendix, we have

$$R_{mk}(\mathbf{p}) \geq R_{mk}^{(i)}(\mathbf{p}) \triangleq \frac{B}{\ln 2}\left[\mathscr{G}_{mk}^{(i)}(\mathbf{p}) - \kappa \mathscr{W}_{mk}^{(i)}(\mathbf{p})\right] \tag{2.21}$$

under the trusted regions

$$q_m(\mathbf{p}) + p_{mk} \leq 2(q_m(\mathbf{p}^{(i)}) + p_{mk}^{(i)}), \forall m, k, \tag{2.22}$$

$$\frac{q_m(\mathbf{p}) + p_{mk}}{q_m(\mathbf{p}^{(i)}) + p_{mk}^{(i)}} \leq 2\frac{q_m(\mathbf{p})}{q_m(\mathbf{p}^{(i)})}, \forall m, k, \tag{2.23}$$

where $\mathscr{G}_{mk}^{(i)}(\mathbf{p})$, and $\mathscr{W}_{mk}^{(i)}(\mathbf{p})$ are defined as in Appendix A, $\kappa = \frac{Q^{-1}(\varepsilon)}{\sqrt{N}}$. As a result, we innerly approximate constraint (2.17e) as

$$R_{mk}^{(i)}(\mathbf{p}) \geq R_{\min}, \forall m,k. \tag{2.24}$$

Convexify of (2.17d): By introducing variables $\mathbf{r} \triangleq \{r_{mk}\}_{\forall m,k}$ that satisfy $r_{mk} \geq 1/R_{mk}$, $\forall m,k$. We can equivalently express (2.17d) as follows:

$$\frac{\theta}{2}\alpha_{mk}^{(i+1)}C_{mk}\left(f_{mk}^{\mathrm{loc}} - \hat{f}_{mk}^{\mathrm{loc}}\right)^2 + (1-\alpha_{mk}^{(i+1)})D_{mk}p_{mk}r_{mk} \leq E_m^{\max}, \forall m,k, \tag{2.25a}$$

$$\frac{1}{R_{mk}^{(i)}(\mathbf{p})} \leq r_{mk}. \forall m,k \tag{2.25b}$$

The constraint (2.25b) is now convex while (2.25a) is still non-convex so we apply the following inequality:

$$xy \leq \frac{1}{2}\left(\frac{\bar{y}}{\bar{x}}x^2 + \frac{\bar{x}}{\bar{y}}y^2\right), \tag{2.26}$$

with $x = p_{mk}$, $\bar{x} = p_{mk}^{(i)}$, $y = r_{mk}$, $\bar{y} = r_{mk}^{(i)}$ to approximate (2.25a) as

$$\frac{\theta}{2}\alpha_{mk}^{(i+1)}C_{mk}(f_{mk}^{\mathrm{loc}} - \hat{f}_{mk}^{\mathrm{loc}})^2 + (1-\alpha_{mk}^{(i+1)})D_{mk}\frac{1}{2}\left(\frac{r_{mk}^{(i)}}{p_{mk}^{(i)}}p_{mk}^2 + \frac{p_{mk}^{(i)}}{r_{mk}^{(i)}}r_{mk}^2\right) \leq E_m^{\max}, \forall m,k, \tag{2.27}$$

which is now a convex constraint.

Convexify of (2.18e) By using $\mathbf{r}$ defined in (2.25b) and (2.18e) can be convexified as

$$t_{cm} \geq (1-\alpha_{mk}^{(i+1)})D_{mk}r_{mk}, \forall m,k, \tag{2.28}$$

Based on the above developments, we solve the following approximate convex program of (2.19) at iteration ith:

$$\underset{\mathbf{p},\mathbf{f},\mathbf{t}|\alpha^{(i+1)}}{\text{minimize}} \quad \underset{\forall m,k}{\max}\{\tau_{mk}(\mathbf{t})\}, \tag{2.29a}$$

$$\text{s.t.} \quad (2.17e) - (2.17g), (2.18c) - (2.18f), (2.22),$$
$$(2.23), (2.24), (2.25b), (2.27), \text{ and } (2.28). \tag{2.29b}$$

This is a convex program so that we can solve it efficiently with the CVX package. For complexity analysis, the convex problem (2.29) comprises $11KM_k + 2K$ linear or quadratic constraints and $5KM_k + K$ scalar decision variables, which leads to the per-iteration computational complexity of $\mathcal{O}\left(\sqrt{11KM_k + 2K}(5KM_k + K)\right)$ [55, Section 6].

Algorithm 1: AO-IA-based algorithm for solving (2.18)

1: **Input**: Set $i = 0$ and randomly choose initial feasible points $\mathscr{S}_1^{(0)}$ and $\mathscr{S}_2^{(0)}$ to constraints in (2.30), (2.29)
    Set the tolerance $\varepsilon = 10^{-3}$ and the maximum number of iterations $I_{\max} = 20$.
2: **Repeat**
3:    Solve problem (2.29) with given $\mathscr{S}_2^{(i)}$ to obtain the optimal solution of $(\mathbf{p}^*, \mathbf{f}^*, \mathbf{r}^*)$ and update $(\mathscr{S}_1^{(i+1)} := (\mathbf{p}^*, \mathbf{f}^*, \mathbf{r}^*)$;
4:    Solve problem (2.30) with given $\mathscr{S}_1^{(i+1)}$ to obtain the optimal solution of (α^*) and update $\mathscr{S}_2^{(i+1)} := (\alpha^*)$;
5:    Set $i := i + 1$;
6: **Until** Convergence or $i > I^{\max}$.
7: **Output**: $\{\alpha^*, \beta^*, \mathbf{p}^*, \mathbf{f}^*\}$ and $\max\{\tau_{mk}(\mathbf{t})\}_{\forall m,k}$.

2.3.2 Task offloading optimisation

In this subsection, we solve (2.18) for given $(\mathbf{p}^{(i)}, \mathbf{f}^{(i)})$ to obtain the next optimal values of offloading policies $(\alpha^{(i+1)})$. The observed optimisation is now given by

$$\underset{\alpha, \mathbf{t}|\mathbf{p}^{(i)}, \mathbf{f}^{(i)}}{\text{minimize}} \quad \max_{\forall m,k}\{\tau_{mk}(\mathbf{t})\}, \tag{2.30a}$$

$$\text{s.t.} \quad (2.17\text{d}) - (2.17\text{g}), (2.18\text{c}) - (2.18\text{f}). \tag{2.30b}$$

This is obviously a convex program and can be solved effectively with standard solvers such as CVX [56]. The per-iteration of solving this convex program is $\mathcal{O}\left(\sqrt{7KM_k} + 2K.4KM_k\right)$.

2.3.3 Proposed algorithm

Let us denote $\mathscr{S}_1^{(i)} \triangleq (\mathbf{p}^{(i)}, \mathbf{f}^{(i)}, \mathbf{r}^{(i)})$ and $\mathscr{S}_2^{(i)} \triangleq (\alpha^{(i)})$, and at the ith iteration, respectively. We now proceed by proposing Algorithm 1 to solve the problem (2.18).

2.4 Numerical simulations

This section expresses the performance metrics and impacts studied through numerical simulations.

2.4.1 Simulations setup

To understand the performance and resolution of the problem (2.17) for minimising the e2e latency, we relied on numerical simulations. The values of parameters used for attaining these results are summarised in Table 2.1.

Table 2.1 Simulation parameters [41,46,48,49]

Parameters	Value
Number of antennas	$L = 8$
Maximum transmit power	$P_m^{\max} = 23$ dBm
Bandwidth	$B = 5$ MHz
Transmission duration URLLC	$\delta = 0.02$ ms
Decoding error probability	$\varepsilon_{mk} = 10^{-5}$
Noise spectral density	-174 dBm/Hz
Maximum blocklength	$N = \tau B = 200$
Number of UEs	$M = \{4, 5, 6\}$
Number of ESs	$K = 2$
Maximum UEs' processing rate	$F_{\max}^{\text{loc}} = 1.5$ GHz
Maximum processing rate	$F_{\max}^{\text{es}} = 3$ GHz
Input data size	$D_{mk} = 100$ KB
Required computation resource	$C_{mk} = 1200 \times 10^6$ cycles
Total delay requirement	$T_m^{\max} = 2$ s
Minimum data rate	$R_{\min}^{ul} = 1$ Mbps
Maximum energy consumption UE	$E_m^{\max} = 0.5$ Joule
Maximum power consumption UE	$P_k^{\max} = 5$ W
Effective capacitance coefficient	$\theta_m = 10^{-27}$ Watt s^3/cycle^3

2.4.2 Results and discussion

2.4.2.1 Convergence of the proposed algorithm

Figure 2.3 clearly demonstrates the convergence of the proposed algorithm in reducing the worst-case e2e latency. In particular, with the model of $M_k = 4$ UEs, $K = 2$ UAVs, the optimising process converges at six iterations. The figure additionally illustrates the impact of the required computation resource on the e2e latency. Unsurprisingly, when the required computation resource increases, the e2e latency of computational tasks gradually increases. For instance, the observed scenario witnesses a considerable rise in the worst-case e2e latency from 0.6s to approximate 0.87s when C_m rises to 1100 megacycles.

2.4.2.2 Impact of required computation resource

Figure 2.4 plots the impacts of required computing resources of the tasks (C_m) in the e2e latency of UEs under the proposed algorithm and other benchmark schemes. Particularly, when the tasks are more complicated which require more computation resources, the e2e latency gradually increases. For instance, in the model of $M_k = 4, K = 2$ as observed in Figure 2.4, when C_m rises from 800 to 1200 megacycles, the worst-case latency obtained by using Algorithm 1 increases by approximate 300 ms. In addition, Figure 2.4 clearly demonstrates that our proposed algorithm is far better than all benchmark schemes. These results prove that joint optimisation of both communication and computation resources significantly improves the performance of MEC-based systems in reducing the e2e latency of UEs.

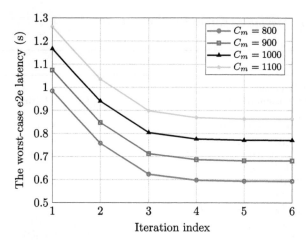

Figure 2.3 Convergence of the proposed algorithm with different values of required computation resource (C_m) in the scenarios of $M_k = 4, K = 2$. ©IET 2022. Reprinted with permission from [1].

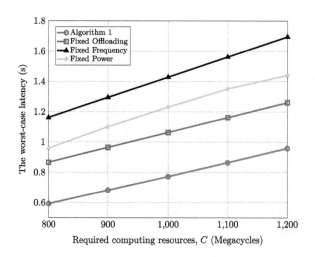

Figure 2.4 The worst-case latency among different values of required computation resource ($C_m \triangleq C$) in the scenarios of $M_k = 4, K = 2$. ©IET 2022. Reprinted with permission from [1].

2.4.2.3 Impact of UE transmit power budget

To investigate the impact of UEs transmit power in reducing the e2e latency, we have conducted experiments among different values of UE power budget in three models of $M_k = \{4, 5, 6\}$ and $K = 2$ UAVs. Figure 2.5 clearly states that when the transmit power budget of UEs increases, the worst-case e2e latency of UEs gradually

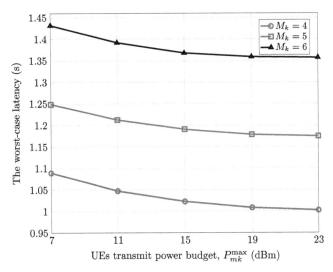

Figure 2.5 The worst-case latency among different values of UE transmit power budget ($P_{mk}^{\max}$) in the scenarios of $M_k = \{4, 5, 6\}$, $K = 2$. ©IET 2022. Reprinted with permission from [1].

reduces. For instance, the model of $M_k = 4$ UEs, $K = 2$ UAVs witnesses a considerable decline in latency by around 100 ms when the maximum of UEs transmit power reaches 23 dBm.

2.4.2.4 Impact of UEs' processing rate

To demonstrate the impact of UEs' processing rate on obtaining the minimised latency and adjusting optimal offloading decisions, Figure 2.6 presents the numerical results of experiments with a range of values of $F_{\max}^{\text{loc}}$. Unsurprisingly, when the processing capacity of UEs increases from 1 to 1.3 GHz, the e2e latency of UEs significantly reduces in both models. This is because the UEs are more powerful in processing tasks locally. Additionally, due to the constraints on UEs energy consumption in (2.17d), the average offloading portions of UEs have to increase to satisfy the energy budget when UEs have higher processing rate.

2.4.2.5 Impact of ESs' processing rate

Figure 2.7 illustrates the impact of the total computation resource of ESs, $F_{\max}^{\text{es}}$, on the e2e latency of computational tasks coming from UEs. The figure clearly shows that the more powerful the ESs are, the less e2e latency can be achieved for both examined scenarios of $M_k = \{4, 6\}$ UEs, $K = 2$ UAVs. Figure 2.7 also indicates under the same computation resource budget of ESs, the model that has more UEs ($M_k = 6$) obtains higher e2e latency than that in the smaller size model ($M_k = 4$). These results definitely demonstrate the effectiveness of the proposed offloading design.

42 Digital twins for 6G

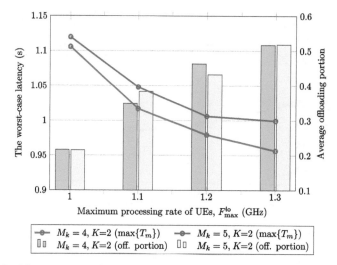

Figure 2.6 The worst-case latency among the maximum values of UEs' processing rate (F_{max}^{loc}) in the scenarios of $M_k = \{4, 5\}, K = 2$. ©IET 2022. Reprinted with permission from [1].

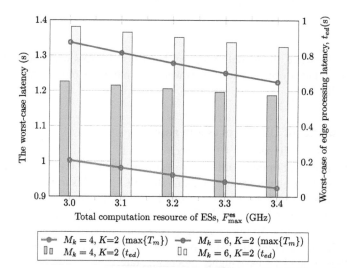

Figure 2.7 The worst-case latency among different values of total ESs' processing rate (F_{max}^{es}) in the scenarios of $M_k = \{4, 6\}, K = 2, P_k^{max} = 8$ W. ©IET 2022. Reprinted with permission from [1].

2.5 Conclusion

This chapter has investigated the DT-assisted UAV edge networks with URLLC for mission-critical applications. The addressed problem has jointly considered both communication and computation perspectives in minimising the e2e latency. The AO-based solution has been proposed to optimise task offloading portions and resource allocation in the system. Extensive simulations were conducted to demonstrate the impacts of various involved parameters to confirm the effectiveness of the proposed solution.

Appendix A: Approximations of G_{mk} and W_{mk} in (2.21)

We first rewrite the SINR of UE (m, k) as $\gamma_{mk}(\mathbf{p}) = p_{mk}/q_{mk}(\mathbf{p})$. By applying the inequality [57, Eq. (72)] for $x = p_{mk}$, $y = q_{mk}(\mathbf{p})$, $\bar{x} = p_{mk}^{(i)}$, and $\bar{y} = q_{mk}(\mathbf{p}^{(i)})$, we have

$$G_{mk}(\mathbf{p}) \geq a_{mk}^{(i)} - \frac{b_{mk}^{(i)}}{p_{mk}} - c_{mk}^{(i)} q_{mk}(\mathbf{p}) \triangleq \mathscr{G}_{mk}^{(i)}(\mathbf{p}) \tag{A.1}$$

where $a_{mk}^{(i)} = \ln(1 + \frac{p_{mk}^{(i)}}{q_{mk}^{(i)}(\mathbf{p})}) + 2\frac{p_{mk}^{(i)}}{p_{mk}^{(i)}+q_m^{(i)}(\mathbf{p})}$, $b_{mk}^{(i)} = \frac{(p_{mk}^{(i)})^2}{p_{mk}^{(i)}+q_m^{(i)}(\mathbf{p})}$ and $c_{mk}^{(i)} = \frac{p_{mk}^{(i)}}{(q_{mk}^{(i)}(\mathbf{p})+p_{mk}^{(i)})q_{mk}^{(i)}(\mathbf{p})}$.

To find an upper bounding convex function approximation of $W_{mk}(\mathbf{p})$, we apply the inequality [57, Eq. (75)] for $x = 1 - 1/(1 + \gamma_{mk}(\mathbf{p}))^2$ and $\bar{x} = 1 - 1/(1 + \gamma_{mk}(\mathbf{p}^{(i)}))^2$, yielding

$$W_{mk}(\mathbf{p}, \boldsymbol{\pi}_{mk}^{(i)}) \leq = d_{mk}^{(i)} - e_{mk}^{(i)} \frac{q_{mk}^2(\mathbf{p})}{(q_{mk}(\mathbf{p}) + p_{mk})^2} \tag{A.2}$$

where

$$d_{mk}^{(i)} = 0.5\sqrt{V_{mk}(\mathbf{p}^{(i)})} + 0.5/\sqrt{V_{mk}(\mathbf{p}^{(i)})} \tag{A.3}$$

$$e_{mk}^{(i)} = 0.5/\sqrt{V_{mk}(\mathbf{p}^{(i)})}. \tag{A.4}$$

The function $\frac{q_{mk}^2(\mathbf{p})}{(q_{mk}(\mathbf{p})+p_{mk})^2}$ in (A.2) is still not convex [57], which can be further approximated by using inequalities [57, Eq. (77)] and [57, Eq. (76)] as

$$\frac{q_{mk}^2(\mathbf{p})}{q_{mk}(\mathbf{p}) + p_{mk}} \frac{1}{q_{mk}(\mathbf{p}) + p_{mk}} \geq \frac{2}{q_{mk}(\mathbf{p}^{(i)}) + p_{mk}^{(i)}}$$
$$\times \left(\frac{2q_{mk}(\mathbf{p}^{(i)})q_{mk}(\mathbf{p})}{q_{mk}(\mathbf{p}^{(i)}) + p_{mk}^{(i)}} - \frac{q_m^2(p_m^{(i)})}{(q_m(p_m^{(i)}) + p_m^{(i)})^2}(q_{mk}(\mathbf{p}) + p_{mk}) \right) - \frac{q_{mk}^2(\mathbf{p})}{(q_{mk}^{(i)}(\mathbf{p}) + p_{mk}^{(i)})^2} \tag{A.5}$$

over the trusted regions defined in (2.22) and (2.23). By substituting this result to (A.2) yields

$$W_{mk}(\mathbf{p}, \boldsymbol{\pi}_{mk}^{(i)}) \leq \mathscr{W}_m^{(i)}(\mathbf{p})$$
$$\triangleq d_{mk}^{(i)} - \frac{2e_{mk}^{(i)}}{q_{mk}(\mathbf{p}^{(i)}) + p_{mk}^{(i)}} \left(2f_{mk}^{(i)} q_{mk}(\mathbf{p}) - (f_{mk}^{(i)})^2 (q_{mk}(\mathbf{p}) + p_{mk}) \right) + \frac{(f_{mk}^{(i)})^2}{q_{mk}^2(\mathbf{p}^{(i)})} q_{mk}^2(\mathbf{p}), \tag{A.6}$$

where $f_{mk}^{(i)} \triangleq \frac{q_{mk}(\mathbf{p}^{(i)})}{q_{mk}(\mathbf{p}^{(i)}) + p_{mk}^{(i)}}$.

References

[1] Y. Li, D. V. Huynh, T. Do-Duy, E. Garcia-Palacios, and T. Q. Duong, "Unmanned aerial vehicle-aided edge networks with ultra-reliable low-latency communications: a digital twin approach," *IET Signal Process.*, vol. 16, no. 8, pp. 897–908, 2022.

[2] Y. Wu, K. Zhang, and Y. Zhang, "Digital twin networks: a survey," *IEEE Internet Things J.*, vol. 8, no. 18, pp. 13789–13804, 2021.

[3] T. Liu, L. Tang, W. Wang, Q. Chen, and X. Zeng, "Digital twin assisted task offloading based on edge collaboration in the digital twin edge network," *IEEE Internet Things J.*, vol. 4662, no. 2, pp. 1427–1444, 2022, doi: 10.1109/JIOT.2021.3086961.

[4] W. Sun, H. Zhang, R. Wang, and Y. Zhang, "Reducing offloading latency for digital twin edge networks in 6G," *IEEE Trans. Veh. Technol.*, vol. 69, no. 10, pp. 2240–12251, 2020.

[5] J. Lopez, J. E. Rubio, and C. Alcaraz, "Digital twins for intelligent authorization in the B5G-enabled smart grid," *IEEE Wireless Commun. Mag.*, vol. 28, no. 2, pp. 48–55, 2021.

[6] D. V. Huynh, V.-D. Nguyen, V. Sharma, O. A. Dobre, and T. Q. Duong, "Digital twin empowered ultra-reliable and low-latency communications-based edge networks in industrial IoT environment," in *Proceedings of the IEEE International Conference on Communications (ICC'22)*, Seoul, Korea, May 16–20, 2022.

[7] M.-N. Nguyen, L. D. Nguyen, T. Q. Duong, and H. D. Tuan, "Real-time optimal resource allocation for embedded UAV communication systems," *IEEE Wireless Commun. Lett.*, vol. 8, no. 1, pp. 225–228, 2018.

[8] L. D. Nguyen, A. Kortun, and T. Q. Duong, "An introduction of real-time embedded optimisation programming for UAV systems under disaster communication," *EAI Endorsed Trans. Ind. Netw. Intell. Syst.*, vol. 5, no. 17, p. e5, 2018.

[9] Y. Shin Kang, B. Jin Park, C. Sun Yoo, Y. Shin Kim, and S.-O. Koo, "Ground test results of flight control system for the Smart UAV," in *Proceedings of the International Conference on Control, Automation and Systems*, Gyeonggi-do, Korea, October 2010, pp. 2533–2536.

[10] Y. Shin Kang, B. Jin Park, A. Cho, C. Sun Yoo, and S.-O. Koo, "Flight test of flight control performance for airplane mode of Smart UAV," in *Proceedings of the 12th International Conference on Control, Automation and Systems*, Guangzhou, China, December 2012, pp. 1738–1741.

[11] C.-S. Yoo, A. Cho, B.-J. Park, Y. shin Kang, S.-W. Shim, and I.-H. Lee, "Collision avoidance of Smart UAV in multiple intruders," in *Proceedings of the 12th International Conference on Control, Automation and Systems*, Guangzhou, China, December 2012, pp. 443–447.

[12] M. Hua, L. Yang, Q. Wu, C. Pan, C. Li, and A. L. Swindlehurst, "UAV-assisted Intelligent reflecting surface symbiotic radio system," *IEEE Trans. Wireless Commun.*, vol. 20, pp. 5769–5785, 2021.

[13] J. Xu, Q. Guo, L. Xiao, Z. Li, and G. Zhang, "Autonomous decision-making method for combat mission of UAV based on Deep Reinforcement Learning," in *Proceedings of the IEEE 4th Advanced Information Technology, Electronic and Automation Control Conference (IAEAC)*, vol. 1, Chengdu, China, Dec. 2019, pp. 538–544.

[14] Y. Wang, H. Wang, and X. We, "Energy-efficient UAV deployment and task scheduling in multi-UAV edge computing," in *Proceedings of the International Conference on Wireless Communications and Signal Processing (WCSP)*, Wuhan Hubei, China, June 2020, pp. 1147–1152.

[15] L. D. Nguyen, K. K. Nguyen, A. Kortun, and T. Q. Duong, "Real-time deployment and resource allocation for distributed UAV systems in disaster relief," in *Proceedings of the IEEE 20th International Workshop on Signal Processing Advances in Wireless Communications (SPAWC)*, Cannes, France, July 2019, pp. 1–5.

[16] J. Ousingsawat and M. E. Campbell, "Optimal cooperative reconnaissance using multiple vehicles," *J. Guid. Control Dyn.*, vol. 30, no. 1, pp. 122–132, 2007.

[17] S. Li, H. Tang, S. He, Y. Shu, T. Mao, J. Li, and Z. Xu, "Unsupervised detection of earthquake-triggered roof-holes from UAV images using joint color and shape features," *IEEE Geosci. Remote Sens. Lett.*, vol. 12, no. 9, pp. 1823–1827, 2015.

[18] F. Mohammed, A. Idries, N. Mohamed, J. Al-Jaroodi, and I. Jawhar, "UAVs for smart cities: Opportunities and challenges," in *Proceedings of the International Conference on Unmanned Aircraft Systems (ICUAS)*, Orlando, FL, May 2014, pp. 267–273.

[19] X. Yang, D. Lin, F. Zhang, T. Song, and T. Jiang, "High accuracy active stand-off target geolocation using UAV platform," in *Proceedings of the IEEE International Conference on Signal, Information and Data Processing (ICSIDP)*, Chongqing, China, December 2019, pp. 1–4.

[20] S. Waharte and N. Trigoni, "Supporting search and rescue operations with UAVs," in *Proceedings of the International Conference on Emerging Security Technologies*, Canterbury, UK, September 2010, pp. 142–147.

[21] S. Wang, Y. Han, J. Chen, Z. Zhang, G. Wang, and N. Du, "A Deep-learning-based sea search and rescue algorithm by UAV remote sensing," in *Proceedings of the IEEE CSAA Guidance, Navigation and Control Conference (CGNCC)*, Xiamen, China, August 2018, pp. 1–5.

[22] T. Do-Duy, L. D. Nguyen, T. Q. Duong, S. R. Khosravirad, and H. Claussen, "Joint optimisation of real-time deployment and resource allocation for UAV-aided disaster emergency communications," *IEEE J. Sel. Areas Commun.*, vol. 30, pp. 3411–3424, 2021.

[23] G. J. Sutton, J. Zeng, R. P. Liu, *et al.*, "Enabling technologies for ultra-reliable and low latency communications: From PHY and MAC layer perspectives," *IEEE Commun. Surveys Tuts.*, vol. 21, no. 3, pp. 2488–2524, 2019.

[24] T. Yoshizawa, S. B. M. Baskaran, and A. Kunz, "Overview of 5G URLLC system and security aspects in 3GPP," in *Proceedings of the IEEE Conference on Standards for Communications and Networking (CSCN)*, Granada, Spain, October 2019, pp. 1–5.

[25] G. Durisi, T. Koch, and P. Popovski, "Toward massive, ultra-reliable, and low-latency wireless communication with short packets," *Proc. IEEE*, vol. 104, no. 9, pp. 1711–1726, 2016.

[26] Y. Wu, J. Shi, K. Ni, L. Qian, W. Zhu, Z. Shi, and L. Meng, "Secrecy-based delay-aware computation offloading via Mobile Edge Computing for Internet of Things," *IEEE Internet Things J.*, vol. 6, no. 3, pp. 4201–4213, 2018.

[27] N. Mu, S. Gong, W. Sun, and Q. Gan, "The 5G MEC applications in smart manufacturing," in *Proceedings of the IEEE International Conference on Edge Computing (EDGE)*, Virtual Conference, October 2020, pp. 45–48.

[28] A. Rafiq, W. Ping, W. Min, S. H. Hong, and N. N. Josbert, "Optimizing energy consumption and latency based on computation offloading and cell association in MEC enabled Industrial IoT environment," in *Proceedings of the 6th International Conference on Intelligent Computing and Signal Processing (ICSP)*, Xi'an, China, April 2021, pp. 10–14.

[29] A. A. Kherani, G. Shukla, S. Sanadhya, *et al.*, "Development of MEC system for indigenous 5G test-bed," in *Proceedings of the International Conference on COMmunication Systems and NETworkS (COMSNETS)*, Bangalore, India, January 2021, pp. 131–133.

[30] J. Li, H. Gao, T. Lv, and Y. Lu, "Deep Reinforcement Learning based computation offloading and resource allocation for MEC," in *Proceedings of the IEEE Wireless Communications and Networking Conference (WCNC)*, Barcelona, Spain, April 2018, pp. 1–6.

[31] J. Wu, Z. Cao, Y. Zhang, and X. Zhang, "Edge-cloud collaborative computation offloading model based on improved partical swarm optimization in MEC," in *Proceedings of the IEEE 25th International Conference on Parallel and Distributed Systems (ICPADS)*, Tianjin, China, December 2019, pp. 959–962.

[32] İ. Altin and M. Akar, "Novel OMA and hybrid NOMA schemes for MEC offloading," in *Proceedings of the IEEE International Black Sea Conference on Communications and Networking (BlackSeaCom)*, Odessa, Ukraine, May 2020, pp. 1–5.

[33] E. Gonzalez-Sosa, I. Frontelo-Benito, R. Kachach, P. Perez, J. J. Ruiz, and A. Villegas, "Audience meter: a use case of deploying machine learning algorithms over 5G networks with MEC," in *Proceedings of the IEEE International Conference on Consumer Electronics (ICCE)*, Phu Quoc Island, Vietnam, January 2020, pp. 1–2.

[34] Y. Yang, Y. Hu, and M. C. Gursoy, "Reliability-optimal designs in MEC networks with finite blocklength codes and outdated CSI," in *Proceedings of the 17th International Symposium on Wireless Communication Systems (ISWCS)*, Virtual Conference, September 2021, pp. 1–6.

[35] G. Tan, H. Zhang, and S. Zhou, "Resource allocation in MEC-enabled vehicular networks: a deep reinforcement learning approach," in *Proceedings of*

the *IEEE Conference on Computer Communications Workshops (INFOCOM WKSHPS)*, Virtual conference, July 2020, pp. 406–411.

[36] S. Lee, S. Lee, and M.-K. Shin, "Low cost MEC server placement and association in 5G networks," in *Proceedings of the International Conference on Information and Communication Technology Convergence (ICTC)*, Jeju-si and Jeju-do, South Korea, October 2019, pp. 879–882.

[37] Z. Jia, D. Li, W. Zhang, and L. Pang, "5G MEC gateway system design and application in industrial communication," in *Proceedings of the 2nd World Symposium on Artificial Intelligence (WSAI)*, Guangzhou, China, June 2020, pp. 5–10.

[38] M. Hao, D. Ye, S. Wang, B. Tan, and R. Yu, "URLLC resource slicing and scheduling in 5G vehicular edge computing," in *Proceedings of the IEEE 93rd Vehicular Technology Conference (VTC2021-Spring)*, Guangzhou, China, June 2021, pp. 1–5.

[39] C. Zhan, H. Hu, Z. Liu, Z. Wang, and S. Mao, "Multi-UAV-enabled mobile-edge computing for time-constrained IoT applications," *IEEE Internet Things J.*, vol. 8, no. 20, pp. 15553–15567, 2021.

[40] Y. K. Tun, Y. M. Park, N. H. Tran, W. Saad, S. R. Pandey, and C. S. Hong, "Energy-efficient resource management in UAV-assisted Mobile Edge Computing," *IEEE Commun. Lett.*, vol. 25, no. 1, pp. 249–253, 2021.

[41] A. A. Nasir, "Latency optimization of UAV-enabled MEC system for virtual reality applications under Rician fading channels," *IEEE Wireless Commun. Lett.*, vol. 10, no. 8, pp. 1633–1637, 2021.

[42] M. Bennis, M. Debbah, and H. V. Poor, "Ultrareliable and low-latency wireless communication: Tail, risk, and scale," *Proc. IEEE*, vol. 106, no. 10, pp. 1834–1853, 2018.

[43] M. S. Elbamby, C. Perfecto, C.-F. Liu, J. Park, S. Samarakoon, X. Chen, and M. Bennis, "Wireless edge computing with latency and reliability guarantees," *Proc. IEEE*, vol. 107, no. 8, pp. 1717–1737, 2019.

[44] Y. Pan, K. Wang, C. Pan, H. Zhu, and J. Wang, "UAV-assisted and intelligent reflecting surfaces-supported terahertz communications," *IEEE Wireless Commun. Lett.*, vol. 10, no. 6, pp. 1256–1260, 2021.

[45] H. Ren, C. Pan, Y. Deng, M. Elkashlan, and A. Nallanathan, "Joint pilot and payload power allocation for massive-MIMO-enabled URLLC IIoT networks," *IEEE J. Sel. Areas Commun.*, vol. 38, no. 5, pp. 816–830, 2020.

[46] C. She, C. Yang, and T. Q. S. Quek, "Radio resource management for ultra-reliable and low-latency communications," *IEEE Commun. Mag.*, vol. 55, no. 6, pp. 72–78, 2017.

[47] Q. Liu, T. Han, and N. Ansari, "Joint radio and computation resource management for low latency mobile edge computing," in *Proceedings of the IEEE Global Communications Conference, GLOBECOM 2018*, Abu Dhabi, United Arab Emirates, 2018.

[48] J. Wang, D. Feng, S. Zhang, A. Liu, and X.-G. Xia, "Joint computation offloading and resource allocation for MEC-enabled IoT systems with imperfect CSI," *IEEE Internet Things J.*, vol. 8, no. 5, pp. 3462–3475, 2021.

[49] C.-F. Liu, M. Bennis, M. Debbah, and H. V. Poor, "Dynamic task offloading and resource allocation for ultra-reliable low-latency edge computing," *IEEE Trans. Commun.*, vol. 67, no. 6, pp. 4132–4150, 2019.

[50] T. Do-Duy, L. D. Nguyen, T. Q. Duong, S. R. Khosravirad, and H. Claussen, "Joint optimisation of real-time deployment and resource allocation for UAV-aided disaster emergency communications," *IEEE J. Sel. Areas Commun.*, vol. 39, no. 11, pp. 3411–3424, 2021.

[51] L. D. Nguyen, K. K. Nguyen, A. Kortun, and T. Q. Duong, "Real-time deployment and resource allocation for distributed UAV systems in disaster relief," in *Proceedings of the IEEE 20th International Workshop on Signal Processing Advances in Wireless Communications (SPAWC)*, Cannes, France, 2019, pp. 1–5.

[52] T. Q. Duong, L. D. Nguyen, and L. K. Nguyen, "Practical optimisation of path planning and completion time of data collection for UAV-enabled disaster communications," in *Proceedings of the 15th International Wireless Communications & Mobile Computing Conference (IWCMC)*, Tangier, Morocco, 2019, pp. 372–377.

[53] B. R. Marks and G. P. Wright, "A general inner approximation algorithm for nonconvex mathematical programs," *Oper. Res.*, vol. 26, no. 4, pp. 681–683, 1978.

[54] A. Beck, A. Ben-Tal, and L. Tetruashvili, "A sequential parametric convex approximation method with applications to nonconvex truss topology design problems," *J. Global Optim.*, vol. 47, no. 1, pp. 29–51, 2010.

[55] A. Ben-Tal and A. Nemirovski, *Lectures on Modern Convex Optimization*. Philadelphia: MPS-SIAM Series on Optimization, SIAM, 2001.

[56] M. Grant and S. Boyd, "CVX: MATLAB software for disciplined convex programming, version 2.1," http://cvxr.com/cvx, March 2014.

[57] A. A. Nasir, H. D. Tuan, H. Nguyen, M. Debbah, and H. V. Poor, "Resource allocation and beamforming design in the short blocklength regime for URLLC," *IEEE Trans. Wireless Commun.*, vol. 20, no. 2, pp. 1321–1335, 2021.

Chapter 3
AI-enabled data management for digital twin networks

Elif Ak[1], Gökhan Yurdakul[2], Ahmed Al-Dubai[3] and Berk Canberk[3]

As we have discussed in previous chapters, digital twins establish contextual relationships with the surrounding entities, providing a holistic view of interconnected systems and environments. With the growing complexity and abundance of data generated by digital twins, effective data management strategies have become paramount. This chapter delves into AI-enabled data management for digital twins, exploring how artificial intelligence techniques empower the collecting, storing, integrating, analyzing, and utilizing of diverse and voluminous data within the digital twin ecosystem. We will see how 6G and IoT networks can unlock valuable insights and optimize operational processes in many aspects by leveraging AI.

3.1 Introduction

Unlike when it first came out, the digital twin not only does deal with descriptive three-dimensional data any longer but also utilizes physical entities with their functions, behaviors, its telemetry data that is generated, and all contextual relations with surrounding entities. Therefore, specialized data management for the digital twin networks (DTNs) is an inevitable research era that should be carefully studied.

3.1.1 Importance of data management in DTNs

First and foremost, data is one of the most essential things for the digital twin. The integration of sensors, the establishment of connections between physical and virtual entities, and the bidirectional flow of information between the virtual and physical realms all rely on careful handling and management of data. Even a single malfunctioning sensor or a disruption in the interconnected links within the digital twin can severely impede the overall digital twin data (DTD) pipeline.

---
[1]Department of Computer Engineering, Istanbul Technical University, Turkey
[2]BTS Group, Istanbul, Turkey
[3]School of Computing, Engineering and The Build Environment, Edinburgh Napier University, UK

On the other hand, AI studies lie on two main assumptions (i) data has already been collected and stored somewhere for preprocessing and analyzing purposes; and (ii) timeliness, synchronization, and velocity of data are not considered in the training step. However, it is challenging to collect comprehensive data for fully training a diagnosis model. In addition, the current environment where data is collected is continuously changing, which results in a previously trained model mostly not applicable to current conditions. Therefore, transfer learning or partial train approaches should be investigated instead of retraining the model from scratch when the DTD matters. Herein, considering basic digital twin requirements, we have listed eight essential data characteristics of a digital twin to manage and leverage AI capabilities thoroughly.

3.1.1.1 Eight essential data characteristics of the digital twin

1. **Heterogeneous**: Digital twins integrate data from various sources and sensors, both internal and external, into the system. This data can have different formats, resolutions, and types, ranging from structured data (e.g., sensor readings and numerical values) to unstructured data (e.g., images, text, and video). Integration and harmonization of heterogeneous data are essential for a comprehensive digital twin representation.
2. **Multi-dimensional**: Digital twins capture and represent a wide range of information about the physical object or system they emulate. This includes not only basic attributes like dimensions, location, and physical properties but also more complex and dynamic data such as real-time sensor readings, operational parameters, and historical performance data.
3. **Real-time**: Real-time data transmission from the original object or product to its digital twin gets enabled when the real objects or products are linked or connected with the digital twin model. This linking and connectivity enable the model to simulate the physical system in real-time and vice versa. Connectivity is an essential characteristic of digital twins, making them different from digital models and shadows.
4. **Interconnected**: Digital twins connect multiple systems, networks, and devices. They are often part of larger ecosystems and enable data exchange and integration between external data sources. This interconnectedness allows digital twins to reach a broader context and facilitate collaborative decision-making.
5. **Time-series**: Digital twins can store time-stamped data in real time. Analysis of such trends, patterns, and historical behavior can be invaluable for identifying anomalies and enabling predictive analytics. This makes them perfect for observing physical counterparts over time.
6. **High volume and velocity**: Digital twins process large volumes of data due to their physical counterparts, which feed them continuously with real-time data. Such volume of incoming data requires efficient data management techniques to handle the velocity.
7. **Contextuality**: Digital twins are good at capturing contextual information by transforming physical objects into digital space. Transformation can include environmental conditions, user interactions, operational context, and contextual

metadata. Contextual data provides a richer understanding of the digital twin's behavior and aids in accurate analysis and decision-making.
8. **Security and privacy-sensitive**: Digital twins handle sensitive data, such as proprietary information, personal data, or operational details. Ensuring data security, privacy, and compliance with regulations is crucial to protect the integrity and confidentiality of the digital twin and its associated data.

Understanding and effectively managing these data characteristics is essential for deriving meaningful insights, performing data analytics, and leveraging the full potential of digital twins. Advanced data management techniques, such as data integration, preprocessing, storage, analysis, and security measures, play a vital role in harnessing the power of data within the digital twin ecosystem. We will examine these techniques in Chapter 8.2 to explore how to use data to obtain meaningful results.

3.1.2 Explanation of AI's role in data management for DTNs

The convergence of AI and digital twin technology has initiated a revolution in data management across various industries, and this is particularly evident in the context of 6G networks. Digital twins, essentially virtual replicas of physical systems, rely heavily on the continual analysis and management of data to offer valuable insights and facilitate decision-making. Beyond any doubt, AI plays a crucial role in digital twins since it automates data processing, enhances data quality, provides predictive modeling, supports real-time decision-making, and conserves privacy.

AI for data acquisition, preprocessing, modeling, and data storage: AI algorithms can be used to intelligently gather data from various sources, including sensors, IoT devices, and human inputs. Machine-learning techniques such as anomaly detection can identify and isolate noisy or faulty data at the source, streamlining the acquisition process. The cleaning, normalization, and transformation of raw data are also vital for accurate modeling [1]. AI-driven techniques, including automatic data imputation and normalization algorithms, can handle these tasks more efficiently and with greater accuracy than manual processes. Moreover, AI-enabled modeling involves the utilization of complex machine-learning algorithms to build representations of physical systems. These models can simulate, predict, and analyze various scenarios, providing invaluable insights for decision-makers. AI also optimizes data storage by determining the most appropriate storage methods, considering the type of data, required access speed, and security considerations [2]. For instance, real-time data might be stored in a Time-Series Database like InfluxDB, while large-scale data can be managed through cloud-based solutions. In Section 3.2, we will see various techniques from scratch, highlighting the twinning process.

AI for data analysis and predictive modeling: AI applies complex algorithms to discover patterns, trends, and hidden insights within the data. Techniques such as clustering, classification, and regression models enable the understanding of complex relationships between variables. It enables predictive modeling by using historical data to forecast future outcomes. Whether predicting equipment failure or optimizing energy consumption, predictive models are at the core of intelligent decision-making within DTNs. For instance, in a 6G network environment, predictive models can be used for network congestion prediction, anomaly detection, or predicting device failure, thus ensuring network resilience and reliability [3]. As a second example, AI algorithms facilitate immediate responses to network changes in near-real-time. This is also important in 6G use cases such as autonomous vehicles and ultra reliable low latency communications (URLLC), where a delay in data processing can lead to critical failures [4]. Data analysis and predictive modeling approaches will be presented in Section 3.3.

AI for ethical considerations and security aspects: AI must be designed and implemented with fairness, transparency, and accountability in mind. Ethical AI ensures that models are free from biases and that their decisions can be explained and justified. AI also should enhance security by implementing intelligent monitoring and intrusion detection systems. Machine-learning models detect unusual patterns and potential security breaches, triggering immediate responses. With the sensitive nature of some data, AI is employed to develop and enforce privacy-preserving techniques, like differential privacy, ensuring that individual data points cannot be reverse-engineered. Finally, ethical considerations, security aspects, and privacy-preserving techniques are discussed in Section 3.4.

3.1.3 Challenges in data management

A continuous flow of high-quality data, privacy and security regulations, domain expertise, interoperability, and ethical considerations are crucial to leveraging the full potential of Digital Twin technologies. However, the diverse and voluminous nature of data sources poses a challenging obstacle to effective analysis and integration for DTNs. In this section, we delve into the multifaceted challenges faced in digital twins data management, and then throughout this chapter, we will see how AI can be used and which techniques are proper to mitigate these challenges.

Data variety: In digital twin environments, there are lots of data sources to collect data. These can include images, text, and structured, unstructured, or semi-structured data. This diversity presents a significant challenge as it demands robust and effective analysis techniques. Integrating and managing this disparate data requires robust ontologies and semantic integration methodologies that can accommodate varying levels of granularity and complexity [5].

Data mining: Data mining, the method of extracting significant patterns from diverse data sources, is essential for enhancing the virtual worlds of digital twins. In DT implementations, data are often large-scale and real-time. Extracting useful patterns and maintaining data standards become complex due to the multifaceted nature of this data. Applying traditional data mining techniques may prove insufficient, necessitating advanced machine learning and big data analytics approaches [6].

Data uncertainty: Data uncertainty poses a significant challenge in digital twin [7]. Variability in environments and physical components complicates the assumption of uniform data distribution. This complexity arises from varying data transmission and reception speeds, missing data, inaccurate measurements, and other factors. Navigating this ambiguity requires innovative data processing and modeling techniques. Proper handling of data uncertainty in digital twins can lead to more robust insights and informed decision-making [8].

Useful data: Ensuring the continuous flow of high-quality data is vital for a digital twin's operation. Insufficient or unreliable data can hinder its functioning. The amount and quality of signals from IoT devices play a critical role. To ensure the appropriate data is obtained and effectively utilized, detailed planning and analysis of device usage are essential. Leveraging predictive analytics and real-time monitoring can further enhance the functionality of a Digital Twin [9].

Privacy and security: Privacy and security are paramount for digital twins, especially in industrial contexts. The vast data consumption and potential risk to vital system information necessitate adherence to prevailing security and privacy regulations. By prioritizing security and privacy, industries can enhance trust and alleviate concerns surrounding digital twin technologies [10].

Domain knowledge expertise: Building and managing a digital twin requires substantial domain expertise. Lack of proper knowledge can lead to mistakes and suboptimal performance. To mitigate this risk, comprehensive training, collaboration with domain experts, and continuous learning are necessary. A robust understanding of both technological and domain-specific aspects is vital to maximize the effectiveness of digital twin applications [11].

Interoperability: Interoperability, or the ability of different systems and devices to work together seamlessly, is another challenge. In a complex DTN, disparate systems must communicate and exchange information effectively. This requires standardized protocols and data formats, along with robust middleware solutions to facilitate integration and interoperability [12].

Ethical considerations: Ethical considerations are an emerging challenge in DTNs. Concerns about how data is used, potential biases in modeling, and implications for workers' rights and privacy must be carefully considered [13]. A transparent and responsible approach to data management, in line with ethical guidelines and societal norms, is essential to maintain public trust and adhere to legal obligations. In Section 3.4, we will see ethical challenges and possible mitigation approaches in detail.

3.1.4 Three states of DTD

The story of the existence of the data in the digital twin covers the duration from the data collection to the transfer to the digital twin model and then processing and finally producing the output. The data management for DTNs covers the handling of the data taking into account eight essential characteristics of the DTD mentioned in Section 3.1.1

Definition 1. *DTD*: *In a nutshell, DTD refers to the collection of information and data associated with a digital twin. This data encompasses a wide range of information collected, generated, stored, processed, or analyzed within the digital twin ecosystem.*

DTD serves as the foundation for analysis, simulation, predictive modeling, decision-making, and other applications that leverage the digital twin's capabilities to monitor, simulate, optimize, and manage the physical entity or system it represents.

In the digital twin ecosystem (regardless of which environment, system, or physical entity is digitalized), the DTD is presented in three states: input data, digital data, and output data.

Input data: In the context of a digital twin, input data represents the data collected from the physical counterpart of the digital twin through the IoT or 6G network stack. It is the real-time data that mirrors the current status, position, or working condition of the physical entity. This data might include sensor readings, status updates, or any other relevant information that the physical object produces.

Physically, this input data resides in a data lake or similar storage system and should be readily accessible for the digital twin components for processing. Besides the digital twin's process model, which simulates the physical object's operations, other digital twin services, such as analytics, also utilize this input data. It's worth noting that the "Input Data" box in the given context is a logical representation, serving as an entry point for physical data into the digital twin system. We will see the details of how input data is collected and being a member of digital data later in Section 3.2.2.

Digital data: Digital data, also referred to as a *digital model* in the context of digital twins, is the computational or process model that simulates the physical object's operations. It receives input data, processes it, and produces output data that symbolically represents the physical object's behavior or output.

Let us assume we are modeling a digital twin for a 6G base station to configure signal strength and channel operations automatically. A digital model could simulate different states of a 6G base station, with each state determined by the input data (e.g., current configurations of the antenna). Then, the digital model results in "virtual output," which is the calculation of the input data values in the digital twin environment. For example, the total number of connected users to the base station is calculated based on the input provided and the digital model (data).

Digital data is emulated through finite-state machines, discrete-time simulations, or 3D modeling software according to the physical entity desired to be modeled. We will see some discussions on data integration with the digital twin model and storage issues through Section 3.2.5.

Output data: Output data refers to the data generated by the digital twin as a result of processing the input data through the digital model. This data could be diverse, representing various aspects of the digital twin and the physical object it mirrors. Examples include status updates of the object, outputs from virtual sensors, results from machine learning models, alert data containing reasons for alerts and timestamps, and more.

Output data is stored in a data store or similar system, making it available for services such as visualization, analytics, and other downstream applications. For instance, machine-learning models might use this data for forecasting or anomaly detection, or a notification service might use the data to send alerts when certain conditions are met. The output data essentially provides insights, recommendations, and actionable information based on the simulation of the physical object in the digital twin. AI-enabled data analysis and interpretation techniques will be discussed to produce output data in Section 3.3.

3.2 The twinning process: AI-driven data acquisition, preprocessing, modeling, and data storage for DTNs

We have seen *Eight Essential Data Characteristics of the Digital Twin* in Section 3.1.1, which is also similar to essential 3V's of Big Data (Volume, Velocity, and Variety) [14], which both of them require various specific procedures like data acquisition, preprocessing, analysis, and storage techniques. So someone might ask *why we cannot directly apply the same techniques used in Big Data projects in the DTNs*. The crucial consideration comes up with the data synchronization issues when we need to collect data for the digital twin. In other words, virtual representations might retrieve data from physical entities through APIs, direct connectivity, sensors, network devices, or even edge computing. However, it should be ensured that the data synchronization mechanisms consider a unified digital twin architecture for analysis and modeling of the twinned object, environment, or system.

Definition 2. *Synchronization and twinning process: The goal of the synchronization is to twin the digital twin's state with the real-world counterpart, ensuring that the digital twin accurately reflects the current state of the physical entity it represents. In order to achieve such synchronization, various series of software and communications procedures are necessary, which is also known as* twinning processes.

The twinning process encompasses various stages such as data collection, transmission, integration, processing, cleaning, and storage. Consequently, in this chapter, we first examine data acquisition aspects, answer essential questions, and then see some basic techniques for collecting data, some data preprocessing methods for cleaning, filtering, and transforming raw data, and finally, data storage methods for digital twins.

3.2.1 Understanding the data acquisition

Data acquisition is the linchpin in creating and maintaining DTNs. The process involves collecting real-time or bulk data from various sources, including IoT devices, sensors, network devices, and user-generated inputs. This data, a perfect reflection of its physical counterpart, is what breathes life into digital twins, making them dynamic entities capable of replicating the behaviors of their real-world equivalents.

In the beginning step towards building the digital twin for any system, network, or thing, we should understand physical entity, data, and what we need to collect. We have already seen the data characteristics of the digital twin in Section 3.1, but some knowledge is still required, and some questions exist that we need to ask ourselves. Considering the data characteristics of the digital twin, we will categorize the data acquisition aspect into three categories: physical features, telemetry data, and contextual relations.

3.2.1.1 Physical features

Physical features refer to the intrinsic properties and characteristics of the physical entity or system that the digital twin represents. These features encompass the object or system's structural, geometric, and material aspects. When acquiring data related to physical features, you should ask the following questions.

1. What are the dimensions, shapes, locations, and configurations of the physical entity? In the case of 6G networks, the physical entities could include network infrastructure components such as base stations, antennas, routers, or IoT devices. Correspondingly, their physical features include size, form, connectivity interfaces, and technical features. Also, each connected device is located in a specific place which helps to understand the connected network in the digital twin.

2. What is the composition of the entity or are there any components contained within the physical object? For example, in the backbone of the 6G networks, many components exist which might seem like a single system in the digital twin, like "core networks." Routers, switches, network servers, and controllers are example components in the core network of the 6G. Moreover, understanding the composition of the physical entities can help to assess factors such as endurance. For instance, considering the material composition of IoT devices can determine their resistance to weather conditions or interference.

3. Are there any unique identifiers or serial numbers associated with the physical entity? Assigning unique identifiers or serial numbers to network devices or IoT devices is crucial for tracking, inventory management, and ensuring proper identification during data collection, configuration, or troubleshooting processes.

3.2.1.2 Telemetry data

Telemetry data encompasses real-time and bulk information on various aspects such as network traffic, latency, bandwidth utilization, and signal quality. This data is integral for monitoring and optimizing the performance of physical entities within the network, enabling the digital twin to emulate the real-world system precisely. By using specialized techniques and trigger mechanisms to acquire telemetry data, insights can be gained into the functional characteristics and requirements of the physical entity, supporting more effective decision-making.

1. What is the data to be monitored from the physical object? For 6G networks, data that could be collected from entities might include network traffic, latency, bandwidth utilization, signal quality, or device connectivity status.

2. Is the data generated in real-time or produced in bulk? Network performance and traffic monitoring metrics such as latency, packet loss, throughput, signal strength, bandwidth utilization, or network congestion level are considered real-time data. Moreover, considering the 6G network operations, service layer agreements (SLA) could be inserted into the digital twin system in bulk, unlike real-time. Therefore, the way data is generated should be handled differently in both the data acquisition and storage steps, which we will see in the database operations in Section 3.2.5 later.

3. If specific data are not available directly from the physical entity, but it is required to be measured, how can they be derived? Let us consider a situation where the physical entity in the 6G network is a base station, and we require information about the signal propagation characteristics in its vicinity, such as path loss or signal coverage maps. In this circumstance, we need to refer to derived features if the direct measurements of path loss or signal coverage are unavailable through the physical entity, such as a base station. Using radio frequency (RF) propagation mathematical models, path loss can be estimated based on distance, frequency, antenna characteristics, and environmental conditions.

4. What is the trigger mechanism to send data from the physical entity? If time matters, what is the sampling rate or frequency at which the data should be captured? The trigger mechanism and sampling rate for data acquisition from the physical entity in 6G networks can vary depending on the specific requirements, use cases, and characteristics of the physical entity.

Definition 3. *Event-based trigger mechanism: The data transmission from the physical entity to the digital twin can be triggered by specific events or conditions. For example, data may be sent when a threshold is exceeded, a status change occurs, or an anomaly is detected. Events could be pre-defined thresholds, sensor readings outside a certain range, or specific triggers configured for monitoring critical parameters.*

Definition 4. *Time-based trigger mechanism: Time-based: data transmission can be scheduled at regular time intervals. In this case, the physical entity sends data to the digital twin at fixed time points, regardless of specific events or conditions. Time-based triggers are helpful when capturing periodic measurements or maintaining a consistent data update frequency. If time-based data acquisition is decided to be used, then the sampling rate should be chosen considering the following criteria:*

- **Real-time monitoring**: *If real-time monitoring is required, the data should be captured at a high sampling rate to ensure timely updates and responsiveness. For example, in 6G networks, real-time monitoring of network performance may necessitate capturing data at sub-second intervals to capture rapid fluctuations.*
- **Granularity and precision**: *The required level of granularity and precision in the data affect the sampling rate. If fine-grained details are essential, a higher sampling rate is typically necessary. For instance, capturing fine-scale variations in temperature or signal strength may require higher-frequency measurements.*
- **Bandwidth and storage constraints**: *Considerations of bandwidth limitations and storage capacities should also be considered. Balancing the need for frequent data capture with the available resources is essential to ensure efficient data transmission, storage, and processing.*

3.2.1.3 Contextual relations

The contextual understanding of a physical object in a digital twin system plays a vital role in accurately replicating its behavior, interconnections, and functions within a digital environment. By addressing the below questions, anyone can create a robust and precise digital representation that mirrors the complexities and nuances of real-world systems.

1. What is the function of the physical object? The function of the physical object in a digital twin system, such as an IoT device or a 6G network component, often corresponds to specific roles or tasks within the network or system. For example, a base station in a 6G network serves to facilitate wireless communication between mobile devices and the network infrastructure, managing resources and routing data.

2. What is the input, process, and output implemented when the object operates? The input, process, and output of a physical object can vary based on the specific entity and function. For a network router in a 6G network, the input may include incoming data packets, the process would involve routing algorithms to determine the optimal path, and the output would be the data packets sent to the next destination in the network.

3. Can the operation of the physical object be modeled as a state machine? Many physical objects in digital twin systems can be modeled as a state machine. In a state machine representation, different operating conditions or "states" are identified, and specific triggers or conditions define transitions between states. For example, a network switch might have states such as "Idle," "Processing," and "Error," with transitions based on network load or error conditions.

4. How long will the physical object be in a certain state? The duration a physical object remains in a particular state depending on various factors such as the operational context, system dynamics, and external objects. For instance, a sensor in "Sleep" mode might remain in that state until activated by a particular event, while a base station in "High Traffic" mode might revert to "Normal" mode when the network load decreases.

5. What is the overall ecosystem to which the physical object belongs? The overall ecosystem for a physical object in a digital twin context often includes various interconnected components, systems, and stakeholders. For a 6G network, this ecosystem might encompass network operators, service providers, hardware manufacturers, regulators, end-users, and other entities interconnected through various technologies and protocols.

6. Are there relationships (e.g., a parent–child relationship) with other objects and potentially other digital twins? Relationships such as parent–child or peer-to-peer can exist among physical objects within digital twin systems. For example, in a network hierarchy, a core router (parent) may manage multiple edge routers (children), or IoT devices might have peer-to-peer relationships within a mesh network.

7. Are there components within the physical object that need to be modeled as digital twins also? It is common to model individual components within a physical object as separate digital twins, especially in complex systems. For example, within a 6G base station, different subsystems, such as the antenna array, power management, and signal processing units, might be modeled as individual digital twins. This granularity enables more detailed monitoring, analysis, and control of the overall system.

3.2.2 Techniques for collecting data from various sources in DTNs

In the previous section, we divided data acquisition into three categories and answered some questions to better understand various aspects of data acquisition through 6G and IoT networks. When the techniques come to the scene for collecting data from various sources, we continue to consider the way of acquisition and characteristic features of DTD. Therefore, these techniques would differ from standard data collection techniques required in many big data projects. This sub-chapter delves into techniques for collecting data from various sources in the context of 6G and IoT for DTNs.

3.2.2.1 Sensor-based data collection

A primary data source for digital twins in 6G networks is sensors embedded within physical entities [15]. These sensors collect data about various parameters, such as temperature, pressure, speed, humidity, and motion. In a 5G and beyond context, sensors can be embedded within the network or IoT devices to collect real-time data about network performance and health. For instance, sensors can measure signal strength, packet loss rate, bandwidth utilization, and other key network metrics. The creation and maintenance of a DTN necessitate the collection and exchange of vast amounts of data between a physical entity and its digital counterpart. This data exchange is facilitated by a range of technologies which can be categorized into two levels: (i) the hardware technology they have and (ii) the communication technology they use.

The emergence of 6G networks and the increasing demand for sensor-based data collection in digital twin and IoT applications necessitate powerful single-board computers (SBCs). These SBCs are the backbone of data acquisition and processing, enabling real-time monitoring, analysis, and synchronization between physical and virtual environments. The following selected SBCs offer a balance of computational power, connectivity options, and support for AI frameworks, enabling efficient data collection and analysis.

- **Raspberry Pi 4:** Raspberry Pi 4 is a highly popular single-board computer due to its affordability, widespread support, and high-performance capabilities. In the most recent version, with up to 8GB LPDDR4 RAM, a quad-core Cortex-A72 (ARM v8) 64-bit SoC, and support for dual-band wireless networking, it is robust enough to handle demanding tasks in a Digital Twin or IoT environment. The in-built Bluetooth 5.0 provides solid connectivity for IoT devices, and the Gigabit Ethernet enables fast data transfer, which is crucial in a 6G network environment. Furthermore, the large community and vast array of support for Raspberry Pi devices make this a robust and versatile choice for these applications [16–18].

- **NVIDIA Jetson family**: Designed specifically for AI and machine-learning applications, the NVIDIA Jetson Nano is an exceptional SBC for DTNs in the context of 6G. The Jetson Nano, combined with its support for AI frameworks like TensorFlow, PyTorch, and Caffe, empowers researchers to implement sophisticated AI algorithms for real-time data analysis and decision-making. Furthermore, the Jetson Nano's Ethernet, Wi-Fi, and Bluetooth connectivity options facilitate seamless integration into 6G-enabled IoT networks, enhancing sensor data collection capabilities.

  Besides Jetson Nano, NVIDIA Volta™ is equipped with a GPU with 48 Tensor Cores and 8GB of LPDDR4x RAM. These specifications make the Jetson Xavier NX capable of processing high volumes of data locally, which is essential for digital twin edge processing [19]. It is also capable of supporting a wide range of IoT applications, and the high-speed processing power it provides can meet the high data rate requirements of 6G networks [20].

- **Adafruit ESP32:** Adafruit ESP32 can be used to collect data from various sensors connected to physical entities such as temperature, humidity, and vibration and send this information to the digital twin model for real-time monitoring. Adafruit ESP32's low cost and accessibility make it an attractive option for prototyping and small-scale deployments. With a variety of GPIO pins and support for I2C, SPI, etc., the Adafruit ESP32 allows for flexible sensor integration, where different types of sensors like accelerometers, gyroscopes, or environmental sensors mirror the physical system in the digital twin accurately. Moreover, its ability to run lightweight machine learning models directly on the device makes it suitable for edge computing where latency and bandwidth are critical [21]. With on-board processing capabilities, the Adafruit ESP32 can perform real-time analysis, reducing the need to send data to a central server, thus minimizing latency in critical applications.

- **BeagleBone Board**: It is quite similar to the Raspberry Pi but has some features that make it more suitable for industrial applications. It has 4GB of onboard flash storage, which makes it more reliable in environments where the SBC may be turned off unexpectedly. It also includes a real-time clock on the board, useful for IoT applications that need to track time accurately and require sensitive synchronization [22,23].

- **Google Coral Dev Board**: The Coral Dev Board from Google is designed with edge AI applications in mind. It features Google's Edge TPU, a system on chip (SoC) designed to run TensorFlow Lite ML models at the edge, thus offering fast, localized data processing which is essential for real-time Digital Twin simulations and IoT devices. Furthermore, with Wi-Fi and Bluetooth connectivity, it can facilitate interconnected devices in 6G networks.

Besides the variety of SBCs, sensor-based data collection comprises diverse communication protocols to manage and transport data through DTNs. At the network level, communication protocols regulate the transmission of data from the SBC to the cloud-based digital twin model. The appropriate protocols to be used in DTNs should optimize data transmission in terms of speed, reliability, and energy efficiency. Here are the most prominent protocols:

- **Message queuing telemetry transport (MQTT)**: A lightweight, publish-subscribe network protocol that transports messages between devices. It is designed for remote locations where a small code footprint is required and network bandwidth is at a premium.
- **Long-range wide-area network (LoRaWAN)**: A media access control (MAC) protocol designed for large-scale public networks with a single operator. It provides low-power, mobile, and secure bi-directional communication in IoT, M2M, and smart city and industrial applications.
- **ZigBee**: A high-level communication protocol that provides a small, low-power solution for efficiently transmitting data over a long range. It is based on the IEEE 802.15.4 standard for wireless personal networks.
- **Wi-Fi**: Based on IEEE 802.11 standard, it offers high-speed and reliable wireless data transfer. The Wi-Fi communication protocol is commonly used where power is not a concern and high data rates are needed.

On the other hand, integrating these protocols through 6G networks in the context of DTNs involves the confluence of data from different sources into a unified, standard format that the digital twin can understand and process. This requires using middleware like MQTT broker or The Things Network (TTN) services to manage the flow of data across the network and facilitate interoperability between various sensor data and the digital twin model.

3.2.2.2 Network topology-based data collection

In the 5G and beyond, network devices are also valuable data sources. Data about network performance, including metrics such as network latency, jitter, packet loss, and bandwidth utilization, provide valuable insights into network health and performance. In a digital twin scenario, this data can help in proactive network management and fault prediction for 5G and 6G networks [24]. To achieve this, a suite of protocols and modeling languages have been developed and are widely used. The most common ones include SNMP, YANG, NETCONF, RESTCONF, and Open Daylight.

- **Simple network management protocol (SNMP)**: SNMP is a widely used protocol designed for network management. It allows for the monitoring, configuration, and control of network devices like switches, routers, servers, printers, and other devices on an IP network. SNMP operates in a manager–agent model, where the manager is the network management system, and the agents are the network devices being managed. It collects data by **polling** the agents at regular intervals and retrieving operational statistics and status information, which can then be used to build the digital twin's network topology and operational status. The relation between SNMP agent and SNMP manager through TRAP receiver via MIB Database can be seen in Figure 3.1.
- **Yet another next generation (YANG)**: YANG is a data modeling language used for the configuration and status data of network devices, defining how to manage the network elements (devices) and the data they hold. It provides a structured and efficient way to model the operations, capabilities, and

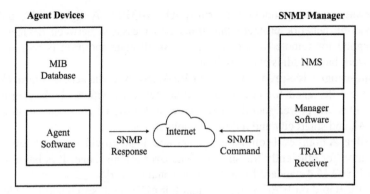

Figure 3.1   SNMP agent and manager

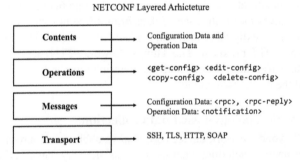

Figure 3.2   NETCONF-layered architecture and examples

features of a network device, enabling the network management system to understand the data it reads and manipulates. In the context of DTNs, the YANG model facilitates a clear representation of the network components and their relationships.

- **Network configuration protocol (NETCONF)**: NETCONF is a network management protocol with an XML-based data encoding, and it provides a set of rules to install, manipulate, and delete the configuration of network devices as seen in Figure 3.2. It enables network devices to inform state and configuration data through a structured and machine-readable interface. NETCONF also provides detailed transaction management, configuration data stores, and capability discovery, unlike SNMP protocol. NETCONF protocol is used with YANG data model to pull, push, and update both operational and configurational data. For an industrial IoT (IIoT), NETCONF can be used to collect data, ensuring synchronization and completeness. For example, DT-enabled fine-grained IIoT air quality monitoring system using YANG and NETCONF is presented [25]. AI algorithms can also analyze these configurations to optimize network performance or predict possible failures.

- **Representational state transfer configuration protocol (RESTCONF)**: RESTCONF provides an HTTP-based protocol that leverages REST principles to access data defined in YANG, as in NETCONF protocol. It allows configuration data, state data, and operations to be accessed using standard HTTP methods like GET, POST, PUT, DELETE, etc. In a smart city digital twin, RESTCONF can be employed to dynamically access and manipulate data related to traffic systems, weather sensors, and public services where HTTP-based communication and APIs exist. For example, in Figure 3.3, K-means clustering ML model is used with real-time data to make intelligent decisions to cluster and to assign Internet access accordingly [26].
- **Internet protocol flow information export (IPFIX)**: Flow data from routers, probes, and other devices that allow flow-based processing can be exported via the IPFIX protocol, sometimes referred to as "NetFlow v10." It is frequently used for gathering thorough data on network traffic flows, allowing for comprehensive network monitoring and analysis. In a DTN of a large data center, IPFIX can be utilized to continuously monitor network traffic between servers, switches, and routers [27]. AI algorithms can analyze this traffic data to detect anomalies, predict potential bottlenecks, and suggest load-balancing strategies to ensure optimal performance.
- **gRPC network management interface (gNMI)**: gRPC (gRPC remote procedure calls) is a transport protocol used by the network management protocol gNMI. It has features for configuration, state retrieval, and RPC operations and is built to meet contemporary management requirements. In a digital twin environment representing an IoT ecosystem, gNMI can be leveraged to enable efficient and scalable communication between different IoT devices and their corresponding digital representations. AI models use gNMI to dynamically adapt device configurations based on environmental conditions or user preferences [28].

3.2.2.3 Edge computing and distributed data collection

Edge computing refers to the decentralization of computing power, bringing computation closer to data sources such as IoT devices. In the context of digital twin technology, edge computing minimizes latency, reduces bandwidth use, and enables real-time processing. Moreover, it also facilitates real-time simulation and analysis. Similar to edge computing, the TinyML approach focuses on deploying machine-learning algorithms on edge physical devices close to the sensor to provide energy-efficient hardware and optimized process.

- **Real-time processing at edge**: Digital twins, or virtual replicas of physical objects or processes, require real-time data to function effectively. Edge computing allows data to be processed close to the source, offering low latency and quick insights [29]. This is crucial for scenarios where immediate response and adaptation are needed, such as industrial automation. Moreover, implementing TinyML algorithms directly on edge devices to process data in real time provides

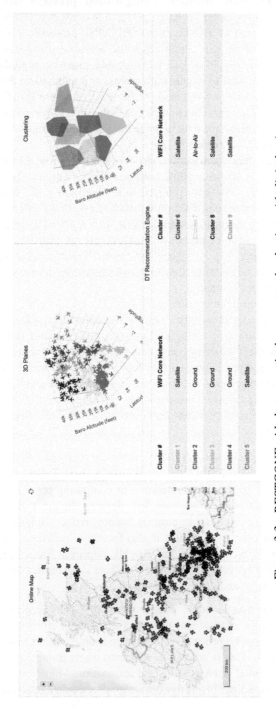

Figure 3.3　*RESTCONF-aided aeronautical core network selection with digital twin*

feature extraction, anomaly detection, or predictive maintenance at edge devices instead of overwhelming twinning processes at the cloud.
- **TinyML and model training at edge**: TinyML allows even small devices to carry out complex analytics, adapt to changes, and make decisions independently. This leads to more dynamic and responsive digital twins. Moreover, TinyML focuses on deploying machine learning models on small, power-efficient hardware like microcontrollers offering intelligent processing with minimal energy consumption. TinyML conducts training, model deployment, and model updating at edge utilizing tools like TensorFlow [30]. In this way, it converts the trained model to TensorFlow Lite for Microcontrollers to run on an edge device.
- **Deep transfer learning**: Deep transfer learning is another key technique applied in data collection. By leveraging a pre-trained model on a large-scale dataset, transfer learning is used to perform learning tasks on a different but related problem, saving computation time and resources. In the context of 6G and IoT, deep transfer learning is also used in fault diagnosis and detection, where data from various sources is collected and processed [31].
- **Data privacy and security measures**: Localized processing enhances the security and privacy of sensitive information, as less data is transmitted over potentially vulnerable networks. This approach also supports to minimize data transfer to centralized systems, reducing latency and network congestion while enabling real-time decision-making and immediate insights. There are different approaches exist for data privacy and security, including (i) encryption techniques to encrypt sensitive data while transmitting between edge devices and central systems; and (ii) authentication and authorization strategies to implement access controls ensuring that only authorized devices and systems can communicate with the digital twin [32].

3.2.3 Data preprocessing methods for cleaning, filtering, and transforming raw data

DTNs are sophisticated digital representations that mirror real-world systems, assets, and processes in real-time. Given their reliance on large volumes of data, the quality of this data is critical to the accuracy, performance, and utility of these digital twins. This necessitates a robust data preprocessing strategy encompassing cleaning, filtering, and transformation of raw data. This section explores various data preprocessing techniques to enhance the data quality for DTN.

3.2.3.1 Data cleaning

Data cleaning is essential to rectify "dirty" data into a consumable format. In the context of DTN, data may be imperfect due to sensor inaccuracies, transmission errors, system glitches, or various other inconsistencies. The process primarily involves handling missing data and managing outliers.

Handling missing data Data collected in DTN often encounter missing values due to sensor malfunction, loss of connectivity, or data entry errors. Such missing data can impair the functionality of digital twins. Techniques for addressing missing data include:

1. Ignoring or deleting: When missing data is minimal or specific records have a significant proportion of missing values, ignoring such data points may be appropriate. However, it should be used judiciously to prevent bias.
2. Manual imputation: Although labor-intensive, certain contexts might require manually filling in missing values for critical parameters.
3. Global constant imputation: Missing values can be replaced with a standard global constant like "NA" or "0" when the prediction of these values is complex.
4. Mean/median imputation: For numerical data, missing values can be replaced with the mean or median of the attribute.
5. Forward or backward fill: This method is valuable in time-series data where the previous or next value can serve as an estimate for the missing value.
6. Data mining algorithms: More sophisticated techniques like regression, interpolation, or machine-learning algorithms can be employed to predict missing values based on other available data.

Handling noise and outliers Noisy data and outliers can distort the model's predictive ability in a DTN. Noise can be managed using binning techniques, which sort data into "bins" or "buckets," followed by smoothing the bins using bin means, medians, or boundaries. Outliers can be detected using visual analysis, boxplots, or clustering. Once detected, they may be smoothed, corrected, or removed altogether to maintain data integrity.

Data integration In DTN, data often comes from diverse sources, ranging from IoT sensors to databases. Integrating these sources into a coherent dataset is a challenging but crucial step. It requires addressing schema integration issues, managing data value conflicts, and eliminating redundant data.

Data transformation Transforming data is crucial to make it suitable for use within a DTN. It includes normalization, aggregation, or other transformation techniques as per the dataset's needs. For example, it is converting all measurements to a common unit or scale or transforming certain continuous variables into categorical variables for easier handling (known as discretization).

Data reduction In situations where the data volume is large, it can be beneficial to reduce it to a more manageable size, while still maintaining its representational quality. Data reduction techniques could include dimensionality reduction (selecting a subset of the original variables), numerosity reduction (choosing alternative, smaller forms of data representation), or data compression. This results in faster processing times and lower storage requirements.

In summary, data preprocessing is a critical aspect of managing and maintaining DTNs. It involves a series of steps, including data cleaning, integration, transformation, and reduction, which together contribute to the overall quality and utility of the data used within the digital twins. Employing robust data preprocessing techniques will ensure that the DTN accurately mirrors its real-world counterpart and can provide reliable, insightful outputs.

AI-enabled data management for digital twin networks 67

3.2.4 Digital twin ontology and data modeling

As discussed in Section 3.2.1, each physical object has properties, relationships, and components. Since we have been exploring digital twin modeling in the context of 6G and IoT networks, it is crucial to capture not only the properties and relationships of the physical object but also telemetry data, component information, and any other relevant data necessary for simulating and representing the physical object's behavior. Consequently, ontologies and data modeling are central elements in the conception and implementation of data in DTNs.

An *ontology* defines the conceptual framework and a high-level abstract representation that describes the various components, attributes, and relationships between different elements of the physical entity being represented. It establishes a common understanding of the domain and provides a standardized vocabulary for describing and representing the digital twin's elements. Properties of various common physical objects (e.g., room, automobile, thing, etc.) have been defined through ontologies. The product ontology has various attributes, such as manufacturer and model number, along with their data types defined already. Additional data attributes may be added to show telemetry data, components, etc. In a sense, ontology can be considered as an extension of data modeling.

On the other hand, *data modeling* for digital twins involves the design and implementation of models to capture the behavior and states of physical entities or processes. These models provide the necessary predictive, diagnostic, or prescriptive capabilities for a digital twin. Data modeling is a more specific and practical aspect of the digital twin concept. It involves defining a structured representation of the data that is collected, processed, and used by the digital twin. Data models are often implemented in databases and are designed to efficiently store, retrieve, update, and delete data items. In Figure 3.4, the relation between ontology and data modeling as well as the underlying physical entity are shown.

Definition 5. *Ontology versus data model*: *Ontologies are essentially semantic models that allow machines to understand and respond to complex queries better. Data*

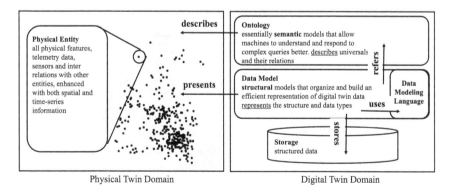

Figure 3.4   The relation among ontology, data modeling, and physical entity in digital twin

models are structural models that organize and build an efficient representation of DTD.

3.2.4.1 Data modeling languages

For modeling the DTD for a physical entity, various tools are available, including lexical data modeling languages, such as digital twin-definition language (DTDL), GraphQL, JSON-LD, and OWL. Additionally, graphical languages, such as unified model language (UML) or entity relationship diagrams (ERD), may be used for modeling the DTD.

- **Digital twin definition language (DTDL):** DTDL is a language designed for describing digital twins as well as the domain-specific information they contain. It enables the interoperability between digital twins and services by defining a standard for the twins' structure, relationship, and interfaces [26].

  A DTDL model for a traffic light system where a smart city project that involves thousands of IoT devices, including traffic lights, weather sensors, and autonomous vehicles, can be defined as below.

```
{
"@context": "dtmi:dtdl:context;2",
"@id": "dtmi:example:TrafficLight;1",
"@type": "Interface",
"displayName": "Traffic Light",
"description": "Digital twin of a city traffic light",
"contents": [
  {
    "@type": "Property",
    "name": "status",
    "schema": "enum",
    "enumValues": [
      {"name": "Red", "value": 1},
      {"name": "Yellow", "value": 2},
      {"name": "Green", "value": 3}
    ]
  },
  {
    "@type": "Telemetry",
    "name": "faultReport",
    "schema": "boolean"
  },
  {
    "@type": "Command",
    "name": "ChangeStatus",
    "commandType": "synchronous",
    "request": {"name": "NewStatus", "schema": "enum"}
  }
]
}
```

- **GraphQL:** GraphQL is an open-source data query and manipulation language that allows clients to define the structure of the data required. It offers a more efficient and powerful alternative to REST and allows for real-time data updates, which are crucial in the dynamic world of digital twins. By using GraphQL, clients can request exactly the data they need. This is especially crucial in IoT and 6G networks, where bandwidth and efficiency are significant considerations. Moreover, GraphQL supports subscriptions that allow real-time updates, which is essential for monitoring critical applications like IoT sensors in healthcare, traffic management, etc. [33]. However, these flexibilities come with complexities that need to be carefully managed, especially concerning performance and security.
- **JSON-LD:** JSON for linked data (JSON-LD) is a lightweight linked data format that helps create structured data over the Internet. It offers an interoperable means to construct, send, and receive data using standard web protocols.
- **Web ontology language (OWL):** OWL is a semantic web language designed to represent the complex real domain of physical entities like things, groups of things, and relations between things. Considering the IoT home use case as a basic OWL definition including superclass, subclasses, object, and data properties as follows:

  Superclass "Device": Every object that falls under the category of IoT devices would belong to a superclass "Device," which is a subclass of *"owl:Thing."*

  Subclasses: Different devices can be subclassed according to their functions or locations, such as "LightingDevice," "TemperatureSensor," and "Security-Camera."

  Object properties: These are relationships that connect instances of classes. In IoT, these might include relations like "connectedTo," "controls," "monitoredBy," etc.

  Individuals: Specific instances of devices like "LivingRoomLight," "KitchenThermostat," etc.

  Data properties: These tie actual data values like "currentTemperature," "lightStatus," etc.

  In this example, we have two specific devices, "LivingRoomLight" and "KitchenThermostat," as instances of "LightingDevice" and "TemperatureSensor," respectively. "KitchenThermostat" controls "LivingRoomLight," an object property. "currentTemperature" is a data property tied to "TemperatureSensor," with a specific value for "KitchenThermostat." This ontology can be extended by adding more classes, relationships, and properties to capture a rich and interconnected network of IoT devices, allowing more sophisticated querying and reasoning about the system's structure and behavior.
- **Unified modeling language (UML) and entity-relationship diagrams (ERDs):** UML is a standardized general-purpose modeling language in the field of software engineering. It provides a set of graphical notation techniques to create visual models of object-oriented software-intensive systems, which can be very useful when designing digital twin systems. Besides the UML, where

many diagrams exist within the modeling language, ERDs are a type of structural diagram used to represent and understand the interrelationships between entities in a database. Both UML and ERD provide an effective graphical representation and a way of modeling the language of the DTD model, facilitating easier design and understanding of the system.

3.2.5 Storage architectures for managing large-scale data in DTNs

DTs can frequently develop independently at a small scale, but larger-scale applications may require a higher level of data management [34]. The integration of AI-driven techniques further adds complexity and necessitates efficient and adaptable data storage methodologies. The cost of data collection and processing, the requirement for standardization across various data sources, and the need for useful, practical data all lead to the challenge [35].

Data storage techniques fall into five main categories as follows, with each providing particular data types and application needs:

1. Real-time series database: Time-series databases are pivotal for managing temporal data like industrial monitoring or business metrics. The selection of an appropriate storage engine is critical for the system when designing a DTD flow that generates temporal data. Another issue is how to query and aggregate a vast volume of sensor data in order to extract valuable information [36].

InfluxDB, a renowned time-series database, offers features like real-time data writing and reading, capable of handling write rates up to one million per second [37]. In the era of IoT and 6G, where latency and instantaneous processing of data is essential, InfluxDB's capacity to manage massive amounts of time-series data is critical. The need to monitor network performance, signal strength, and numerous other parameters in real-time makes this tool indispensable in these contexts.

2. Graph database: Graph databases, specializing in representing relationships between data components, utilize graph models to enhance querying [38]. It highlights links between data components and stores relevant "nodes" in graphs to accelerate context preservation.

An example is the knowledge graph, which portrays the interconnectedness of entities in the world. For complex DTNs, knowledge graphs offer a rich tool to explore network relations across various dimensions, thus forming an integral part of digital twin representation [39].

As one of the knowledge graphs, Neo4j is utilized for complex network mapping and understanding inter-device relationships; and Neo4j allows for robust modeling of IoT networks and 6G infrastructure. By mapping nodes and relationships, insights into network topology and potential vulnerabilities can be identified. Figure 3.5 shows a practical scenario for a digital twin of wind turbines where data is collected through a SCADA unit by utilizing a 5G-Next generation-radio access network (5G-NG-RAN), and finally, temporal convolution network (TCN) followed by a non-parametric k-nearest neighbor (kNN) regression for AI predictive modeling and Neo4j for graph storage and analysis [40].

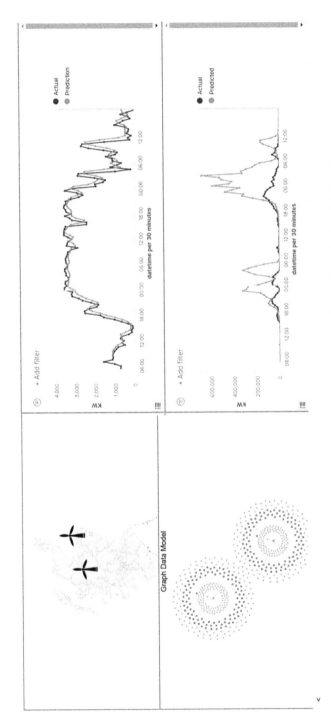

Figure 3.5 Graph database usage for wind turbines digital twin with Neo4j

3. **Key-value database:** Key-value databases, a subset of non-relational databases, use unique keys with corresponding values. They are suitable for storing log files from connected cyber-physical systems (CPSs) [38]. As demonstrated in Redis, it is perfect for storing large-scale, high-speed data in IoT; and it enables rapid data lookup using keys, supporting real-time analytics and caching in IoT applications [41].

4. **Document database:** A document database is a type of database that organizes data into collections and stores data like individual pieces of information in formats resembling documents. It is highly adaptable for storing different types of data because each of these components, documents, can have its own unique format.

Document databases like MongoDB organize data into collections, storing them in document-like formats. This flexibility makes them highly adaptable to varied data types. In IoT, the data's heterogeneity requires adaptable storage solutions. MongoDB's document-centric approach allows flexibility and performance, supporting varied sensor data and logs [42].

5. **Cloud-based storage:** Cloud-based storage solutions, employing services from providers like Amazon Web Services, Google Cloud, or Microsoft Azure, have emerged as a versatile choice for handling DTD.

The appeal lies in the reduced technological complexity and the increased adaptability to the rapidly changing digital landscape. By leveraging distributed computing and advanced security measures, cloud storage ensures scalability and reliability for large-scale digital twin applications [43].

With 6G's ultra-low latency requirements, edge computing can bring computation and data storage closer to the location where it's needed, minimizing delays. Platforms like Amazon Greengrass or Azure IoT Edge offer such capabilities [44]. On the other hand, distributed cloud storage across multiple regions ensures accessibility, redundancy, and compliance with local data regulations. Tools like Amazon S3 or Google Cloud Storage provide these functionalities [45] and give promising results for digital twin storage requirements.

3.3 AI-enabled data analysis and interpretation

Digital twin technology represents a bridge between the physical and digital world by creating a virtual representation of a real-world object or system. The introduction of AI into this paradigm has significantly expanded its capabilities, enabling complex data analysis, what–if scenarios, and prescriptive analysis. Considering the AI analysis for digital twin in IoT and 6G networks, broadly, there are four categories as follows: prescriptive analytics, what–if analytics, spatial analytics, and streaming and time series analytics. In this section, we will see AI analysis approaches with appropriate AI and ML models and specific use cases.

3.3.1 Overview of AI algorithms for data analytics in DTNs

Definition 6. *Prescriptive analytics Prescriptive analytics is used to analyze which steps to take for a specific situation. It is often described as a combination of* *descriptive and predictive analysis. When used in commercial applications, prescriptive analytics helps decipher large amounts of information to obtain more precise conclusions.*

Prescriptive analysis in IoT and 6G networks is utilized in the optimization of processes, interoperable decision making, and enhancing human decision-making. Multi-objective optimization algorithms like pareto optimization [46] adjust conflicting goals, such as minimizing costs while maximizing quality in industrial processes. Algorithms like particle swarm optimization [47] are also used for large-scale optimization, such as network routing in 6G networks. Graph neural networks (GNNs) model the interconnectedness of various IoT devices, and GNNs can provide insightful analysis for system-wide optimizations enhancing interoperable decision-making. Finally, explainable AI (XAI) models ensure that complex AI-driven decisions in critical areas like healthcare IoT or autonomous vehicles in 6G networks are interpretable and transparent to human users.

Definition 7. *What–if analytic What–if analysis empowers decision-makers to simulate various scenarios and their outcomes. For a digital twin of IoT and 6G networks, for example, AI models and tests how changes in device configuration or a network parameter might affect the final product [48].*

What–if analytic models different conditions and parameters, and AI algorithms can forecast potential failures, delays, or inefficiencies, allowing proactive adjustments. What–if analysis fosters a culture of experimentation without risk, enabling experts to explore new strategies, configurations, or architectures. Recurrent neural networks (RNNs) analyze time-series data in IoT and 6G networks and evaluate different conditions predicting network performance and performing outstanding results. Bayesian networks are also useful for modeling uncertainties and dependencies among various IoT components in different scenarios.

Definition 8. *Spatial analytics This method is used to analyze location-based IoT data and applications. Spatial analytics deciphers various geographic patterns, determining any type of spatial relationship between various physical objects. Parking applications, smart cars, and crop planning are all examples of applications that benefit from spatial analytics.*

Graph neural networks (GNNs) present an excellent approach to analyzing spatial features in 6G and IoT ecosystems. These networks are particularly effective when the data is structured in graphs, as they can capture complex patterns and dependencies between nodes. It also leverages the positions, including devices, their relations, and even user mobility [49].

Definition 9. *Streaming and time series analytics Streaming analytics, sometimes referred to as event stream processing, facilitates the analysis of massive "in-motion" data sets. These real-time data streams can be analyzed to detect emergency or urgent situations, facilitating an immediate response. The types of IoT data that benefit from streaming analytics include those used in traffic analysis, air trafficking, and the tracking of financial transactions. Time series analytics is based on time-based data, and data are analyzed to reveal any anomalies, patterns, or trends. Two* systems that greatly benefit from time series analytics are health-monitoring and weather-monitoring systems.

In the context of a digital twin of a city's 6G networks, the streaming analysis provides real-time guidance on traffic management and load balancing based on various factors like weather, events, or accidents. Edge AI models, which we discussed in Section 3.2.2, utilize lightweight deep learning models that run on edge devices in IoT, enabling real-time analysis and decision-making, like adjusting traffic signals based on congestion. Moreover, federated learning approach in 6G networks allows for decentralized model training across devices in real-time with streaming data, preserving privacy while enabling personalized services.

3.4 Ethical considerations and security aspects

DTNs, when combined with AI-enabled data management, have the potential to revolutionize industries and the academic era. On the other hand, DTs require vast amounts of data to operate, which brings forth discussions about the ethics and privacy of the individuals from whom the data is being collected. Ethical considerations involve ensuring that the data collected is obtained with proper consent and used responsibly. Security concerns also arise, as the data must be protected from unauthorized access and potential misuse.

3.4.1 Ethical considerations of using AI in data management

In terms of efficiency, accuracy, and innovation, artificial intelligence in data management provides appealing possibilities. To achieve ethical compliance, however, its integration necessitates careful investigation. The below approaches all take distinct perspectives on ethical considerations [50].

- **Fairness:** Biases in the training data for the AI can lead to discriminatory outcomes. To combat this, training data should be collected from diverse sources.
- **Transparency:** The AI's decision-making process can be complex and unexplainable, which causes trust issues. It is recommended to increase information disclosure by those creating or using AI systems to achieve greater transparency.
- **Privacy:** Depending on the service being provided, some personal data that can lead to the behavioral analysis of the user might be collected. Security measures should be taken to protect sensitive information.
- **Ownership:** Ethical ownership involves defining rights, transparent distribution, respecting data rights, and navigating shared data contributions as well as fostering fair agreements and collaborative governance models.
- **Non-maleficence and beneficence:** Non-maleficence ensures that AI systems and the data they process do not cause harm, which includes identifying and mitigating potential risks and negative impacts, as well as preventing discrimination, bias, and unintended results while beneficence involves actively promoting the well-being and positive outcomes of individuals and society.

- **Accountability:** Accountability in AI-enabled data management involves individuals taking responsibility for AI systems' behavior, outcomes, and consequences. It involves identifying who is responsible, acknowledging errors, mitigating biases, and ensuring AI aligns with human values and norms.
- **Trust:** Trust in AI is based on transparency, responsibility, and ethical design. Trust is built when AI procedures are explainable, accountable for faults, and aligned with human values.
- **Human autonomy and consent:** Human autonomy and control in AI-enabled data management involve preserving individuals' decision-making power and influence over AI systems. This ethical principle ensures that AI improves human capabilities by enabling individuals to understand, intervene, and modify decisions. It emphasizes transparency, explainability, and human agency preservation in the face of technological advancements.

3.4.2 Security aspects, including data privacy and protection

The significance of a digital twin is immense when utilized for objects, products, or processes. It holds trust and drives critical decisions for real-world counterparts. Consequently, addressing data security and privacy concerns becomes essential [51].

Cyberattacks may tamper with the data being sent from the IOT sensors on the physical twin. This could lead to differences between the twins, which can cause potential car accidents in the vehicular digital twins, incorrect treatment for patients in the healthcare digital twins, or financial losses in the industrial digital twins. Additionally, if there is a leak in the data that is collected to develop digital twins, sensitive information such as people's routines and health problems can be exposed, this is where the General Data Protection Regulation (GDPR) steps in, setting guidelines to ensure ethical data handling [52].

3.4.3 Privacy-preserving techniques for protecting sensitive data in DTNs

DTNs deal with large-scale data, some of which can be highly sensitive. Several methods can be employed to maintain privacy:

- **Federated learning:** Raw data is processed locally instead of being sent to a central server, keeping the private data secure [53].
- **Blockchain:** In addition to being decentralized, blockchain technology improves data privacy by offering a tamper-proof and transparent ledger for recording data transactions [54].
- **Edge computing:** As the data is processed near the data source, private information does not arrive on central servers [53].
- **Authentication and role-based permissions:** Authentication ensures that only authorized personnel can access and manipulate data, preventing unauthorized access, whereas role-based permissions entail assigning specific roles (such as administrators, analysts, or technicians) to individuals and granting them access only to the data required for their roles [55].

- **Homomorphic encryption:** Enables computations on encrypted data to be carried out without the need for decryption. This method maintains data confidentiality throughout the processing [55].
- **Secure multi-party computation (SMPC):** Prevents data leakage during data sharing between digital twins (inter-twin communication) [55].
- **Data minimization and masking:** By ensuring that only necessary data is kept, data minimization and masking reduce the chance of data exposure. Data masking replaces private information with fictitious information that is structurally similar, maintaining data integrity while ensuring confidentiality [56].
- **Data anonymization:** Data anonymization involves stripping the data of identifiable information while still keeping it helpful for training [57].
- **Differential privacy:** Involves adding controlled noise to data, thus making it harder to single out identifiable data while still enabling meaningful analysis [58].
- **Privacy by design:** Privacy by design entails incorporating measures to ensure the security and protection of sensitive information at every stage of the design and development process.

3.5 Conclusion and future directions

3.5.1 Summary of the chapter

This chapter on AI-enabled data management for DTNs presented a comprehensive overview of the fascinating convergence between AI and digital twin technologies.

In Section 8.1, the importance of data management in DTNs was underscored, detailing AI's role, challenges, and the three states of DTD. The detailed twinning process, involving AI-driven data acquisition, preprocessing, modeling, and data storage for DTNs, is elaborated in Section 8.2.

Section 8.3 delved into AI-enabled data analysis and interpretation, highlighting various algorithms, techniques for extracting insights, and the overall data flow. Finally, Section 8.4 presented the ethical considerations and security aspects, exploring the critical themes of ethics, privacy, and security in this rapidly evolving field.

3.5.2 Remaining challenges and open research questions in AI-enabled data management for DTNs

The relationship between artificial intelligence, DTNs, and data management is an engaging area with various opportunities. But navigating this complicated environment necessitates a careful and sophisticated approach to collecting, processing, and analyzing the DTD. Thus, there are still challenges and opportunities for further research. First of all, integrating data from various sources requires standards that ensure seamless *interaction, interpretation, and interoperability*. Future work can focus on developing universally accepted protocols and methodologies to leverage *standardization* on digital twin. Second, with the exponential growth of data, particularly in the context of IoT and the emerging 6G network, *scalable solutions* need to be devised to manage large-scale data efficiently. Moreover, as AI models become more complex, ensuring they adhere to *ethical principles* is a critical area of ongoing concern. The creation of ethical compliance frameworks tailored to AI-enabled DTNs will be a vital research avenue. Furthermore, *security* in the age of quantum computing and increasingly sophisticated cyber threats demands continuous innovation. Research into quantum-resistant cryptographic methods and AI-driven security protocols could be pioneering. Similarly, AI's energy consumption, especially in large-scale DTNs, presents a *sustainability* challenge. Exploring *energy-efficient* algorithms and green computing strategies remains an essential future direction. Last but not least, ensuring that AI models are *transparent, interpretable*, and aligning with human values and expectations is an area ripe for exploration. Balancing the need for *personalized* services with *privacy* preservation is a fragile task. Research into federated learning, differential privacy, and other privacy-preserving techniques specific to digital twin applications will be increasingly important.

In conclusion, AI-enabled data management for DTNs is a vibrant, multifaceted field with deep potential to reshape industries and societies. However, it is a journey with complexity, requiring collaboration across disciplines and various ethical landscapes. The future of this field seems to be an exciting intersection of technology, ethics, creativity, and human endeavor, representing a transformative era in the history of technological advancement.

References

[1] E. Sisinni, A. Saifullah, S. Han, U. Jennehag, and M. Gidlund, "Industrial Internet of Things: challenges, opportunities, and directions," *IEEE Transactions on Industrial Informatics*, vol. 14, no. 11, pp. 4724–4734, 2018.

[2] M. Usama, J. Qadir, A. Raza, *et al.*, "Unsupervised machine learning for networking: techniques, applications and research challenges," *IEEE Access*, vol. 7, pp. 65579–65615, 2019.

[3] M. Chen, S. Mao, and Y. Liu, "Big data: a survey," *Mobile Networks and Applications*, vol. 19, pp. 171–209, 2014. Available: https://doi.org/10.1007/s11036-013-0489-0.

[4] T. B. Ahammed, R. Patgiri, and S. Nayak, "A vision on the artificial intelligence for 6g communication," *ICT Express*, vol. 9, no. 2, pp. 197–210, 2023.

[5] I. Horrocks, M. Giese, E. Kharlamov, and A. Waaler, "Using semantic technology to tame the data variety challenge," *IEEE Internet Computing*, vol. 20, no. 6, pp. 62–66, 2016.

[6] Y. Pan and L. Zhang, "A bim-data mining integrated digital twin framework for advanced project management," *Automation in Construction*, vol. 124, p. 103564, 2021.

[7] M. Liu, S. Fang, H. Dong, and C. Xu, "Review of digital twin about concepts, technologies, and industrial applications," *Journal of Manufacturing Systems*, vol. 58, pp. 346–361, 2021.

[8] J. Ríos, G. Staudter, M. Weber, R. Anderl, and A. Bernard, "Uncertainty of data and the digital twin: a review," *International Journal of Product Lifecycle Management*, vol. 12, no. 4, pp. 329–358, 2020.

[9] A. Fuller, Z. Fan, C. Day, and C. Barlow, "Digital twin: enabling technologies, challenges and open research," *IEEE Access*, vol. 8, pp. 108952–108971, 2020.

[10] Y. Wu, K. Zhang, and Y. Zhang, "Digital twin networks: a survey," *IEEE Internet of Things Journal*, vol. 8, no. 18, pp. 13789–13804, 2021.

[11] M. Zhang, F. Tao, B. Huang, A. Liu, L. Wang, N. Anwer, and A. Nee, "Digital twin data: methods and key technologies," *Digital Twin*, vol. 1, p. 2, 2022.

[12] V. Piroumian, "Digital twins: universal interoperability for the digital age," *Computer*, vol. 54, no. 1, pp. 61–69, 2021.

[13] D. de Kerckhove, "The personal digital twin, ethical considerations," *Philosophical Transactions of the Royal Society A*, vol. 379, no. 2207, p. 20200367, 2021.

[14] D. Laney and M. Beyer, *The Importance of "Big Data": A Definition*. Gartner, 6 2012. Available: https://www.gartner.com/en/documents/2057415.

[15] F. Tao, H. Zhang, A. Liu, and A. Y. C. Nee, "Digital twin in industry: State-of-the-art," *IEEE Transactions on Industrial Informatics*, vol. 15, no. 4, pp. 2405–2415, 2019.

[16] A. To, M. Liu, M. Hazeeq Bin Muhammad Hairul, J. G. Davis, J. S. Lee, H. Hesse, and H. D. Nguyen, "Drone-based AI and 3D reconstruction for digital twin augmentation," in *International Conference on Human–Computer Interaction*. Springer, 2021, pp. 511–529.

[17] D. Lehner, J. Pfeiffer, E.-F. Tinsel, et al., "Digital twin platforms: requirements, capabilities, and future prospects," *IEEE Software*, vol. 39, no. 2, pp. 53–61, 2021.

[18] S. H. Khajavi, N. H. Motlagh, A. Jaribion, L. C. Werner, and J. Holmström, "Digital twin: vision, benefits, boundaries, and creation for buildings," *IEEE Access*, vol. 7, pp. 147406–147419, 2019.

[19] G. C. Deac, C. N. Deac, C. L. Popa, M. Ghinea, and C. E. Cotet, "Machine vision in manufacturing processes and the digital twin of manufacturing architectures," in *Proceedings of the 28th DAAAM International Symposium*, 2017, pp. 0733–0736.

[20] O. El Marai, T. Taleb, and J. Song, "Roads infrastructure digital twin: a step toward smarter cities realization," *IEEE Network*, vol. 35, no. 2, pp. 136–143, 2020.

[21] A. Arsiwala, F. Elghaish, and M. Zoher, "Digital twin with machine learning for predictive monitoring of CO_2 equivalent from existing buildings," *Energy and Buildings*, vol. 284, p. 112851, 2023.

[22] R. Lynn, M. Sati, T. Tucker, J. Rossignac, C. Saldana, and T. Kurfess, "Realization of the 5-axis machine tool digital twin using direct servo control from cam," *National Institute of Standards and Technology (NIST) Model-Based Enterprise Summit*, 2018.

[23] K. Chaiprabha and R. Chancharoen, "A deep trajectory controller for a mechanical linear stage using digital twin concept," in *Actuators*, vol. 12, no. 2. MDPI, 2023, p. 91.

[24] R. Minerva, G. M. Lee, and N. Crespi, "Digital twin in the IoT context: A survey on technical features, scenarios, and architectural models," *Proceedings of the IEEE*, vol. 108, no. 10, pp. 1785–1824, 2020.

[25] Y. Yigit, K. Huseynov, H. Ahmadi, and B. Canberk, "YA-DA: Yang-based data model for fine-grained IIoT air quality monitoring," in *2022 IEEE Globecom Workshops (GC Wkshps)*, 2022, pp. 438–443.

[26] T. Bilen, E. Ak, B. Bal, and B. Canberk, "A proof of concept on digital twin-controlled wifi core network selection for in-flight connectivity," *IEEE Communications Standards Magazine*, vol. 6, no. 3, pp. 60–68, 2022.

[27] R. Hofstede, P. Čeleda, B. Trammell, I. Drago, R. Sadre, A. Sperotto, and A. Pras, "Flow monitoring explained: from packet capture to data analysis with NetFlow and IPFIX," *IEEE Communications Surveys & Tutorials*, vol. 16, no. 4, pp. 2037–2064, 2014.

[28] Y. Yigit, B. Bal, A. Karameseoglu, T. Q. Duong, and B. Canberk, "Digital twin-enabled intelligent DDoS detection mechanism for autonomous core networks," *IEEE Communications Standards Magazine*, vol. 6, no. 3, pp. 38–44, 2022.

[29] A. Masaracchia, V. Sharma, B. Canberk, O. A. Dobre, and T. Q. Duong, "Digital twin for 6G: taxonomy, research challenges, and the road ahead," *IEEE Open Journal of the Communications Society*, vol. 3, pp. 2137–2150, 2022.

[30] P. P. Ray, "A review on TinyML: state-of-the-art and prospects," *Journal of King Saud University—Computer and Information Sciences*, vol. 34, no. 4, pp. 1595–1623, 2022.

[31] C. Tan, F. Sun, T. Kong, W. Zhang, C. Yang, and C. Liu, "A survey on deep transfer learning," in *Artificial Neural Networks and Machine Learning—ICANN 2018: 27th International Conference on Artificial Neural Networks, Rhodes, Greece, October 4–7, 2018, Proceedings, Part III 27*. New York, NY: Springer, 2018, pp. 270–279.

[32] M. Mukherjee, R. Matam, L. Shu, L. Maglaras, M. A. Ferrag, N. Choudhury, and V. Kumar, "Security and privacy in fog computing: challenges," *IEEE Access*, vol. 5, pp. 19293–19304, 2017.

[33] C. Yang, X. Tu, J. Autiosalo, R. Ala-Laurinaho, J. Mattila, P. Salminen, and K. Tammi, "Extended reality application framework for a digital-twin-based smart crane," *Applied Sciences*, vol. 12, no. 12, p. 6030, 2022.

[34] S. Singh, E. Shehab, N. Higgins, *et al.*, "Data management for developing digital twin ontology model," *Proceedings of the Institution of Mechanical Engineers, Part B: Journal of Engineering Manufacture*, vol. 235, no. 14, pp. 2323–2337, 2021.

[35] M. Attaran and B. G. Celik, "Digital twin: benefits, use cases, challenges, and opportunities," *Decision Analytics Journal*, vol. 6, p. 100165, 2023. Available: https://www.sciencedirect.com/science/article/pii/S277266222300005X.

[36] V. Damjanovic-Behrendt and W. Behrendt, "An open source approach to the design and implementation of digital twins for smart manufacturing," *International Journal of Computer Integrated Manufacturing*, vol. 32, no. 4–5, pp. 366–384, 2019.

[37] V. Kamath, J. Morgan, and M. I. Ali, "Industrial IoT and digital twins for a smart factory: an open source toolkit for application design and benchmarking," in *2020 Global Internet of Things Summit (GIoTS)*, 2020, pp. 1–6.

[38] Y. Lu, C. Liu, I. Kevin, K. Wang, H. Huang, and X. Xu, "Digital twin-driven smart manufacturing: connotation, reference model, applications and research issues," *Robotics and Computer-integrated Manufacturing*, vol. 61, p. 101837, 2020.

[39] Y. Zhu, D. Chen, C. Zhou, L. Lu, and X. Duan, "A knowledge graph based construction method for digital twin network," in *2021 IEEE 1st International Conference on Digital Twins and Parallel Intelligence (DTPI)*, 2021, pp. 362–365.

[40] M. Fahim, V. Sharma, T.-V. Cao, B. Canberk, and T. Q. Duong, "Machine learning-based digital twin for predictive modeling in wind turbines," *IEEE Access*, vol. 10, pp. 14184–14194, 2022.

[41] P. Muñoz, J. Troya, and A. Vallecillo, "Using UML and OCL models to realize high-level digital twins," in *2021 ACM/IEEE International Conference on Model Driven Engineering Languages and Systems Companion (MODELS-C)*, 2021, pp. 212–220.

[42] Z. Wang, R. Gupta, K. Han, H. Wang, A. Ganlath, N. Ammar, and P. Tiwari, "Mobility digital twin: concept, architecture, case study, and future challenges," *IEEE Internet of Things Journal*, vol. 9, no. 18, pp. 17452–17467, 2022.

[43] M. Mashaly, "Connecting the twins: a review on digital twin technology and its networking requirements," *Procedia Computer Science*, vol. 184, pp. 299–305, 2021.

[44] W. Shi, J. Cao, Q. Zhang, Y. Li, and L. Xu, "Edge computing: vision and challenges," *IEEE Internet of Things Journal*, vol. 3, no. 5, pp. 637–646, 2016.

[45] N. Bessis and C. Dobre, *Big data and Internet of Things: a roadmap for smart environments*. New York, NY: Springer, 2014, vol. 546.

[46] C. Zhan, H. Hu, X. Sui, Z. Liu, and D. Niyato, "Completion time and energy optimization in the UAV-enabled mobile-edge computing system," *IEEE Internet of Things Journal*, vol. 7, no. 8, pp. 7808–7822, 2020.

[47] N. Potu, C. Jatoth, and P. Parvataneni, "Optimizing resource scheduling based on extended particle swarm optimization in fog computing environments," *Concurrency and Computation: Practice and Experience*, vol. 33, no. 23, p. e6163, 2021.

[48] E. Ak, K. Duran, O. A. Dobre, T. Q. Duong, and B. Canberk, "T6CONF: digital twin networking framework for IPv6-enabled net-zero smart cities," *IEEE Communications Magazine*, vol. 61, no. 3, pp. 36–42, 2023.

[49] E. Ak and B. Canberk, "Forecasting quality of service for next-generation data-driven WiFi6 campus networks," *IEEE Transactions on Network and Service Management*, vol. 18, no. 4, pp. 4744–4755, 2021.

[50] A. Jobin, M. Ienca, and E. Vayena, "The global landscape of AI ethics guidelines," *Nature machine intelligence*, vol. 1, no. 9, pp. 389–399, 2019.

[51] Y. Wang, Z. Su, S. Guo, M. Dai, T. H. Luan, and Y. Liu, "A survey on digital twins: architecture, enabling technologies, security and privacy, and future prospects," *IEEE Internet of Things Journal*, pp. 1–1, 2023.

[52] C. Alcaraz and J. Lopez, "Digital twin: a comprehensive survey of security threats," *IEEE Communications Surveys & Tutorials*, vol. 24, no. 3, pp. 1475–1503, 2022.

[53] C. He, T. H. Luan, R. Lu, Z. Su, and M. Dong, "Security and privacy in vehicular digital twin networks: challenges and solutions," *IEEE Wireless Communications*, vol. 30, pp. 1–8, 2022.

[54] K. Liu, Z. Yan, X. Liang, R. Kantola, and C. Hu, "A survey on blockchain-enabled federated learning and its prospects with digital twin," *Digital Communications and Networks*, vol. 1, pp. 1–24, 2022, https://doi.org/10.1016/j.dcan.2022.08.001.

[55] J. Domingo-Ferrer, O. Farras, J. Ribes-González, and D. Sánchez, "Privacy-preserving cloud computing on sensitive data: a survey of methods, products and challenges," *Computer Communications*, vol. 140, pp. 38–60, 2019.

[56] R. Mukta, H.-y. Paik, Q. Lu, and S. S. Kanhere, "A survey of data minimisation techniques in blockchain-based healthcare," *Computer Networks*, vol. 205, p. 108766, 2022.

[57] S. Sengupta, V. Kaulgud, and V. S. Sharma, "Cloud computing security–trends and research directions," in *2011 IEEE World Congress on Services*. IEEE, 2011, pp. 524–531.

[58] G. Li, T. H. Luan, X. Li, J. Zheng, C. Lai, Z. Su, and K. Zhang, "Breaking down data sharing barrier of smart city: a digital twin approach," *IEEE Network*, pp. 1–9, 2023.

Chapter 4
AI-based traffic analysis in digital twin networks

Sarah Al-Shareeda[1], Khayal Huseynov[2], Lal Verda Cakir[3], Craig Thomson[3], Mehmet Ozdem[4] and Berk Canberk[3]

In today's networked world, digital twin networks (DTNs) are revolutionizing how we understand and optimize physical networks. These networks, also known as "digital twin networks (DTNs)" or "networks digital twins (NDTs)," encompass many physical networks, from cellular and wireless to optical and satellite. They leverage computational power and AI capabilities to provide virtual representations, leading to highly refined recommendations for real-world network challenges. Within DTNs, tasks include network performance enhancement, latency optimization, energy efficiency, and more. To achieve these goals, DTNs utilize AI tools such as machine learning (ML), deep learning (DL), reinforcement learning (RL), federated learning (FL), and graph-based approaches. However, data quality, scalability, interpretability, and security challenges necessitate strategies prioritizing transparency, fairness, privacy, and accountability. This chapter delves into the world of AI-driven traffic analysis within DTNs. It explores DTNs' development efforts, tasks, AI models, and challenges while offering insights into how AI can enhance these dynamic networks. Through this journey, readers will gain a deeper understanding of the pivotal role AI plays in the ever-evolving landscape of networked systems.

4.1 DTNs ecosystem

DTNs establish a parallel digital domain that mirrors a wide spectrum of physical networks encompassing diverse realms such as wireless networks, mobile networks 4G/5G/5G+/6G, optical networks, underwater, ground, campus, vehicular, airborne, satellite networks, and even space networks. By seamlessly integrating with and harnessing the computational power and AI capabilities intrinsic to this expansive landscape, DTNs transcend conventional boundaries [1]. This empowerment allows for the rapid execution of simulations and predictive processes within the mirrored environment, ultimately formulating highly refined feedback recommendations

[1] Artificial Intelligence and Data Engineering Department, Istanbul Technical University, Turkey
[2] BTS Group, Istanbul, Turkey
[3] School of Computing, Engineering, and The Built Environment, Edinburgh Napier University, UK
[4] Turk Telecom, Istanbul, Turkey

tailored to the difficulties of the associated real-world network. These refined recommendations, shaped by the marriage of computational power and domain-specific expertise, are then seamlessly disseminated to the actual entities within the physical network. This entire process manifests as a self-sustaining closed-loop data transmission paradigm, where the DTN continuously learns, evolves, and adapts based on the outcomes of its recommendations in the real world. The architecture of the DTN is partitioned into three layers (sometimes two of these three layers are further split into two sublayers) [2–5], as depicted in Figure 4.1:

- Physical layer: This layer aggregates and preprocesses data from the physical network. The linkage between the physical layer of the digital twin (DT) and the corresponding physical network is referred to as intra-twin communication.
- Virtual layer: The virtual layer of the DT crafts a virtual emulation of the underlying physical network, therein orchestrating the analysis and computation of the collected cleaned data through the integration of AI-driven ML/DL capabilities.
- Service or decision layer: The service or decision layer leverages the insights from the virtual layer to engender well-informed decisions and recommendations for optimizations within the physical realm via intra-twin communication.

Intrinsically intertwined within the DTN fabric, virtual entities use inter-twin links to establish intercommunication. The virtual twin architecture capitalizes on simulation, analysis, and optimization capacities to exhibit tangible enhancements across diverse aspects of the physical network, encompassing connectivity, robust communication, security protocols, scalability, Quality of Service (QoS) considerations, routing efficiency, and data-driven decision-making mechanisms. Comprehensive surveys have exhibited DTNs' multifaceted dimensions, potential, and transformative impact across various sectors. Wu *et al.* [6] unveil a panoramic view of the emerging concept of DTNs, delving into their foundational features, definitions, technologies, challenges, and potential application scenarios. A parallel work of Tang *et al.* [7] navigates the convergence of DT technology and 6G wireless systems, probing the realm of DT edge networks. Their investigation involves integrating DT technology with mobile/multi-access edge computing to fortify network performance, security, and cost-effectiveness. Similarly, Kuruvatti *et al.* [8] analyze existing literature, tracing the deployment of DT technology in the context of 6G, and engage in discussions regarding use cases, standards development, and the trajectory of future research. Mashaly [9] emphasizes the pivotal role of latency, efficiency, and security in effectuating successful DTN implementations. In a quest for novel techniques, McManus *et al.* [10] explore emergent strategies that facilitate data-driven control in new environments, encompassing SLAM-based sensing, network softwarization, RL, and collaborative testing techniques that pave the way for robust DTN constructions. Kroyer and Holzinger [11] dissect DT technology, plumbing its functionalities, core objectives, and technical prerequisites to align capabilities with specific application domains.

Transitioning toward Industry 4.0 integration, Zeb *et al.* [12] delve into the relationship between DT technology and Industry 4.0 Internet of Things (IoT) networks.

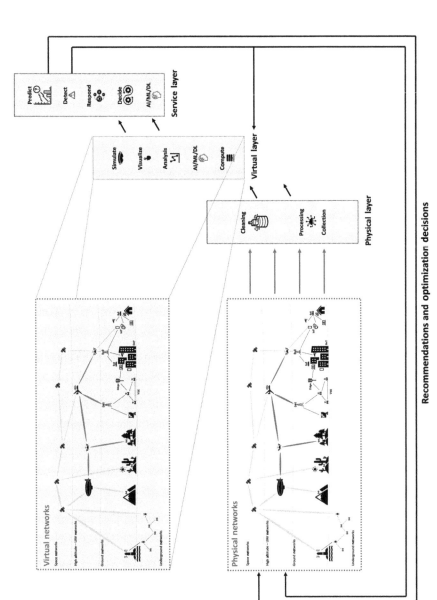

Figure 4.1 DTNs descriptive ecosystem

Their survey underscores the role of DTs in optimizing industrial processes, accentuating their interaction with IoT, cloud computing, ML, and advanced analytics. Similarly, Hakiri et al. [13] showcase the power of such virtual twin networks to revolutionize industries by interconnecting products, machinery, and human endeavors. Their analysis addresses challenges and underscores how DTs boost maintainability, flexibility, and responsiveness within the Industrial IoT. On the other hand, Su et al. [14] traverse the evolving landscape of the Internet of DTs, scrutinizing its architecture, technologies, security, and privacy concerns. Concurrently, Khan et al. [15] survey the application of FL in DT-enabled vehicular networks, exploring the fusion of FL and DT modeling to meet diverse latency, reliability, and experiential quality requirements. In addition to the survey above studies, numerous endeavors have been put forth to establish and fabricate DTs for a diverse array of network domains, encompassing cellular networks, wireless networks, optical networks, satellite and aeronautic networks, vehicular networks, and industrial IoT networks. The subsequent section provides an extensive overview of these concerted initiatives in constructing DTNs.

4.2 DTNs development efforts: literature review

4.2.1 Networks in general

Many efforts have been made to develop the DTNs and explore their potential. Among significant contributions, Larsson [16] explores network virtualization integration, focusing on scalability using domain-specific languages and providing insights into potential virtualization technologies within the DTN landscape. Addressing the challenges posed by emerging technologies, Hui et al. [17] propose DTNs as tools for "What-if" evaluations, addressing emerging technology challenges by leveraging data-driven methods. Similarly, Kislyakov [18] contributes a theoretical framework for defining DTNs within Industry 4.0 contexts, establishing clear communication models, and defining the roles of various models. Exploring the intricacies of handling diverse network data, Yang et al. [19] develop a sophisticated DTN system capable of managing multi-source heterogeneous network data, facilitating a holistic view of network performance. Innovating the architecture of future networks, Tan et al. [20] propose a cyber DT-based framework that supports intelligent services and resource management across both cyber and physical domains, heralding a new era of network adaptability. Additionally, Zhu et al. [21] introduce a comprehensive DTN architecture, enriching network capabilities through multi-layer integration and creating a unified framework for enhanced connectivity.

Contributing to the deployment methodology of DTs, Luan et al. [22] highlight the transformative potential of DTNs in leveraging computing capabilities. Hamzaoui and Julien [23] present a structured approach that addresses DTs' interactional and systemic aspects, harmonizing the development of DT technology across various fields. On the other hand, Chen et al. [24] delve into the classification of DTN models, paving the way for orchestrated simulations and optimized performance verification. Building on this classification framework, Szanto et al. [25] categorize DTs based on application roles, demonstrating the power of bi-directional communication through real-time protocols. These diverse contributions underscore the profound potential of DTNs to revolutionize network paradigms, redefine connectivity, and bridge the gap between physical and digital realms.

4.2.2 Cellular networks: 5G and beyond

Efforts in developing DTNs for mobile networks are proliferating, driven by a collective aspiration to enhance management practices and elevate operational efficiency. The investigations within this domain have yielded notable works. In the context of 5G, Tao *et al.* [26] delve into the role of DTNs as a catalyst for advancements, traversing architectural nuances and transformative technologies. Likewise, Seilov *et al.* [27] meticulously tailor DTNs to intricacies within the telecom landscape, enhancing the lifecycle with streamlined processes and vigilant traffic monitoring. Likewise, Tao *et al.* [28] introduce a thought-provoking paradigm, suggesting the data-driven modeling of 5G core networks through DT frameworks. Other works navigate the intricate terrain of optimizing mobile networks, weaving a synergy between NDTs and AI/ML [29–31].

In the landscape of 6G, Lu *et al.* [32] ingeniously embed DT technology within the edge networks, harnessing AI to orchestrate optimal network efficiency. Lin *et al.* [33] pivot toward real-time DTNs, outlining pragmatic implementations alongside AI/ML-driven optimization strategies. The endeavor of Duong *et al.* [34] embarks on a journey through multi-tier computing within the 6G realm, with AI, to boost the metaverse's architectural foundations. Guo *et al.* [35] spotlight the application of DT technology in 6G networks, specifically emphasizing heightening the QoS for mobile devices and applications. Masaracchia *et al.* [36] reveal the potential of DT collaboration for fortifying intelligent and resilient radio access networks (RANs) in the 6G landscape.

Further studies encompass diverse explorations, each carving a distinct niche within the expansive landscape of DTNs for mobile networks. Collectively, these endeavors resonate as a testament to the potential of DTNs specifically calibrated to harmonize with the intricacies of mobile networks [37–44].

4.2.3 Wireless networks

Efforts to develop DTNs for wireless networks have also been explored, aiming to organize data flows and enhance connectivity. Almeida *et al.* [45] introduce DTs to wireless networks, blending simulation and experimentation. Their position-based ML propagation loss model enhances ns-3 simulations, estimating propagation loss and significantly advancing network evaluation. Bariah *et al.* [46] envision DTs converging with smart city applications, particularly in wireless network domains. Their comprehensive representation of wireless network elements integrates AI for training, reasoning, and decision-making, amplifying wireless technologies in smarter, more sustainable smart cities. These explorations highlight the fusion of DT technologies with modern communication approaches in wireless networks. Further use cases are included in Section 4.3.

4.2.4 Optical networks

The integration of DTNs into optical networks is rapidly gaining attention, fueled by the capabilities of AI-powered virtual counterparts. Chen *et al.* [47] lead the way by deploying DTNs using ML techniques. This pioneering effort showcases the transformative potential of combining DTNs with AI, reshaping network operations. Zhuge [48] focuses on DTs for self-driving optical networks, skillfully integrating AI to comprehensively assess these networks' lifecycles. This work underscores the ongoing pursuit of adaptable and intelligent optical networks. Kuang *et al.* [49] delve into the evolution of intelligent optical infrastructures guided by purpose-built algorithms and AI-infused DT models. Likewise, Solmaz *et al.* [50] highlight DTs' potential in photonics-based architectures, envisioning energy-efficient smart campuses and emphasizing their diverse capabilities.

Janz *et al.* [51] shed light on AI-empowered optical transmission performance assessment. This entails automating tasks such as provisioning and risk mapping, and enhancing network efficiency through AI interventions. Wang *et al.* [52] disrupt the optical communication landscape with AI-driven DTs designed to tackle evolving challenges in efficient system management. Zhuge *et al.* [53] offer a comprehensive tutorial on AI-driven modeling, telemetry, and self-learning, enhancing DT capabilities to address intricate aspects like optical transmission impairments. Mello *et al.* [54] explore NDTs in data-driven optical networks, focusing on intent-based allocation strategies to improve efficiency and reliability. Authors in [55,56] adeptly integrate DT technology with optical wireless communication networks, leveraging AI to enhance reliability and network performance. Their efforts advance misconfiguration detection and Quality of Transmission (QoT) estimation. Similarly, Vilalta *et al.* [57] introduce innovative architectural frameworks for DT optical networks, guiding precise network design and optimization.

4.2.5 Satellite and aeronautic networks

Within non-terrestrial networks such as satellite and aeronautic networks, creating aeronautic DTNs has been presented by Chang *et al.* [58] in 2022. They carve a distinct mark by conceiving a multi-fidelity simulator designed for wireless unmanned aerial vehicle (UAV) networks. In response to the challenges of AI/ML-driven control, their amalgamation of two popular simulators orchestrates synchronized simulations. In parallel, Bilen *et al.* [59] navigate the intricacies of aeronautical landscapes, enhancing the selection of wireless core networks in dynamically evolving aero-settings as proof of DTN adaptability. Confronting data collection challenges, Moorthy *et al.* [60] steer a multi-fidelity simulator toward wireless UAV networks. The synergy of other simulators ignites a symphony of coordinated simulations, laying the bedrock for DT-infused advancements in UAV applications. The work of Brunelli *et al.* [61] tailors a DT model for a 3D urban air mobility interwoven with dynamic links guided by heuristic cost considerations. Zhou *et al.* [62] and Al-Hraishawi *et al.* [63] lead with developing a hierarchical DTN tailored to the distinctive satellite communication requirements. This endeavor showcases DT technology's ability to overcome challenges, spanning design, emulation, deployment, and maintenance. The hierarchical DTN assumes a dual role, seamlessly intertwining communication and networking twins.

4.2.6 Vehicular networks

In this area of research, there are many efforts to harness DTNs' potential to enhance vehicular networks' performance and efficiency. Palmieri *et al.* [64] contribute with a method that develops accurate digital-twin models for multi-agent vehicular networks. The approach involves simulating components across modeling languages and assessing the impact of network delay using AI techniques. In real-time traffic monitoring, Fennell [65] explores Apache Kafka's potential as a communication link between motorway sensors and a DT. The research aims to establish a communication architecture with improved availability, throughput, and low latency, investigating network traffic analysis's impact on latency and throughput. Wang and Chen [66] introduce a 5G-based framework for driverless tracked vehicles in the context of the Internet of Vehicles (IoV). Vehicle-to-Everything (V2X) (Vehicle-to-Everything) communication comes to the forefront as Wagner *et al.* [67] delve into the interaction between traffic light controllers and road vehicles. Investigating state-of-the-art V2X communication technologies, the study develops a traffic control system adhering to V2X protocols. These are a few of the many efforts in this domain; other use cases are highlighted in Section 4.3.

4.2.7 Industrial IoT networks

Innovative steps in constructing DTNs have been taken for the Industrial IoT (IIoT) networks. Kherbache *et al.* [68] propose a comprehensive architecture for IIoT, introducing an NDT tailored for closed-loop network management. Guimaraes *et al.* [69] pave the way for automated IoT instrumentation networks through DTs. Isah *et al.* [70] present a data-driven DTN architecture that bridges the gap between the physical and digital realms. Kherbache *et al.* [4] design an NDT to enhance IIoT network management and optimization. Jagannath *et al.* [71] delve into real-time modeling using DT frameworks within the IoT landscape. Rizwan *et al.* [72] combine FL and DT technology to empower IoT networks. Hakiri *et al.* [73] embark on developing the Hyper-5G project's NDT, a platform geared toward replicating IoT networks for experimentation, especially in evaluating novel IoT services. These endeavors reshape the fabric of IoT networks, seamlessly integrating the physical and digital dimensions for enhanced connectivity and insightful management.

4.3 Key tasks in DTNs analysis

The traffic observed within the physical network signifies a complex web of information exchanges and interactions among the tangible components of the network. Following the initial phases of filtering, refining, and preprocessing at the physical layer of the DT, this traffic seamlessly traverses to the virtual layer of the twin architecture. This particular layer provides a versatile foundation that adeptly handles the

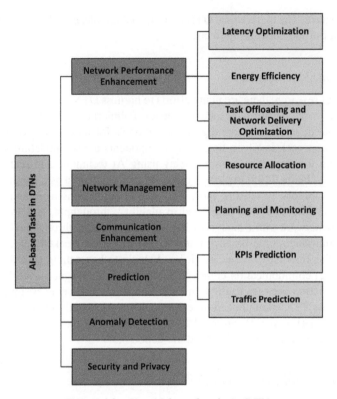

Figure 4.2 Key AI-based tasks in DTNs

following various facets, Figure 4.2, of network enhancement, resource management, communication optimization, predictive analysis, anomaly detection, and security and privacy assurance:

1. Network performance enhancement: A pivotal role of the DTN is optimizing the network's performance. It maintains a constant vigil over data exchanges, ensuring the network operates at its best efficiency. This proactive stance allows for the early identification and swift resolution of emerging performance issues, safeguarding network operations' uninterrupted and efficient flow; Section 4.3.1 covers the related AI-based works.
2. Network management: Efficiently overseeing network resources is a keystone of DTN functionality. It encompasses the sound allocation of resources, dynamic load balancing, and adept network configuration to guarantee seamless and reliable network communication; Section 4.3.2 addresses this key task of DTN.
3. Communication enhancement: DTN tirelessly endeavors to augment communication within the network. It strives to streamline data transmission, minimize latency, and ensure uninterrupted connectivity. This is quintessential for perpetuating efficient data exchange amongst network constituents and users; Section 4.3.3 unravels the literature and uses AI-based tools and techniques.

4. Prediction analysis: Armed with historical data, the DTN's predictive capabilities forecast forthcoming network trends and potential disruptions. This foresight empowers preemptive interventions, whether bolstering resources in anticipation of increased demand or proactively addressing impending issues before they impact network performance; Section 4.3.4 covers the related AI-based works.
5. Anomaly detection: DTN's strength lies in rapidly identifying anomalies and irregularities within the network. Regarding security breaches, technical glitches, or atypical network behavior, DTN should swiftly alert administrators, enabling them to take immediate remedial action; Section 4.3.5 covers the related AI-based works.
6. Security and privacy: Network security and data privacy are non-negotiable. DTN methodically scrutinizes data flows, unearths vulnerabilities, and fortifies network defenses against potential threats. Moreover, it ensures strict adherence to data privacy regulations, safeguarding sensitive information from unauthorized access; Section 4.3.6 covers the related AI-based works.

Harmoniously executed tasks empower DTN to offer a holistic suite of analytical capabilities, fostering proactive traffic analysis and optimization. To facilitate the effective execution of these tasks, advanced AI-driven models and analytical tools are strategically integrated throughout the DTN architecture. This collaborative endeavor culminates in elevated network efficiency and heightened reliability from the initial stages of data aggregation to the culminating decision-making phase. It significantly contributes to the overall optimization of network operations.

Key tasks in DTNs analysis

- **Network performance enhancement**: Ensuring best network operation.
- **Network management**: Efficiently allocating network resources.
- **Communication enhancement**: Optimizing data exchange and connectivity.
- **Prediction analysis**: Forecasting future network trends and disruptions.
- **Anomaly and fault detection**: Swiftly identifying irregularities and threats.
- **Security and privacy analysis**: Protecting network integrity and data privacy.

4.3.1 AI-based network performance enhancement

A substantial body of academic research on enhancing network performance through AI exists. This body of research delves into incorporating DTNs to enhance the effectiveness of different types of networks. These studies collectively examine the confluence of AI, ML, and DL techniques to tackle challenges in network operation and ultimately achieve improved performance. Notably, the research and literature address these primary facets of network performance:

4.3.1.1 Latency optimization

In the pursuit of optimizing network performance by reducing latency, several research endeavors have emerged. Saravanan *et al.* [74] employ innovative approaches grounded in DTs to enhance mobile edge computing within the dynamic 6G network environment. Leveraging AI techniques, specifically the Lyapunov approach and Actor-Critic (A3C) learning, they address the intricate challenges arising from user mobility and the volatile nature of edge computing environments. Yang et al. [75] introduce a flow emulation framework tailored for delay–tolerant networks. They emphasize accurately replicating network traffic patterns, offering insights into optimizing latency. Along the same lines, Van Huynh et al. [76] take on the noteworthy challenge of achieving fairness-aware latency minimization in DT-aided edge computing. In [77,78], the same group of authors concentrates on minimizing latency in computation offloading, particularly within IIoT environments, utilizing DT wireless edge networks. Wang et al. [79] present an innovative time-sensitive networking (TSN) scheduling technique applied in delay–tolerant networks; their research centers on optimizing data transmission and scheduling within intricate network structures. Tang et al. [2] endeavor to attain differentiated service level agreements (SLAs), low latency, and deterministic bandwidth in complex network architectures by deploying DTNs. Likewise, Ferriol-Galmes et al. [80] introduce TwinNet, a cutting-edge graph neural network (GNN)-based model designed to estimate QoS metrics accurately. Meanwhile, Duong et al. [81] shift their research focus toward comprehensive end-to-end latency minimization within DT-aided offloading scenarios in UAV networks. Their considerations extend to ultra-reliable low latency communication (URLLC) and UAVs. Jiang et al. [82] employ the proximal policy optimization (PPO) AI approach to optimize resource-intensive task computing in dynamic UAV networks. In parallel, Li et al. [83] delve into intelligent task offloading within UAV-enabled mobile edge computing (MEC) environments, utilizing DTs. They employ the double deep Q-network (DDQN) algorithm for optimization. Moreover, Wang et al. [84] delve into the intricate realm of computation offloading decisions for UAVs, particularly in disaster scenarios. Their research leverages the upper confidence bound (UCB)-based stable matching algorithm to enhance decision-making processes, ultimately optimizing latency and network performance (Table 4.1).

Table 4.1 Latency optimization works

Work	Utilized AI tools
Saravanan et al. [74]	Lyapunov approach and A3C learning
Yang et al. [75]	Emulation framework
Van Huynh et al. [76–78]	N/A
Wang et al. [79]	Time-sensitive scheduling
Tang et al. [2]	N/A
Ferriol-Galmes et al. [80]	TwinNet (GNN)
Duong et al. [81]	N/A
Jiang et al. [82]	PPO
Li et al. [83]	DDQN
Wang et al. [84]	UCB-based stable matching algorithm

Table 4.2   Energy efficiency enhancement works

Work	Used AI tool
Zhao et al. [85]	Association scheme
Chen et al. [86]	A3C method
Shui et al. [87]	DQN
Liu et al. [88]	ML and DRL

4.3.1.2   Energy efficiency

Several noteworthy studies have emerged in the context of network energy efficiency enhancement. Zhao et al. [85] focus on optimizing energy consumption and computational overhead in 6G networks through an innovative DT-edge association scheme. Chen et al. [86] explore the integration of MEC and DTs to minimize energy consumption, utilizing the asynchronous advantage A3C method for optimization. Shui et al. [87] tackle reducing energy usage in cell-free networks, maintaining URLLC requirements, and employing AI optimization methods such as DQN and DT. In IoT networks, Liu et al. [88] address energy efficiency issues, leveraging AI, ML, and deep RL (DRL) to optimize energy consumption through resource prediction and migration strategies. These studies collectively contribute to pursuing more sustainable and efficient network infrastructures (Table 4.2).

4.3.1.3   Task offloading and content delivery optimization

In the rapidly evolving realm of the Internet of Everything (IoE), driven by advancements in 5G/6G and AI technologies, Yi et al. [89] create a highly efficient content delivery system specially designed for the IoE environment, and they have employed DT technology for this purpose. Their primary goal is to fine-tune content delivery using advanced AI techniques like long short-term memory (LSTM) and RL. Similarly, Gao et al. [90] are tackling the impending challenges posed by the massive content delivery demands expected in the 5G and 6G communication era. Their strategy is to integrate DT technology into edge networks, enabling precise measurement of Quality-of-Decision (QoD) and Quality-of-Experience (QoE) metrics. These metrics, in turn, inform content delivery strategies, catering to both machine-centric and user-centric communication scenarios. Moving forward, Guemes-Palau et al. [91] delve into the complexity of network management, where they emphasize the use of DTNs and optimization techniques. Their approach involves harnessing the power of DRL and evolutionary strategies (ES) to optimize network performance. In a related endeavor, Lin et al. [92] introduce a novel packet-action sequence model, a tool for accurately modeling network behavior within DTNs. Their work focuses on understanding the dynamic nature of network behaviors, highlighting the crucial role of AI in comprehending and optimizing these intricate patterns.

Shifting gears slightly, Ursu et al. [93] have directed their attention toward optimizing cloud network functions, specifically HTTP load balancing. Their research takes place within Kubernetes (K8s) cluster environments. It enhances the modeling

Table 4.3 Task offloading and content delivery optimization works

Work	Used AI tool
Yi et al. [89]	LSTM and RL
Gao et al. [90]	QoD and QoE metrics
Guemes-Palau et al. [91]	DRL and ES
Lin et al. [92]	Packet-action sequence model
Ursu et al. [93]	ML load balancing
Zhao et al. [94]	Genetic algorithm
Sun et al. [95]	FL
Han et al. [96]	PL
Liu et al. [97]	Decision tree and double DL

and optimization of network function behaviors using DTs and ML techniques. Further up in the skies, Zhao et al. [94] tackle challenges related to low-orbit satellite networks integrated with terrestrial networks using DTs. They employ genetic algorithm techniques to optimize network performance, reduce handover frequency, enhance data delivery, and improve network efficiency. Sun et al. [95] aim to elevate air–ground network performance using FL. Their approach involves collaborative model training without the need to share sensitive data. Dynamic DTs are integrated to provide real-time reflections on network status, contributing to this ambitious enhancement effort. Lastly, Han et al. [96] present a pioneering Polymorphic Learning (PL) framework for DTNs. Their emphasis is on ensuring secure and personalized services. Their contribution lies in optimizing learning algorithms tailored to individualized service needs. Lastly, Liu et al. [97] present a pioneering DT-assisted scheme thoughtfully tailored to the emerging landscape of 6G networks. Their efforts revolve around the intelligent offloading of tasks from mobile users to cooperative mobile-edge servers, with the overarching aim of optimizing the utilization of network resources. The shrewd application of AI techniques fortifies this endeavor, including the Decision Tree algorithm and double DL, employed to fine-tune task-offloading decisions, ultimately ushering in enhanced network performance (Table 4.3).

4.3.2 AI-based network management
4.3.2.1 Planning and monitoring
Regarding planning and placement, Qu et al. [98] introduce a decision tree framework for intelligent network management, primarily emphasizing network optimization and management strategies. Similarly, Zhao et al. [99] propose a network planning system based on DTNs to enhance efficiency through architectural design. Corici and Magedanz [100] explore the potential of DTs in optimizing 5G network management. In a related context, Bilen et al. [101] propose a DT framework for managing aeronautical ad-hoc networks within the context of 5G and beyond. Chukhno et al. [102] focus on network management by dynamically placing social DTs within advanced 5G+/6G networks. Additionally, they aim to enhance data

exchange efficiency and service discovery, improving operational efficiency by optimizing IoT DT placement within the network. In a similar context, Xiao et al. [103] present an evolutionary framework to optimize server layout within DTNs. In vehicular networks, Zhao et al. [104] propose a DT architecture to enhance vehicular network routing policies, incorporating policy learning and real-time adaptation using DT representations. Fu et al. [105] concentrate on improving decision-making in connected autonomous vehicles through a DT-assisted framework, employing hierarchical multi-agent RL to enhance vehicle collaboration and decision-making efficiency. Al-Hamid and Al-Anbuky [106] delve into the modeling and analysis of dynamic groupings in vehicular networks, aiming to assess the performance of dynamic vehicle groups, particularly during the self-healing and self-formation phases.

Monitoring and analysis are critical aspects in the evolving network management landscape. Jiao et al. [107] explore DT management in tower networking, focusing on system analysis and design methods. Ren et al. [108] present a comprehensive scheme for managing SLA quality in a two-level Cloud RAN, employing AI techniques such as GraphSAGE, Deep Double Q-Network (DDQN), and Bayesian convolutional neural network (BCNN). Rosello et al. [109] introduce the NDT as an experimental and verification framework for 6G technology. Kherbache et al. [110] emphasize crafting a DTN tailored for the IIoT with a primary goal of real-time intelligent management. Wei et al. [111] introduce the concept of a DTN to foster innovation in Industry 4.0, specifically highlighting data-driven routing within the DTN framework. Raj et al. [112] develop a DTN architecture tailored for software-defined networking (SDN)-based networks, allowing for monitoring and verification without disrupting the live system. This architecture incorporates knowledge graph (KG) and template-based contextual information (Table 4.4).

Table 4.4 Planning and monitoring works

Work	Used AI tools
Qu et al. [98]	Decision tree
Zhao et al. [99]	N/A
Corici and Magedanz [100]	N/A
Bilen et al. [101]	N/A
Chukhno et al. [102]	N/A
Xiao et al. [103]	ES
Zhao et al. [104]	Policy learning
Fu et al. [105]	Hierarchical multi-agent RL
Al-Hamid and Al-Anbuky [106]	N/A
Jiao et al. [107]	N/A
Ren et al. [108]	GraphSAGE, DDQN, and BCNN
Rosello et al. [109]	N/A
Kherbache et al. [110]	N/A
Wei et al. [111]	N/A
Raj et al. [112]	KG and template-based contextual information

4.3.2.2 Resource allocation

In the realm of network management through efficient resource allocation, various cutting-edge approaches have emerged. For cellular networks, Sun et al. [113] initiate this narrative by introducing a DTN designed for 5G technology. Their work focuses on upholding stringent SLA standards, a crucial aspect of network performance. This approach ultimately boosts network efficiency by facilitating comprehensive mapping and management. Baranda et al. [114] introduce an AI/ML platform integrated into the 5G workflow for scaling based on real-time data metrics, catering to SLA management in a DT service scenario. Zhou et al. [115] introduce a resource management scheme that uses DT technology for 6G. In [116], Huang et al. focus on developing a collective RL method to efficiently allocate resources in real time and adapt to varying service demands. Tao et al. [117] delve into resource management within DTNs for 6G service requests. They focus on enhancing service response using a software-defined DTN architecture and Proximal Policy Optimization DRL (PPO-DRL). Duran et al. [118] aim to enhance core network management efficiency by integrating intelligent methods into topology discovery processes. The method reduces complexity and resource consumption, mainly employing multilayer perceptron (MLP) for visit decision recommendations. Vila et al. [40] introduce an NDT architecture tailored for RANs within 5G networks. Their approach leverages RL to train and optimize RAN operations through the NDT framework.

Su et al. [119] shift the focus toward addressing compute-intensive applications within MEC environments. They introduce a DT-based task offloading scheme, leveraging the potent double DQN (DDQN) approach. This novel strategy optimizes resource allocation, enhancing network performance. Also, for MEC, Dai et al. [120] bring forward a DTN-assisted system designed to optimize service placement and workload distribution. Their primary focus revolves around the strategic placement of services and efficient distribution of workloads within these systems. He et al. [121] introduce a hierarchical FL framework, seamlessly integrating DT and MEC into cellular networks. This approach optimizes resource allocation and network performance, featuring DRL. Guo et al. [122] introduce a network sensing edge deployment optimization mechanism for DT systems. The mechanism enhances management efficiency by optimizing edge deployment based on network state, employing an activity estimation model and a chaotic particle swarm optimization (PSO) algorithm. Merging DT and MEC concepts, Yuan et al. [123] use DNN for task offloading with A3C, aiming to optimize traffic in pursuit of reduced inference latency. Luo et al. [124] harness DT technology to boost efficiency in wireless communication networks. Their central focus lies in optimizing resource allocation through distributed DRL techniques. Wieme et al. [125] optimize Bluetooth mesh networks with DT technology, emphasizing efficient relay selection. Their work underscores the pivotal role of AI-driven optimization techniques in enhancing network behavior and configuration.

For SDNs, Naeem et al. [126] propose a DT-enabled deep distributional Q-network (DDQN) framework to optimize resource allocation and network slicing policies. Abdel-Basset et al. [127] delve into slicing-enabled communication networks, aiming for optimal resource allocation while maintaining QoS. Their solution relies on FL and differential privacy techniques. Likewise, Hong et al. [128] present the NetGraph DT platform for intelligent data center network management, and, in [129], a platform combining SDN and DT technology for autonomous network management is introduced. Similarly, Gong et al. [130] propose a holistic network virtualization architecture that combines DT and network slicing. They introduce an environment-aware offloading mechanism based on integrated sensing and communication systems to address computation offloading challenges. The use of AI, including the Shapley-Q value and DDPG algorithm, aids in solving optimization problems related to task scheduling and resource allocation.

In IIoT and IoT networks, Lu et al. [131] merge DTs with edge networks, creating DTNs for IoT optimization. They employ blockchain-empowered FL and RL for traffic pattern optimization and resource allocation. Dai et al. [132] employ DTN and the Lyapunov optimization method with the A3C algorithm to enhance energy efficiency and processing in IIoT systems by optimizing resource allocation. Bellavista et al. [133] propose an application-driven DTN middleware for IIoT. Their focus is on simplifying device interactions and dynamically managing network resources. Luan et al. [134] develop an intelligent industrial system that synergizes AI and DTs, ultimately optimizing network performance in smart sensors for manufacturing. AI-driven load-balancing strategies are among their tools to improve network performance. Guo et al. [135] propose a device-to-device communication-aided DT-edge network for efficient management in 6G IIoT. Tang et al. [136] delve into the realm of Industry 4.0, where they utilize DTN technology to personalize services. AI techniques like multi-agent deep deterministic policy gradients (MA-DDPGs) are instrumental in optimizing resource allocation for personalized IIoT services. Geisler et al. [137] manage overload in IoT mobile networks, specifically focusing on the intricate task of orchestrating signaling traffic.

Morette et al. [138] optimize optical networks by harnessing the power of DT technology, emphasizing the pivotal role of ML techniques in elevating network performance and the strategic allocation of vital resources. In parallel, the authors of [139,140] utilize a DT-based approach at the physical layer of optical networks to evaluate transmission quality and overall network performance. Hao et al. [141] focus on automating the scheduling and maintenance of optical transmission networks. The DTN-based automatic scheduling method uses a modified DDPG algorithm for topology optimization, aiming to reduce maintenance costs and improve resource utilization. Wu et al. [142] introduce a multifactor-associated network topology portrait (NTP) scheme for DTN in optical networks. The scheme aims to optimize dynamic routing computation using different routing algorithms, emphasizing improving efficiency and performance.

Fu et al. [143] aim to enhance air logistics through DTN with AI-driven optimization for satellite and aerial-based networks. The framework incorporates transformer-based information fusion and multi-agent DRL for UAV cooperation in route planning. In emergency communication scenarios, Guo [144] focuses on establishing seamless connections between users and resource-constrained aerial base stations using device-to-device communication and advanced Q-learning techniques. Gong et al. [145] pioneer innovative approaches, employing a blockchain-aided

Stackelberg game model and a Lyapunov stability theory-based model-agnostic meta-learning framework. These methodologies are strategically applied to foster optimal resource allocation and informed decision-making within the intricate landscape of satellite-ground integrated DTNs. Zhang et al. [146] fine-tune dynamic data transmission within the expansive domain of DT services. UAVs take center stage as AI techniques deftly chart optimal flight paths and refine data transmission strategies, effectively elevating network performance and enhancing user satisfaction. Likewise, Gong et al. [145] venture into the dynamic realm of UAV-assisted edge computing systems. Their work, rooted in DRL, is dedicated to the meticulous optimization of service migration processes, promising to augment network efficiency further.

For vehicular networks, Dai and Zhang [147] propose enhancing vehicular edge computing networks using adaptive DT-enabled networks and employing DRL to minimize offloading latency. Li et al. [83] introduce a novel approach for computing resource management of edge servers in vehicular networks. They emphasize constructing tailored two-tier DTs and employing AI for real-time resource allocation. Cazzella et al. [148] contribute a data-driven approach to create a DT for V2X communication scenarios emphasizing addressing high mobility challenges. These diverse approaches illuminate the path toward efficient resource allocation in the ever-evolving network management landscape. Researchers have collectively paved the way for innovative solutions, leveraging DT technology and AI techniques to optimize resource allocation from cellular networks to IoT, optical networks, air logistics, and vehicular networks (Table 4.5).

4.3.3 AI-based communication enhancement

In the realm of communication enhancement through DTN, several notable contributions have been made by researchers employing AI, ML, and DL techniques. Sun et al. [95] introduce the DT edge network concept for mobile offloading decision-making in 6G environments; their contribution utilizes Lyapunov optimization and the A3C DRL algorithm. Wang et al. [149] develop a DTN for UAV swarm-based 5G emergency networks. They leverage DL techniques for swarm deployment under varying conditions to enhance communication efficiency during emergencies. Jian et al. [150] propose DT technology for communication channels beyond 5G and 6G communication systems. Xiang et al. [151] create a 5G wireless network DT system using ray tracing propagation modeling. The system accurately models wireless signals for 5G networks, helping identify weak coverage areas and optimize engineering parameters. In the IIoT context, the authors of [131,152] introduce DT edge networks of FL to optimize communication efficiency and reduce transmission energy costs. Liang et al. [153] address the limitations of traditional wavelength division multiplexing communication networks for power IoT with an electric–elastic optical network architecture. Their work optimizes communication bandwidth scheduling and introduces a UCB automatic routing selection algorithm. Li et al. [154] study a DT-empowered integrated sensing, communication, and computation network, optimizing multi-input multi-output (MIMO) radars and computation offloading energy consumption. They employ the multiagent proximal policy optimization (MAPPO) framework for intelligent offloading decisions.

Table 4.5 Resource allocation works

Work	Used AI tools
Sun et al. [113]	DTN and SLA mapping
Baranda et al. [114]	ML
Zhou et al. [115]	N/A
Huang et al. [116]	Collective RL
Tao et al. [117]	PPO-DRL
Duran et al. [118]	MLP
Vila et al. [40]	RL
Su et al. [119]	DDQN
Dai et al. [120]	N/A
He et al. [121]	Hierarchical FL and DRL
Guo et al. [122]	N/A
Yuan et al. [123]	DNN and A3C
Luo et al. [124]	Distributed DRL
Wieme et al. [125]	N/A
Naeem et al. [126]	DDQN
Abdel-Basset et al. [127]	FL and differential privacy
Hong et al. [128]	NetGraph
Gong et al. [130]	N/A
Lu et al. [131]	Blockchain-empowered FL, and RL
Dai et al. [132]	DTN, Lyapunov optimization, and A3C algorithm
Bellavista et al. [133]	Middleware
Luan et al. [134]	N/A
Guo et al. [135]	N/A
Tang et al. [136]	MADDPG
Geisler et al. [137]	N/A
Morette et al. [138]	ML techniques
Curri et al. [139] and Borraccini et al. [140]	N/A
Hao et al. [141]	Automatic scheduling and DDPG
Wu et al. [142]	Multifactor-associated NTP
Fu et al. [143]	Transformer-based fusion and multi-agent DRL
Guo [144]	Q-learning
Gong et al. [145]	blockchain, Stackelberg model, and Lyapunov stability Meta-learning
Zhang et al. [146]	N/A
Liu et al. [97]	Decision tree and double DL
Dai and Zhang [147]	DRL
Li et al. [83]	N/A
Cazzella et al. [148]	N/A

Regarding V2X communications, Zelenbaba et al. [155] use DTs to evaluate hardware and system performance in wireless vehicular communication links. They create site-specific DTs for assessing link reliability in vehicular communication scenarios. Lv et al. [156] optimize optical wireless communications in intelligent transportation systems (ITS) with a focus on visible light communications (VLC). They propose a carrier-less amplitude/phase modulation scheme and employ convolutional neural networks (CNNs) for feature extraction. Demir et al. [157] apply

Table 4.6 Communication enhancement works

Work	Used AI tools
Sun et al. [95]	Lyapunov optimization and A3C DRL
Wang et al. [149]	DL
Jian et al. [150]	N/A
Xiang et al. [151]	Ray tracing propagation modeling
[131,152]	FL
Liang et al. [153]	UCB automatic routing selection algorithm
Li et al. [154]	MAPPO
Zelenbaba et al. [155]	N/A
Lv et al. [156]	CNN
Demir et al. [157]	N/A
Liu et al. [158]	ML

DTs in connected and autonomous vehicles to improve wireless QoS in non-line-of-sight scenarios. Their methodology predicts and mitigates the impact of obstacles on wireless communication. Liu et al. [158] focus on achieving energy-efficient communication in the IoV within 6G mobile networks. They introduce a DT method combined with ML to model the millimeter-wave channel and optimize energy-efficient communication. These contributions underscore the significant role of AI, ML, and DL techniques within DTNs to enhance communication efficiency and address emerging challenges in various communication scenarios (Table 4.6).

4.3.4 AI-based prediction analysis

DTNs have gained prominence recently for their ability to address various prediction analysis tasks across diverse domains. These tasks often involve using AI, ML, and DL techniques to enhance predictions and optimize network performance. A collection of works that employ DTNs to tackle prediction challenges has been presented. To predict key performance indicators (KPIs), Wang et al. [159] utilize DT technology to manage the complexities of slicing in 5G networks. They create virtual representations of slicing-enabled networks and predict performance changes. They hint at using the GNN model to capture relationships among slices and predict metrics like end-to-end latency. Schippers et al. [160] introduce a novel DT approach to predict KPIs for mission-critical vehicular applications. They employ AI techniques to accurately predict KPIs such as data rate and latency, enhancing QoS predictions for smart city services. Similarly, Baert et al. [161] leverage DT technology to optimize Bluetooth mesh networks for IoT applications, using AI-driven DT to predict and optimize end-to-end latencies, packet delivery ratios, and path distributions. Ferriol-Galmes et al. [162] enhance network modeling using GNNs and DL techniques, modeling computer networks at the fine-grained flow level to predict per-flow KPIs accurately. Saravanan et al. [163] build a scalable NDT to predict per-path mean delay in large-scale communication networks, exploring various neural network (NN) architectures, including GNNs, to enhance prediction accuracy.

Padmapriya and Srivenkatesh [164] focus on enhancing the functionality and maintenance of smart home gadgets using the deep CNN logistic regression model with DTs to predict gadget functionality performance. Fu *et al.* [165] propose a time delay prediction algorithm for vehicular networks using NNs, aiming to address performance degradation caused by network latency. Li *et al.* [166] propose a learning-based NDT for efficiently estimating wireless network configuration KPIs before physical implementation. He *et al.* [167] optimize wireless networks with massive MIMO technology using a DTN approach, employing a conditional GAN (C-GAN) for accurate KPI predictions and pre-validation.

For traffic prediction, Lai *et al.* [168] develop a DL approach in DTNs, introducing the eConvLSTM model for predicting background traffic matrices in LANs and improving prediction accuracy significantly. Morette *et al.* [138] enhance traffic QoT estimation and prediction in optical NDTs using ML, introducing an NN architecture for accurate predictions across different configurations and devices. Nie *et al.* [169] develop a network traffic prediction algorithm for vehicular networks using deep Q-learning (DQN) and generative adversarial networks (GANs). In terms of ITS traffic prediction, Ji *et al.* [170] predict the spatiotemporal congestion resulting from traffic accidents in urban road networks using a convolutional LSTM (ConvLSTM) network, focusing on macroscopic traffic operation. Xu *et al.* [171] introduce TraffNet, a DL framework for road NDTs that considers the causality of traffic volume from vehicle trajectory data to improve traffic prediction accuracy for just-in-time decision-making. Likewise, Dangana *et al.* [172] develop a prototype system for metamorphic object transportation, focusing on predicting the transportation process of deformable interlinked linear objects.

These studies collectively demonstrate the versatility of DTNs in addressing prediction analysis tasks across different domains and the role of AI, ML, and DL techniques in improving prediction accuracy and network optimization (Table 4.7).

Table 4.7 Prediction analysis works

Work	Used AI techniques
Wang *et al.* [159]	GNN
Schippers *et al.* [160]	N/A
Baert *et al.* [161]	N/A
Ferriol-Galmes *et al.* [162]	GNNs and DL
Saravanan *et al.* [163]	NN and GNNs
Padmapriya and Srivenkatesh [164]	Deep CNN logistic regression model
Fu *et al.* [165]	NN
Li *et al.* [166]	N/A
He *et al.* [167]	C-GAN
Lai *et al.* [168]	eConvLSTM model
Morette *et al.* [138]	ML and NN architecture
Nie *et al.* [169]	DQN and GAN
Ji *et al.* [170]	Conv-LSTM
Xu *et al.* [171]	DL
Dangana *et al.* [172]	N/A

4.3.5 AI-based fault and anomaly detection

In fault and anomaly detection, recent academic literature has witnessed a proliferation of research endeavors that leverage DTNs. Toward fault detection, Zhu et al. [173] focus on 5G networks where the scarcity of fault samples and monitoring data has been a persistent hurdle. They use the Average Wasserstein GAN with Gradient Penalty (AWGAN-GP) to enhance failure discovery and prediction. The XGBoost algorithm is harnessed for real-time fault localization within the physical network. Zheng et al. [5] introduce a fault self-recovery method targeted at 5G networks. Their methodology employs a data governance approach to construct models representing physical network components in a DT. Visual topology technology is deployed to extract Knowledge-as-a-Service (KaaS) capabilities that facilitate call quality tests, fault-propagation chain reasoning, and disaster recovery analysis. Mayer et al. [174] demonstrate an ML-based framework for localizing soft failures in optical transport networks. Their focus is on harnessing SDN and streaming-based telemetry to automate detecting soft failures that may not trigger conventional alarms. Their experimental setup entails an SDN-controlled network with transponders and an optical line system providing telemetry data. Artificial NN ML plays a pivotal role in their failure localization. Similarly, Wang and Chen [175] optimize the backbone optical transport network. Their work introduces the concept of DTs to address scenarios like equipment defects, fault warnings, and trend analysis.

On the other hand, Calvo-Bascones et al. [176] propose an anomaly detection methodology tailored specifically for industrial systems with the concept of a Snitch DT, which models the connections between physical behaviors to detect anomalies. Leveraging spatiotemporal features and quantile regression, they characterize the behavior of individual physical entities. Li et al. [177] aim to combat Internet service quality degradation by proposing a DT system capable of quasi-real-time simulation for comprehensive indicator analysis, anomaly detection, and intelligent operation. In a slightly different vein, Zhu et al. [178] pivot their research toward analyzing the causes and effects of noise and distortion in the DT context, particularly focusing on AI-driven optical networks. Liu et al. [179] contribute to amalgamating DT technology into vehicular edge networks to support cybersecurity and anomaly detection within vehicular networks. Their primary thrust revolves around establishing a distributed trust evaluation system. Kaytaz et al. [180] put forth a pioneering GNN anomaly detection framework tailored to ensuring reliability within multi-dimensional data streams within ITS. Their approach employs unsupervised and supervised ML techniques, including BiDirectional Generative Adversarial Networks (BiGAN), affinity propagation, and Graph CNN, to model data streams and perform anomaly detection (Table 4.8).

4.3.6 AI-based security and privacy preservation

In recent years, DTNs have witnessed significant growth and innovation in addressing security and privacy challenges. This literature review explores the common themes and key contributions across various works that utilize AI, ML, and DL techniques to enhance security and privacy in the context of DTNs.

Table 4.8 Fault and anomaly detection works

Work	Used AI techniques
Zhu et al. [173]	AWGAN-GP and XGBoost
Zheng et al. [5]	Data governance and visual topology
Mayer et al. [174]	ML and ANN
Wang and Chen [175]	N/A
Calvo-Bascones et al. [176]	Snitch DT, spatiotemporal features, and quantile regression
Li et al. [177]	N/A
Zhu et al. [178]	N/A
Liu et al. [179]	Distributed trust evaluation
Kaytaz et al. [180]	GNN, BiGAN, affinity propagation, and graph CNN

Dong et al. [181] introduce a bidirectional mapping between physical and virtual spaces within DT technology. Their dual blockchain framework enhances data security in DT scenarios, emphasizing the importance of trustworthiness. Kumar and Khari [182] propose a DT-based framework for network forensic analysis to detect and prevent cyberattacks. The integration of tools like Nmap and Wireshark suggests AI utilization. Son et al. [183] focus on secure sharing of DT data using cloud computing and blockchain, emphasizing privacy preservation and data security in wireless channels. They employ formal methods like BAN logic and the AVISPA simulation tool for security analysis. Qu et al. [184] address DTN challenges through a blockchain-enabled adaptive asynchronous FL paradigm. Their work integrates AI techniques, blockchain, and consensus algorithms to enhance privacy, security, and reliability in DTNs. Zhan et al. [185] tackle practical DT deployment in cloud-native networks. Su and Qu [186] introduce an intrusion detection method using FL and LSTM models for network traffic analysis within DTNs. Their approach enhances privacy and accuracy and outperforms existing methods in real-time monitoring. Yigit et al. [187] propose an intelligent DDoS detection mechanism for core networks. Bagrodia [188] emphasizes assessing cyber resiliency in defense systems, particularly in complex scenarios. They build a DTN to analyze network operations and vulnerabilities to counter potential threats.

In the rapidly evolving landscape of cellular networks, Lu et al. [189] integrate DTNs and 6G networks to fortify wireless connectivity in IIoT. They use FL and blockchain to enhance reliability, security, and data privacy. They also leverage multiagent RL for optimized edge association, showcasing tangible efficiency improvements and cost reductions. Vakaruk et al. [190] use DTNs and ML tools to train cybersecurity experts. Chen et al. [191] introduce the DT protection function, a security interaction engine that safeguards message confidentiality, integrity, stability, and non-reputation while disseminating policies within DTNs. Wang et al. [192] propose a DT-based approach to autonomously provision security functions for 5G network slices. The core objective is dynamically allocating security capabilities based on KPIs, ensuring real-time network slice security. Ozdogan et al. [193]

leverage ML techniques to optimize 6G network parameters and address privacy concerns. They integrate blockchain and transfer learning (TL) to maintain data security and privacy during network recovery and expansion.

Addressing security and privacy tasks in the context of DTNs within IoT and IIoT environments, Kumar *et al*. [194] integrate blockchain and DL techniques like LSTM Sparse AutoEncoder and Multi-Head Self-Attention (MHSA)-based Bidirectional Gated Recurrent Unit (BiGRU) algorithms to enhance communication security, detect attacks, and improve data privacy. Lv *et al*. [195] integrate quantum communication techniques with DTs to enhance IIoT communication security. They propose a channel encryption scheme that uses quantum communication principles to ensure secure IIoT communication. Grasselli *et al*. [196] enable experimentation in DTNs with network topologies, attacks, and countermeasures without impacting real cyber-physical systems. Feng *et al*. [197] adopt a game theory approach to enhance the security of industrial DTNs. Their study explores vulnerability mining and repair using evolutionary game theory, cooperative strategies for vulnerability patch development, and the integration of differential evolution into the Wolf Colony Algorithm (WCA). Danilczyk *et al*. [198] propose a novel implementation of the SHA-256 hash algorithm for creating a blockchain of sensor readings in a DT environment, focusing on secure two-party communications. Their work ensures data integrity and security through a chained checksum approach. Feng *et al*. [199] involve interference source location schemes, encryption techniques, and filtering methods to reduce interference attacks in IoT-based DT communication networks. Jiang *et al*. [200] propose a blockchain-based DT edge networks framework for secure and flexible DT construction in IoT. Their approach utilizes cooperative FL and blockchain, emphasizing secure model updates and efficient resource allocation. Bozkaya *et al*. [201] also introduce a holistic approach integrating DT and blockchain into edge networks to optimize task scheduling in IoT applications. Their Proof of Evaluation (PoE)-based algorithm employs genetic algorithms to balance energy efficiency and delay reduction while ensuring data integrity. Qian *et al*. [202] focus on constructing a secure DT for the Marine IoT using FL and non-orthogonal multiple access. Kherbache *et al*. [203] address the security and resilience of information and weapon systems against cyber threats, particularly in complex scenarios. Lv and Qiao [204] explore sustainable DTs and context-aware computing to enhance network security in industrial environments. Wang *et al*. [84] offer a comprehensive exploration of security challenges within DT-enabled wireless systems. Their framework utilizes AI techniques like game theory and ML to enhance the system's security against attacks and flawed parameters.

Wu *et al*. [205] explore security concerns and computational intelligence in a drone information system employing DL techniques to fortify the defenses in the evolving landscape of UAV-based networks. Their work focuses on predictive modeling, leveraging an improved LSTM network to analyze data, specifically control signal data, to predict potential attacks on the drone system. They introduce differential privacy frequent subgraphs and utilize DTs technology to map the drone's physical operating environment to maintain data privacy. Likewise, Dai *et al*. [206] leverage DTs to elevate strategy optimization and decision-making within air–ground integrated networks, particularly in urban sensing and disaster relief applications. The study explores an aerial blockchain-based approach to fortify data security within secure federated aerial learning. He *et al.* [207] introduce a Federated Continuous Learning framework with a Stacked Broad Learning System (FCL-SBLS) tailored for intrusion detection systems in edge UAV-based IoT. They aim to enhance security and privacy while training intrusion detection models with distributed UAV data. The proposed framework, assisted by the DTN, facilitates continuous learning and training of models through asynchronous FL and a DDPG-based UAV selection scheme. AI components like FL and DDPG for UAV selection are pivotal in optimizing the IDS model's performance.

In V2X-based networks, several scholarly endeavors tackle the pressing security and privacy issues. Liu *et al.* [179] enhance cybersecurity within vehicular networks by establishing a distributed trust evaluation system aided by DTs. Xu *et al.* [208] introduce a cloud-based DT that synchronizes with the corresponding autonomous vehicle (AV) in real-time, fostering information fusion and computing. Moreover, their work introduces a concrete authentication protocol [209] to ensure secure communication between AVs and DTs, underscoring the significance of communication security and privacy protection. Similarly, Liu *et al.* [210] propose blockchain technology and DTs to fortify IoV security. Their work encompasses creating a secure communication architecture for IoV based on immutable and traceable blockchain data. Moreover, they develop a risk forecast model for IoV node security utilizing Wasserstein Distance GAN (WaGAN) models, showcasing the accelerated learning rates due to Wasserstein distance. Zhang *et al.* [211] propose a blockchain-based smart parking scheme within DT-empowered networks to monitor and predict traffic conditions around parking lots. This scheme leverages blockchain and smart contracts to ensure reliable data storage and correct parking responses while prioritizing privacy protection for drivers. He *et al.* [212] delve into the intricacies of vehicular DTNs within the context of AVs, highlighting security and privacy concerns. Their work primarily focuses on the challenges posed by the collection of sensitive user information and exposure to open network environments. Yang *et al.* [213] introduce a combined multi-armed bandit-based auction incentive mechanism designed to identify the quality of participants in the vehicular edge network without compromising sensitive information. This mechanism aims to enhance the efficiency and accuracy of the DT model within vehicular edge networks. Collectively, these works underscore the transformative potential of integrating DTNs into V2X networks, with AI, ML, and DL techniques serving as integral components for strengthening security, privacy, and network efficiency (Table 4.9).

4.4 Main AI models and tools harnessed by DTNs

In our analysis of the literature reviewed in the previous section, we observed the extensive utilization of AI models, techniques, and tools to address the mentioned tasks within DTNs. Ordered by their frequency of use in contemporary academic research, the following list offers valuable insights into the pivotal technologies that are shaping the future of DTNs. They are mainly ML, DL, RL, FL, and graph-based.

Table 4.9 Security and privacy works with AI tools

Work	Used AI tools
Dong et al. [181]	Dual blockchain
Kumar and Khari [182]	Nmap and Wireshark
Son et al. [183]	BAN logic, AVISPA, AI, and blockchain
Qu et al. [184]	Blockchain, FL, and consensus algorithms
Zhan et al. [185]	N/A
Su and Qu [186]	FL and LSTM
Yigit et al. [187]	N/A
Bagrodia [188]	N/A
Lu et al. [189]	FL, blockchain, and multiagent RL
Vakaruk et al. [190]	ML
Chen et al. [191]	N/A
Wang et al. [192]	N/A
Ozdogan et al. [193]	ML, blockchain, and TL
Kumar et al. [194]	Blockchain and DL
Lv et al. [195]	Quantum
Kumar et al. [194]	Blockchain, LSTM Sparse AutoEncoder, and MHSA-based BiGRU
Lv et al. [195]	Quantum and encryption
Grasselli et al. [196]	N/A
Feng et al. [197]	Game theory, evolutionary game theory, and differential evolution
Danilczyk et al. [198]	SHA-256, blockchain, and chained checksum
Feng et al. [199]	Interference source location schemes and encryption
Jiang et al. [200]	Blockchain and cooperative FL
Bozkaya et al. [201]	blockchain and genetic algorithm
Qian et al. [202]	FL and non-orthogonal multiple access
Kherbache et al. [203]	N/A
Lv and Qiao [204]	N/A
Wang et al. [84]	Game theory and ML
Wu et al. [205]	DL, LSTM, and differential privacy frequent subgraphs
Dai et al. [206]	Blockchain
He et al. [207]	FCL-SBLS, asynchronous FL, and DDPG
Liu et al. [179]	Distributed trust
Xu et al. [208]	authentication
Liu et al. [210]	Blockchain and WaGAN model
Zhang et al. [211]	Blockchain and smart contracts
He et al. [212]	N/A
Yang et al. [213]	Multi-armed bandit-based auction incentive mechanism

4.4.1 ML tools and models

In DTNs, ML tools enable predictive analysis, enhance security and privacy, optimize task offloading, and drive energy efficiency. These tools encompass a range of techniques, from traditional methods to more advanced algorithms, each tailored to address specific challenges within the digital twin (DT) framework [84,88,93,94,163, 165,173,174,190,193,201].

NNs are a fundamental ML tool widely applied in DTNs. Mayer *et al.* [174], Fu *et al.* [165], and Saravanan *et al.* [163] have employed NNs for prediction analysis. These versatile models demonstrate their strength in understanding complex data patterns and temporal dependencies, making them indispensable for accurate predictions.

Neural networks

NNs are models inspired by the human brain's structure and functioning. They consist of interconnected nodes organized in layers to process and learn from data.

Genetic algorithms, as showcased by Bozkaya *et al.* [201] and Zhao *et al.* [94], step into the spotlight with a focus on security, privacy, and task optimization. Genetic algorithms are trained to explore search spaces to find optimal solutions. Bozkaya *et al.* utilize genetic algorithms to strengthen security and privacy measures within DTNs. Zhao *et al.*, on the other hand, leverage genetic algorithms for task offloading and content delivery optimization, optimizing resource allocation and enhancing content delivery efficiency.

Genetic algorithms

Genetic algorithms are optimization algorithms inspired by natural selection and genetics. They are used to find solutions to optimization and search problems. Genetic algorithms maintain a population of potential solutions and evolve them over generations through selection, crossover (recombination), and mutation.

Transfer learning (TL) finds its application in enhancing security and privacy within DTs [193]. TL allows for knowledge transfer from one domain to another, improving the robustness of security measures and safeguarding sensitive information.

Transfer learning

TL is an ML technique in which a pre-trained model, typically trained on a large dataset, is adapted for a different but related task. Instead of training a model from scratch, TL leverages the knowledge and feature representations learned by the pre-trained model. This approach can significantly reduce training time and data requirements for the target task and is especially useful when labeled data is scarce or expensive to obtain.

Finally, the Extreme Gradient Boost (XGBoost) algorithm is harnessed by Zhu *et al.* [173] for prediction analysis. XGBoost excels at handling structured data and enhancing prediction accuracy, making it an ideal choice for data-driven predictions in DT systems.

Extreme Gradient Boost (XGBoost)
XGBoost is a popular ML algorithm that belongs to the ensemble learning category. It is designed to handle various supervised learning tasks, including classification and regression. XGBoost combines the predictions of multiple weak learners into a strong, robust model. It uses a gradient boosting framework, which optimizes model performance by iteratively adding new trees that correct the errors made by previous ones.

4.4.2 DL models and techniques

DL has emerged as an indispensable tool in tackling the multifaceted challenges within DTNs. In our comprehensive literature review, DL is pivotal in enhancing network performance and managing various network aspects, encompassing planning, resource allocation, communication refinement, traffic and KPIs prediction, anomaly detection, and reinforcing network security and privacy [89,96,162,164,170,173,180, 186,194,205,210].

CNNs stand at the forefront of this technological advancement. These specialized models excel at processing grid-like data, particularly images. Their application in DTs, as demonstrated by Padmapriya and Srivenkatesh [164], revolves around prediction analysis. By leveraging the inherent capabilities of CNNs, researchers have successfully harnessed image-based data to make accurate predictions within the DT framework. Delving deeper into image analysis, deep CNNs build upon CNN foundations. They, too, have found a niche in prediction analysis within DTs. This sophisticated extension of CNNs offers enhanced capabilities for processing complex visual data, contributing to more accurate predictions.

Convolutional neural networks (CNNs)
CNNs are specialized NN architectures designed primarily for processing grid-like data, such as images and videos. They employ convolutional layers to learn hierarchical features automatically.

Moving beyond CNN, LSTM networks are pivotal in DTs. LSTM networks, renowned for their capability to capture temporal dependencies and sequences, have been harnessed by multiple researchers. Su and Qu [186], Wu *et al.* [205], and Yi *et al.* [89] employ LSTM to enhance security, privacy, and task optimization within DTNs. LSTM's power in understanding time-series data allows it to detect patterns and anomalies, safeguard sensitive information, and optimize task allocation. Convolutional LSTM represents a hybrid approach, combining the strengths of CNNs and LSTM networks. This amalgamation is particularly advantageous when spatiotemporal relationships must be considered. Ji *et al.* [170] have employed Conv-LSTM for prediction analysis within DTNs. The model's capacity to capture complex spatiotemporal dependencies makes it valuable for making precise predictions regarding future states or events.

Long short-term memory
LSTM is a recurrent NN (RNN) architecture designed to model sequential data. LSTMs have memory cells that can capture and remember information over long sequences, making them suitable for time series analysis tasks.

Convolutional LSTM (C-LSTM)
Convolutional LSTM, or C-LSTM, is a hybrid neural network architecture that combines the spatial processing capabilities of CNNs with the sequential modeling capabilities of LSTM networks. C-LSTMs are employed in tasks involving spatiotemporal data analysis, such as video and gesture recognition.

Generative adversarial networks (GANs) have also made their presence within DTs. They are a class of models known for data generation that have been leveraged by Ferriol-Galmes *et al.* [162] to enhance prediction analysis within DTs. The generative abilities of GANs have been instrumental in creating synthetic data to supplement prediction tasks, ultimately improving the accuracy of predictions. Furthermore, variants of GANs, such as the Bi-directional GAN (BiGANs), with their bi-directional generative capabilities, offer a unique advantage in generative tasks and anomaly detection. Kaytaz *et al.* [180] have successfully implemented BiGANs for fault and anomaly detection within DTs, harnessing their ability to capture complex data relationships and deviations. Another noteworthy variant is the Wasserstein GAN (WaGAN). It has been tailored by Liu *et al.* [210] to support security and privacy measures within DTNs focusing on blockchain-related security enhancements. This highlights the adaptability of GANs and their variants to address specific security challenges. The Augmented Wasserstein GAN with Gradient Penalty (AWGAN-GP) represents a specialized GAN variant designed to generate and maintain data distribution. Zhu *et al.* [173] have harnessed AWGAN-GP for prediction analysis within DTs.

Generative adversarial networks (GANs)
GANs are a class of DL models comprising two NNs, a generator and a discriminator, which engage in a game-theoretic framework. GANs generate synthetic data by learning the underlying distribution.

Wasserstein distance GAN (WaGAN)
WaGAN is a variant of GANs that employs the Wasserstein distance metric to measure the dissimilarity between generated and real data distributions. WaGANs are valued for their stability and improved training in GANs.

Average Wasserstein GAN with Gradient Penalty (AWGAN-GP)
AWGAN-GP is a variant of GANs that combines the Wasserstein GAN with a gradient penalty term in the loss function. This modification enhances training stability and encourages smoother data generation.

4.4.3 RL and optimization techniques

RL introduces dynamic decision-making capabilities to DTNs. These tools are characterized by their ability to make sequential decisions to optimize various aspects, including energy efficiency, latency, security, privacy, and task optimization [74,82,83,86–89,91,96,189,207].

Asynchronous advantage actor-critic (A3C) takes the lead, focusing on enhancing energy efficiency within DTNs [86] to optimize decision-making in environments with multiple agents. It proves invaluable in optimizing resource allocation and reducing energy consumption. Saravanan et al. [74] also employ A3C but specifically emphasize latency optimization. A3C's ability to adapt and learn optimal policies over time minimizes latency within DTs, ensuring real-time responsiveness and efficient data transfer.

Actor-critic
A3C is an RL technique where two distinct components work together. The "actor" is responsible for making decisions or actions, while the "critic" evaluates those actions by estimating their value. A3C aims to optimize policies by concurrently adjusting the decision-making strategy (actor) and value estimation (critic), leading to more effective learning in complex environments.

Extended from RL, deep reinforcement learning (DRL) encompasses a range of techniques that learn optimal policies through interactions with the environment. Liu et al. [88] employ DRL to optimize resource allocation, contributing to energy efficiency enhancements and reduced environmental impact. Another variant is the multi-agent RL. It is a collective approach that enables collaborative learning among multiple agents, enhancing security measures and safeguarding sensitive information. In DTNs, Lu et al. [189] address security and privacy concerns using multi-agent RL techniques.

Hierarchical multi-agent RL
Hierarchical multi-agent RL is an approach that involves multiple agents with varying levels of decision-making authority. It introduces a hierarchical structure where high-level agents make global decisions while lower-level agents handle more detailed tasks.

Similarly, Q-learning has been developed as a model-free RL technique for solving Markov Decision Processes (MDPs). It estimates the value of taking specific actions in different states and iteratively updates these estimates to find an optimal policy. Q-learning is particularly useful for solving problems with uncertain or unknown environment dynamics. Deep Q-network (DQN) is a variant of Q-learning. Shui et al. [87] concentrate on enhancing energy efficiency within DTs using DQNs. DQN's ability to approximate optimal Q-values for actions is harnessed to optimize resource allocation, reduce energy consumption, and boost sustainability in DT systems. Double DQN (DDQN) excels in learning optimal action-value functions. By applying DDQN, Li et al. [83] fine-tuned latency parameters, ensuring that data processing and communication in DTs occur with minimal delay.

Double deep Q-network
DDQN is a DRL. DDQN addresses overestimation issues in Q-learning by employing two separate NNs to estimate action values.

Proximal policy optimization (PPO) is an RL scheme known for its stability and efficiency in training policies. It focuses on iteratively improving policies by making small updates, ensuring that the new policy does not stay consistent with the old one. Jiang et al. [82] utilize PPO to fine-tune latency-related parameters, ensuring efficient task execution and data transfer within the DT ecosystem.

Multi-agent PPO (MA-PPO)
MA-PPO extends the PPO algorithm to scenarios involving multiple interacting agents. MA-PPO enables multiple agents to learn and adapt their policies in a coordinated manner, considering the influence of other agents in the environment.

Deep deterministic policy gradient (DDPG) is an RL method suitable for continuous action spaces. In DTN, DDPG is introduced by He et al. [207] to facilitate secure data sharing and analysis within federated DTNs.

Multi-agent deep deterministic policy gradient (MADDPG)
MADDPG is an extension of the deep deterministic policy gradient (DDPG) algorithm designed for multi-agent RL scenarios. MADDPG allows multiple agents to collaborate and learn in environments where their actions affect the environment and each other.

4.4.4 FL and collaborative learning

FL is an ML approach designed for collaborative training of models across decentralized devices or edge nodes. Instead of sending raw data to a central server, FL allows devices to train models locally and share only model updates with the central server. This approach enhances data privacy and efficiency while enabling collective model improvement across a distributed network [95,184,186,189,200,202,207]. Variants of FL schemes have been used in DTNs. Dynamic FL, exemplified by Sun et al. [95], focuses on task offloading and optimizing content within DTNs. Dynamic FL's adaptive approach optimizes computational tasks and content delivery to meet specific performance requirements. The other variant, the asynchronous FL, optimizes learning by allowing participants to update their models at their own pace, minimizing data exposure. He et al. [207] use asynchronous FL to enhance privacy within federated DTNs. The last variant is cooperative FL, as showcased by Jiang et al. [200]. It fosters collaboration among participating entities while maintaining data isolation. This approach ensures that sensitive information remains confidential, paving the way for secure collaborative model training within DTs. Hierarchical FL (HFL) extends the concept of FL to incorporate hierarchical structures within decentralized networks. HFL introduces layers of communication and coordination, where local devices or nodes collaborate within smaller groups before sharing their updates with higher-level aggregators. This hierarchical approach enhances scalability and adaptability in large-scale federated learning systems. The FL with Secure Bi-Level Optimization (FCL-SBLS) takes a unique approach to FL by incorporating secure bi-level optimization. This method ensures that data privacy and security are paramount while allowing collaborative model training and knowledge sharing within federated DTNs [207].

Stacked broad learning system (FCL-SBLS)
FCL-SBLS is a framework that combines the capabilities of stacked autoencoders and Broad Learning Systems. It is used for feature learning, data representation, and classification tasks. FCL-SBLS leverages FL techniques to extract hierarchical and abstract features from data, enhancing its ability to handle complex patterns.

4.4.5 Graph and network analysis techniques

Graph-based analysis techniques have gained prominence within the context of DTNs. These approaches leverage the inherent interconnectedness of data and devices within the DT ecosystem and many challenges, such as fault and anomaly detection, latency optimization, and prediction analysis [80,159,162,163,180].

GNNs are NN models that process and analyze graph-structured data. GNNs operate by propagating information through nodes and edges of a graph, enabling tasks such as node classification, link prediction, and graph-level predictions. GNNs are proficient at modeling complex relationships and dependencies in graph data. As illustrated by Kaytaz et al. [180], they find their application in fault and anomaly detection within DTs. GNNs enable the identification and mitigation of faults and anomalies, ensuring the robustness and reliability of DT systems. As demonstrated by Ferriol-Galmes et al. [80], by analyzing the network structure and interdependencies among components, GNNs facilitate data transfer and processing optimization, reducing latency and ensuring real-time responsiveness. Prediction analysis, a cornerstone of DTN tasks, relies on the predictive capabilities of GNNs. Wang et al. [159],

Ferriol-Galmes *et al.* [162], and Saravanan *et al.* [163] harness GNNs to make accurate predictions about future states and events. By considering the intricate relationships among data points and devices, GNNs enhance prediction accuracy, enabling informed decision-making within DT systems. Furthermore, graph CNNs have been presented to extend the capabilities of traditional GNNs. Graph CNNs excel in modeling complex data relationships within graph structures, making them particularly suited for identifying and mitigating faults and anomalies in DT systems as employed by Kaytaz *et al.* [180].

Graph CNN
A graph CNN is an NN architecture specialized for processing graph data. Graph CNNs adapt convolutional operations to work on graph-structured data, allowing them to capture local and global patterns within graphs.

4.5 Main challenges in AI-based DTNs

4.5.1 Key challenges

The application of these various AI tools within DTNs is undoubtedly promising. However, several challenges may arise in adopting and effectively utilizing these tools in the context of DTNs and their associated tasks. Some key challenges are [1,214–218]:

- Data quality and availability: AI tools heavily rely on data, and DTNs are no exception. Ensuring high-quality and diverse data is crucial for training accurate models. In DTNs, obtaining real-time, reliable, and comprehensive data can be challenging due to the complexity and dynamic nature of the mirrored physical systems. Inadequate or noisy data can lead to suboptimal model performance.
- Scalability: DTNs often involve vast, complex systems with many interconnected components. Scaling AI models to handle the increasing size and complexity of DTNs can be computationally intensive. A significant challenge is ensuring that AI tools can efficiently operate on large-scale DTs while maintaining real-time performance.
- Interpretability: Understanding and interpreting the decisions made by AI models is crucial, especially in applications like security, privacy, and fault detection within DTNs. Many advanced AI models, particularly DL and RL, are often regarded as "black boxes," making it challenging to explain their decision-making processes, which can be a barrier to trust and accountability.
- Robustness: The resilience of AI models to adversarial attacks and unexpected system behavior is paramount, particularly in security and privacy applications. Ensuring that AI tools can detect and adapt to novel threats or anomalies in the dynamic environment of DTNs is a significant challenge.
- Energy efficiency: While AI tools can contribute to energy efficiency in DTNs, they can also be computationally intensive and power-hungry. Striking

a balance between the computational demands of AI models and the energy-efficient operation of DTs is a challenge, particularly in resource-constrained environments.
- Privacy concerns: Data privacy is critical in FL and collaborative settings. Ensuring that FL and other collaborative AI tools protect sensitive information while allowing for effective model training is a delicate balancing act.
- Latency and real-time processing: Many AI tools, especially RL and FL, require iterative processes that may not align with the real-time demands of DTNs, particularly in latency-sensitive applications. Ensuring that AI tools meet real-time requirements while delivering optimal results is a significant challenge.
- Model complexity: Some AI models can be highly complex and require significant computational resources. In DTNs, especially in edge and resource-constrained environments, deploying and maintaining such complex models can be challenging.
- Model generalization: AI models must be able to generalize from historical data to adapt to new situations or changes in DTNs. Ensuring that models can adapt effectively when faced with previously unseen scenarios is an ongoing challenge.
- Integration and interoperability: DTNs often involve a mix of heterogeneous systems, and integrating AI tools seamlessly into these environments can be complex. Ensuring compatibility, interoperability, and ease of integration with existing DT systems is crucial.

4.5.2 Responsible AI considerations

Addressing these obstacles will foster successful and responsible AI implementation for more efficient systems. To countermeasure the challenges faced in adopting AI across its key tasks, responsible strategies must be devised for successful implementation. These strategies promote fairness, accuracy, transparency, and security [29,218,219].

> **Responsible AI**
> *Responsible AI refers to the ethical and moral considerations associated with the development and deployment of AI systems. It encompasses a set of principles and practices to ensure that AI technologies are used in a way that aligns with human values and respects the rights and well-being of systems.*

4.5.2.1 Focusing on data quality and representation
- **Data preprocessing and bias mitigation**: Employ rigorous data preprocessing techniques to address historical biases in DTN data. Utilize diverse and representative datasets to ensure accurate predictions and avoid reliance on outdated information.
- **Enhanced data quality control**: Implement robust data quality control measures to ensure reliable DT model representation. Source and integrate data from

multiple reliable sources to provide a comprehensive system view. Facilitate informed decision-making and enhance network management capabilities.
- **Integration of simulation and real-time data**: Integrate simulation and real-time data to improve prediction accuracy within DTNs. This integration enhances the DT's ability to make accurate predictions. Incorporate probabilistic modeling to account for unpredictable events and behaviors, resulting in more reliable performance enhancement strategies and recommendations.

4.5.2.2 Focusing on AI model reliability and transparency
- **Transparent AI models**: Prioritize using explainable AI models within DTNs to enhance transparency and interpretability. Provide detailed explanations for model recommendations and strategies to build trust among stakeholders. Transparency supports informed decision-making.
- **Continuous model refinement**: Recognize the complexity of DTNs and the need for ongoing improvement. Continuously refine and optimize AI models to ensure their reliability in dynamic DTN environments. Validate models using real-world data to maintain effective performance.

These strategies are tailored to address the challenges and goals of DTNs. By focusing on data quality, security, privacy, and the reliability and transparency of AI models, these strategies aim to enhance various aspects of DTNs, including performance, network management, communication, prediction accuracy, anomaly detection, and overall security and privacy.

4.6 Conclusion and key points

DTNs overview

- DTNs span a wide spectrum of physical networks, including cellular, wireless, optical, satellite, and more.
- These virtual networks seamlessly integrate computational power and AI capabilities, breaking conventional boundaries.
- Recommendations from DTNs are disseminated to physical network entities.
- This process operates as an autonomous closed-loop data transmission paradigm.
- DTNs continuously learn, evolve, and adapt based on real-world outcomes.

Key tasks in DTNs

- Network performance enhancement, including latency optimization, energy efficiency, and task offloading.
- Network management, including resource allocation planning and monitoring.

- Communication enhancement.
- Prediction analysis for traffic and KPIs.
- Anomaly detection, security, and privacy preservation.

AI tools in DTNs

- ML tools: NNs, genetic algorithms, TL, and XGBoost.
- DL tools: CNNs, LSTMs, and GANs.
- RL tools: A3C, DRL, multi-agent RL, hierarchical RL, Q-learning, DDQN, MA-PPO, and MADDPG.
- FL tools: dynamic FL, asynchronous FL, cooperative FL, and FCL-SBLS.
- Graph-based techniques: GNNs and graph CNNs.

Challenges faced by AI schemes

- Data quality and availability.
- Scalability.
- Interpretability.
- Robustness.
- Privacy concerns.
- Security.
- Energy efficiency.
- Model complexity.
- Generalization.
- Integration.

Responsible AI strategies

- Bias mitigation and fairness.
- Transparency.
- Privacy preservation.
- Accountability.

In essence, DTNs represent a dynamic and evolving landscape where AI technologies, driven by domain expertise, are poised to revolutionize network performance, management, and security across many physical network domains, offering substantial benefits and opportunities for innovation.

References

[1] Rathore MM, Shah SA, Shukla D, et al. The role of AI, machine learning, and big data in digital twinning: A systematic literature review, challenges, and opportunities. *IEEE Access*. 2021;9:32030–32052.

[2] Tang Z, Chen D, Sun T, et al. Intelligent awareness of delay-sensitive Internet traffic in digital twin network. *IEEE Journal of Radio Frequency Identification*. 2022;6:891–895.

[3] Huang B, Wang D, Li H, et al. Network selection and QoS management algorithm for 5G converged shipbuilding network based on digital twin. In: *2022 10th International Conference on Information and Education Technology (ICIET)*. Piscataway, NJ: IEEE; 2022. p. 403–408.

[4] Kherbache M, Maimour M, and Rondeau E. Network digital twin for the Industrial Internet of Things. In: *2022 IEEE 23rd International Symposium on a World of Wireless, Mobile and Multimedia Networks (WoWMoM)*. Piscataway, NJ: IEEE; 2022. p. 573–578.

[5] Zheng Y, Kong H, Wang N, et al. Practice on fifth-generation core (5GC) network fault self-recovery based on a digital twin. *Digital Twin*. 2022;2:18.

[6] Wu Y, Zhang K, and Zhang Y. Digital twin networks: A survey. *IEEE Internet of Things Journal*. 2021;8(18):13789–13804.

[7] Tang F, Chen X, Rodrigues TK, et al. Survey on digital twin edge networks (DITEN) toward 6G. *IEEE Open Journal of the Communications Society*. 2022;3:1360–1381.

[8] Kuruvatti NP, Habibi MA, Partani S, et al. Empowering 6G communication systems with digital twin technology: A comprehensive survey. *IEEE Access*. 2022;10:112158–112186.

[9] Mashaly M. Connecting the twins: A review on digital twin technology and its networking requirements. *Procedia Computer Science*. 2021;184:299–305.

[10] McManus M, Cui Y, Zhang JZ, et al. Digital twin-enabled domain adaptation for zero-touch UAV networks: Survey and challenges. *Computer Networks*. 2023;236:110000.

[11] Kroyer J and Holzinger K. Digital twins of computer networks. *Network*. 2022;41.

[12] Zeb S, Mahmood A, Hassan SA, et al. Industrial digital twins at the nexus of nextG wireless networks and computational intelligence: A survey. *Journal of Network and Computer Applications*. 2022;200:103309.

[13] HAKIRI A, Gokhale A, Ben Yahia S, et al. A comprehensive survey on digital twin for future networks and emerging IoT industry. SSRN 4535810. 2023.

[14] Su Z, Guo S, Dai M, et al. A survey on digital twins: Architecture, enabling technologies, security and privacy, and future prospects. TechRxiv. 2023.

[15] Khan LU, Mustafa E, Shuja J, et al. Federated learning for digital twin-based vehicular networks: Architecture and challenges. *IEEE Wireless Communications*. 2023;31:1–8.

[16] Larsson R. Creating digital twin distributed networks using switches with programmable data plane. Linkoping University; 2021.
[17] Hui Y, Zhao G, Yin Z, *et al.* Digital twin enabled multi-task federated learning in heterogeneous vehicular networks. In: *2022 IEEE 95th Vehicular Technology Conference: (VTC2022-Spring)*. Piscataway, NJ: IEEE; 2022. p. 1–5.
[18] Kislyakov SV. A digital twin model of the smart city communication infrastructure. *International Journal of Embedded and Real-Time Communication Systems (IJERTCS)*. 2022;13(1):1–16.
[19] Yang H, Lü P, Sun T, *et al.* Multi-source heterogeneous data processing technology for digital twin network. In: *2022 IEEE 22nd International Conference on Communication Technology (ICCT)*. Piscataway, NJ: IEEE; 2022. p. 1829–1834.
[20] Tan B, Qian Y, Lu H, *et al.* Toward a future network architecture for intelligence services: A cyber digital twin-based approach. *IEEE Network*. 2021;36(1):98–104.
[21] Zhu Y, Chen D, Zhou C, *et al.* A knowledge graph based construction method for digital twin network. In: *2021 IEEE 1st International Conference on Digital Twins and Parallel Intelligence (DTPI)*. Piscataway, NJ: IEEE; 2021. p. 362–365.
[22] Luan TH, Liu R, Gao L, *et al.* The paradigm of digital twin communications. arXiv preprint arXiv:210507182. 2021.
[23] Hamzaoui MA and Julien N. Social cyber-physical systems and digital twins networks: A perspective about the future digital twin ecosystems. *IFAC-PapersOnLine*. 2022;55(8):31–36.
[24] Chen D, Yang H, Zhou C, *et al.* Classification, building and orchestration management of digital twin network models. In: *2022 IEEE 22nd International Conference on Communication Technology (ICCT)*. Piscataway, NJ: IEEE; 2022. p. 1843–1846.
[25] Szántó N, Horváth I, and Csapó Á. A digital twin case study exhibiting bi-directional communication and control. In: *2022 IEEE 1st International Conference on Internet of Digital Reality (IoD)*. Piscataway, NJ: IEEE; 2022. p. 000105–000106.
[26] Tao S, Cheng Z, Xiao-Dong D, *et al.* Digital twin network (DTN): concepts, architecture, and key technologies. *Acta Automatica Sinica*. 2021;47(3):569–582.
[27] Seilov SZ, Kuzbayev A, Seilov A, *et al.* The concept of building a network of digital twins to increase the efficiency of complex telecommunication systems. *Complexity*. 2021;2021:1–9.
[28] Tao Z, Guo Y, He G, *et al.* Deep learning-based modeling of 5G core control plane for 5G network digital twin. *IEEE Transactions on Cognitive Communications and Networking*. 2024;10(1):238–251.
[29] Jang S, Jeong J, Lee J, *et al.* Digital twin for intelligent network: Data lifecycle, digital replication, and AI-based optimizations. *IEEE Communications Magazine*. 2023;61(11):96–102.

[30] Sanz Rodrigo M, Rivera D, Moreno JI, *et al.* Digital twins for 5G networks: A modeling and deployment methodology. *IEEE Access.* 2023;11: 38112–38126.

[31] Pantovic V, Milovanovic D, Starcevic D, *et al.* 5G mobile networks and digital twins concept: Research challenges in network DT emulation. In: *2022 4th International Conference on Emerging Trends in Electrical, Electronic and Communications Engineering (ELECOM).* Piscataway, NJ: IEEE; 2022. p. 1–4.

[32] Lu Y, Maharjan S, and Zhang Y. Adaptive edge association for wireless digital twin networks in 6G. *IEEE Internet of Things Journal.* 2021;8(22):16219–16230.

[33] Lin X, Kundu L, Dick C, *et al.* 6G digital twin networks: From theory to practice. *IEEE Communications Magazine.* 2023;61(11):72–78.

[34] Duong TQ, Van Huynh D, Khosravirad SR, *et al.* From digital twin to metaverse: The role of 6G ultra-reliable and low-latency communications with multi-tier computing. *IEEE Wireless Communications.* 2023;30(3): 140–146.

[35] Guo Q, Tang F, Rodrigues TK, *et al.* Five disruptive technologies in 6G to support digital twin networks. *IEEE Wireless Communications.* 2024;31(1):149–155.

[36] Masaracchia A, Sharma V, Fahim M, *et al.* Digital twin for open RAN: Toward intelligent and resilient 6G radio access networks. *IEEE Communications Magazine.* 2023;61(11):112–118.

[37] Sheen B, Yang J, Feng X, *et al.* A digital twin for reconfigurable intelligent surface assisted wireless communication. arXiv preprint arXiv:200900454. 2020.

[38] Ahmadi H, Nag A, Khar Z, *et al.* Networked twins and twins of networks: An overview on the relationship between digital twins and 6G. *IEEE Communications Standards Magazine.* 2021;5(4):154–160.

[39] Shu M, Sun W, Zhang J, *et al.* Digital-twin-enabled 6G network autonomy and generative intelligence: Architecture, technologies and applications. *Digital Twin.* 2022;2:16.

[40] Vilà I, Sallent O, and Pérez-Romero J. On the design of a network digital twin for the radio access network in 5G and beyond. *Sensors.* 2023;23(3):1197.

[41] Gong J, Yu Q, Li T, *et al.* Scalable digital twin system for mobile networks with generative AI. In: *Proceedings of the 21st Annual International Conference on Mobile Systems, Applications and Services*; 2023. p. 610–611.

[42] Mirzaei J, Abualhaol I, and Poitau G. Network digital twin for Open RAN: The key enablers, standardization, and use cases. arXiv preprint arXiv:230802644. 2023.

[43] Raza SM, Minerva R, Crespi N, *et al.* Definition of digital twin network data model in the context of edge-cloud continuum. In: *2023 IEEE 9th International Conference on Network Softwarization (NetSoft).* Piscataway, NJ: IEEE; 2023. p. 402–407.

[44] Apostolakis N, Chatzieleftheriou LE, Bega D, *et al.* Digital twins for next-generation mobile networks: Applications and solutions. *IEEE Communications Magazine*. 2023;61(11):80–86.

[45] Almeida EN, Fontes H, Campos R, *et al.* Position-based machine learning propagation loss model enabling fast digital twins of wireless networks in ns-3. In: *Proceedings of the 2023 Workshop on ns-3*; 2023. p. 69–77.

[46] Bariah L, Sari H, and Debbah M. Digital twin-empowered smart cities: A new frontier of wireless networks. TechRxiv. 2022.

[47] Chen H, Xu X, Simsarian JE, *et al.* Digital twin of a network and operating environment using augmented reality. arXiv preprint arXiv:230315221. 2023.

[48] Zhuge Q. AI-driven digital twin for optical networks. In: *European Conference and Exhibition on Optical Communication*. Washington, DC: Optica Publishing Group; 2022. p. Mo3A–1.

[49] Kuang L, Wu J, and Yin S. Construct digital twin models in cyber space for physical objects of intelligent optical network. In: *2022 IEEE International Conferences on Internet of Things (iThings) and IEEE Green Computing & Communications (GreenCom) and IEEE Cyber, Physical & Social Computing (CPSCom) and IEEE Smart Data (SmartData) and IEEE Congress on Cybermatics (Cybermatics)*. Piscataway, NJ: IEEE; 2022. p. 300–305.

[50] Solmaz G, Cirillo F, and Fattore U. Digital twins in the future networking: A use case in IOWN. In: *2022 IEEE 8th World Forum on Internet of Things (WF-IoT)*. Piscataway, NJ: IEEE; 2022. p. 1–8.

[51] Janz C, You Y, Hemmati M, *et al.* Digital twin for the optical network: Key technologies and enabled automation applications. In: *NOMS 2022–2022 IEEE/IFIP Network Operations and Management Symposium*. Piscataway, NJ: IEEE; 2022. p. 1–6.

[52] Wang D, Zhang Z, Zhang M, *et al.* The role of digital twin in optical communication: Fault management, hardware configuration, and transmission simulation. *IEEE Communications Magazine*. 2021;59(1):133–139.

[53] Zhuge Q, Liu X, Zhang Y, *et al.* Building a digital twin for intelligent optical networks. *Journal of Optical Communications and Networking*. 2023;15(8):C242–C262.

[54] Mello DA, Mayer KS, Escallón-Portilla AF, *et al.* When digital twins meet optical networks operations. In: *2023 Optical Fiber Communications Conference and Exhibition (OFC)*. Piscataway, NJ: IEEE; 2023. p. 1–3.

[55] Eldeeb H, Naser S, Bariah L, *et al.* Digital twin-assisted OWC: Towards smart and autonomous networks. TechRxiv. 2022.

[56] Velasco L, Devigili M, and Ruiz M. Applications of digital twin for autonomous zero-touch optical networking. In: *2023 International Conference on Optical Network Design and Modeling (ONDM)*. Piscataway, NJ: IEEE; 2023. p. 1–5.

[57] Vilalta R, Casellas R, Gifre L, *et al.* Architecture to deploy and operate a digital twin optical network. In: *Optical Fiber Communication Conference*. Washington, DC: Optica Publishing Group; 2022. p. W1F–4.

[58] Chang X, Yang C, Wang H, *et al.* KID: Knowledge graph-enabled intent-driven network with digital twin. In: *2022 27th Asia Pacific Conference on Communications (APCC)*. Piscataway, NJ: IEEE; 2022. p. 272–277.

[59] Bilen T, Ak E, Bal B, *et al.* A proof of concept on digital twin-controlled WiFi core network selection for in-flight connectivity. *IEEE Communications Standards Magazine*. 2022;6(3):60–68.

[60] Moorthy SK, Harindranath A, McManus M, *et al.* A middleware for digital twin-enabled flying network simulations using UBSim and UB-ANC. In: *2022 18th International Conference on Distributed Computing in Sensor Systems (DCOSS)*. Piscataway, NJ: IEEE; 2022. p. 322–327.

[61] Brunelli M, Ditta CC, Postorino MN. A framework to develop urban aerial networks by using a digital twin approach. *Drones*. 2022;6(12):387.

[62] Zhou Y, Zhang R, Liu J, *et al.* A hierarchical digital twin network for satellite communication networks. *IEEE Communications Magazine*. 2023;61(11):104–110.

[63] Al-Hraishawi H, Alsenwi M, Lagunas E, *et al.* Digital twin for non-terrestrial networks: Vision, challenges, and enabling technologies. arXiv preprint arXiv:230510273. 2023.

[64] Palmieri M, Quadri C, Fagiolini A, *et al.* Co-simulated digital twin on the network edge: The case of platooning. In: *2022 IEEE 23rd International Symposium on a World of Wireless, Mobile and Multimedia Networks (WoWMoM)*. Piscataway, NJ: IEEE; 2022. p. 613–618.

[65] Fennell C. A communication architecture for transportation digital twins using Apache Kafka. Trinity College Dublin, Ireland. 2022.

[66] Wang J and Chen W. Internet of vehicles and digital twin framework for unmanned tracked vehicles based on 5G communication. In: *International Conference on Signal Processing and Communication Technology (SPCT 2022)*, vol. 12615. Bellingham, WA: SPIE; 2023. p. 203–210.

[67] Wágner T, Ormándi T, Tettamanti T, *et al.* SPaT/MAP V2X communication between traffic light and vehicles and a realization with digital twin. *Computers and Electrical Engineering*. 2023;106:108560.

[68] Kherbache M, Maimour M, and Rondeau E. When digital twin meets network softwarization in the industrial IoT: Real-time requirements case study. *Sensors*. 2021;21(24):8194.

[69] Guimarães KP, Haddad DB, and da Rocha Henriques F. Digital twins-based self-regulation system for instrumentation networks in IoT applications. *International Journal of Engineering and Science*. 2022;12:35–42.

[70] Isah A, Shin H, Aliyu I, *et al.* A data-driven digital twin network architecture in the Industrial Internet of Things (IIoT) applications. arXiv preprint arXiv:231214930. 2023.

[71] Jagannath J, Ramezanpour K, and Jagannath A. Digital twin virtualization with machine learning for IoT and beyond 5G networks: Research directions for security and optimal control. In: *Proceedings of the 2022 ACM Workshop on Wireless Security and Machine Learning*; 2022. p. 81–86.

[72] Rizwan A, Ahmad R, Khan AN, *et al.* Intelligent digital twin for federated learning in AIoT networks. *Internet of Things*. 2023;22:100698.

[73] Hakiri A, Yahia SB, and Aniruddha SG. Hyper-5G: A cross-Atlantic digital twin testbed for next generation 5G IoT networks and beyond. In: *2023 IEEE 26th International Symposium on Real-Time Distributed Computing (ISORC)*. Piscataway, NJ: IEEE; 2023. p. 230–235.

[74] Saravanan J, Rajendran R, Muthu P, et al. Performance analysis of digital twin edge network implementing bandwidth optimization algorithm. *International Journal Of Computing and Digital System*. 2021;12:851–858.

[75] Yang H, Li Y, Yao K, et al. A systematic network traffic emulation framework for digital twin network. In: *2021 IEEE 1st International Conference on Digital Twins and Parallel Intelligence (DTPI)*. Piscataway, NJ: IEEE; 2021. p. 94–97.

[76] Van Huynh D, Nguyen VD, Khosravirad SR, et al. Fairness-aware latency minimisation in digital twin-aided edge computing with ultra-reliable and low-latency communications: A distributed optimisation approach. In: *2022 56th Asilomar Conference on Signals, Systems, and Computers*. Piscataway, NJ: IEEE; 2022. p. 1045–1049.

[77] Van Huynh D, Nguyen VD, Khosravirad SR, et al. URLLC edge networks with joint optimal user association, task offloading and resource allocation: A digital twin approach. *IEEE Transactions on Communications*. 2022;70(11):7669–7682.

[78] Van Huynh D, Nguyen VD, Sharma V, et al. Digital twin empowered ultra-reliable and low-latency communications-based edge networks in industrial IoT environment. In: *ICC 2022-IEEE International Conference on Communications*. Piscataway, NJ: IEEE; 2022. p. 5651–5656.

[79] Wang S, Kua J, Jin J, et al. Optimal graph partitioning for time-sensitive flow scheduling towards digital twin networks. In: *Proceedings of the 1st Workshop on Digital Twin & Edge AI for Industrial IoT*; 2022. p. 7–12.

[80] Ferriol-Galmés M, Suárez-Varela J, Paillissé J, et al. Building a digital twin for network optimization using graph neural networks. *Computer Networks*. 2022;217:109329.

[81] Duong TQ, Van Huynh D, Li Y, et al. Digital twin-enabled 6G aerial edge computing with ultra-reliable and low-latency communications. In: *2022 1st International Conference on 6G Networking (6GNet)*. Piscataway, NJ: IEEE; 2022. p. 1–5.

[82] Jiang L, Mu JH, Zheng H, et al. Virtual-real mapping error aware computing task offloading and adaptive resource optimization in digital twin driven UAV networks. *Journal of Beijing University of Posts and Telecommunications*. 2022;45(6):138.

[83] Li M, Gao J, Zhou C, et al. Digital twin-driven computing resource management for vehicular networks. In: *GLOBECOM 2022–2022 IEEE Global Communications Conference*. Piscataway, NJ: IEEE; 2022. p. 5735–5740.

[84] Wang B, Sun Y, Jung H, et al. Digital twin-enabled computation offloading in UAV-assisted MEC emergency networks. *IEEE Wireless Communications Letters*. 2023;12(9):1588–1592.

[85] Zhao R, Zhang K, and Zhang Y. Energy-efficient edge association in digital twin empowered 6G networks. In: *2022 IEEE 22nd International*

Conference on Communication Technology (ICCT). Piscataway, NJ: IEEE; 2022. p. 869–874.

[86] Chen L, Gu Q, Jiang K, et al. A3C-based and dependency-aware computation offloading and service caching in digital twin edge networks. *IEEE Access*. 2023;11:57564–57573.

[87] Shui T, Hu J, Yang K, et al. Cell-free networking for integrated data and energy transfer: Digital twin based double parameterized DQN for energy sustainability. *IEEE Transactions on Wireless Communications*. 2023;22(11):8035–8049.

[88] Liu Q, Tang L, Wu T, et al. Deep reinforcement learning for resource demand prediction and virtual function network migration in digital twin network. *IEEE Internet of Things Journal*. 2023;10(21):19102–19116.

[89] Yi B, Lv J, Wang X, et al. Digital twin driven and intelligence enabled content delivery in end-edge-cloud collaborative 5G networks. *Digital Communications and Networks*. 2022;2352–8648. https://doi.org/10.1016/j.dcan.2023.01.004.

[90] Gao Y, Liao J, Wei X, et al. Quality-aware massive content delivery in digital twin-enabled edge networks. *China Communications*. 2023;20(2):1–13.

[91] Güemes-Palau C, Almasan P, Xiao S, et al. Accelerating deep reinforcement learning for digital twin network optimization with evolutionary strategies. In: *NOMS 2022–2022 IEEE/IFIP Network Operations and Management Symposium*. Piscataway, NJ: IEEE; 2022. p. 1–5.

[92] Lin G, Gel J, Wu Y, et al. Digital twin networks: Learning dynamic network behaviors from network flows. In: *2022 IEEE Symposium on Computers and Communications (ISCC)*. Piscataway, NJ: IEEE; 2022. p. 1–6.

[93] Ursu RM, Zerwas J, Krämer P, et al. Towards digital network twins: Can we machine learn network function behaviors? In: *2023 IEEE 9th International Conference on Network Softwarization (NetSoft)*. Piscataway, NJ: IEEE; 2023. p. 438–443.

[94] Zhao L, Wang C, Zhao K, et al. INTERLINK: A digital twin-assisted storage strategy for satellite-terrestrial networks. *IEEE Transactions on Aerospace and Electronic Systems*. 2022;58(5):3746–3759.

[95] Sun W, Xu N, Wang L, et al. Dynamic digital twin and federated learning with incentives for air-ground networks. *IEEE Transactions on Network Science and Engineering*. 2020;9(1):321–333.

[96] Han C, Yang T, Li X, et al. Polymorphic learning of heterogeneous resources in digital twin networks. In: *ICC 2022-IEEE International Conference on Communications*. Piscataway, NJ: IEEE; 2022. p. 1185–1189.

[97] Liu T, Tang L, Wang W, et al. Digital-twin-assisted task offloading based on edge collaboration in the digital twin edge network. *IEEE Internet of Things Journal*. 2021;9(2):1427–1444.

[98] Qu W, Li X, and Chen W. A general digital twin framework for intelligent network management. In: *2022 IEEE/CIC International Conference on Communications in China (ICCC)*. Piscataway, NJ: IEEE; 2022. p. 760–765.

[99] Zhao Z, Wang F, Gao Y, et al. Design of a digital twin for spacecraft network system. In: *2022 IEEE 5th International Conference on Electronics*

and *Communication Engineering (ICECE)*. Piscataway, NJ: IEEE; 2022. p. 46–50.
[100] Corici M and Magedanz T. Digital twin for 5G networks. In: *The Digital Twin*. Berlin: Springer; 2023. p. 433–446.
[101] Bilen T, Canberk B, and Duong TQ. Digital twin evolution for hard-to-follow aeronautical ad-hoc networks in beyond 5G. *IEEE Communications Standards Magazine*. 2023;7(1):4–12.
[102] Chukhno O, Chukhno N, Araniti G, *et al.* Placement of social digital twins at the edge for beyond 5G IoT networks. *IEEE Internet of Things Journal*. 2022;9(23):23927–23940.
[103] Xiao L, Han D, Weng TH, *et al.* An evolutive framework for server placement optimization to digital twin networks. *International Journal of Communication Systems*. 2023;36(14):e5553.
[104] Zhao L, Bi Z, Hawbani A, *et al.* ELITE: An intelligent digital twin-based hierarchical routing scheme for softwarized vehicular networks. *IEEE Transactions on Mobile Computing*. 2023;22(9):5231–5247.
[105] Fu X, Yuan Q, Liu S, *et al.* Communication-efficient decision-making of digital twin assisted Internet of vehicles: A hierarchical multi-agent reinforcement learning approach. *China Communications*. 2023;20(3):55–68.
[106] Al-Hamid DZ and Al-Anbuky A. Vehicular intelligence: Towards vehicular network digital-twin. In: *2022 27th Asia Pacific Conference on Communications (APCC)*. Piscataway, NJ: IEEE; 2022. p. 427–432.
[107] Jiao K, Xu H, Ling L, *et al.* A mobile application-based tower network digital twin management. In: *The International Conference on Cyber Security Intelligence and Analytics*. Berlin: Springer; 2023. p. 369–377.
[108] Ren Y, Guo S, Cao B, *et al.* End-to-end network SLA quality assurance for C-RAN: A closed-loop management method based on digital twin network. *IEEE Transactions on Mobile Computing*. 2023. p. 1–18.
[109] Roselló MM, Cancela JV, Quintana I, *et al.* Network digital twin for non-public networks. In: *2023 IEEE 24th International Symposium on a World of Wireless, Mobile and Multimedia Networks (WoWMoM)*. Piscataway, NJ: IEEE; 2023. p. 495–500.
[110] Kherbache M, Maimour M, and Rondeau E. Digital twin network for the IIoT using Eclipse Ditto and Hono. *IFAC-PapersOnLine*. 2022;55(8):37–42.
[111] Wei Z, Wang S, Li D, *et al.* Data-driven routing: A typical application of digital twin network. In: *2021 IEEE 1st International Conference on Digital Twins and Parallel Intelligence (DTPI)*. Piscataway, NJ: IEEE; 2021. p. 1–4.
[112] RAJ DRR, Shaik TA, Hirwe A, *et al.* Building a digital twin network of SDN using knowledge graphs. *IEEE Access*. 2023;11:63092–63106.
[113] Sun X, Zhou C, Duan X, *et al.* A digital twin network solution for end-to-end network service level agreement (SLA) assurance. *Digital Twin*. 2021;1:5.
[114] Baranda J, Mangues-Bafalluy J, Zeydan E, *et al.* AIML-as-a-service for SLA management of a digital twin virtual network service. In: *IEEE INFOCOM 2021-IEEE Conference on Computer Communications Workshops (INFOCOM WKSHPS)*. Piscataway, NJ: IEEE; 2021. p. 1–2.

[115] Zhou C, Gao J, Li M, et al. Digital twin-empowered network planning for multi-tier computing. *Journal of Communications and Information Networks*. 2022;7(3):221–238.

[116] Huang Z, Li D, Cai J, et al. Collective reinforcement learning based resource allocation for digital twin service in 6G networks. *Journal of Network and Computer Applications*. 2023;217:103697.

[117] Tao Y, Wu J, Lin X, et al. DRL-driven digital twin function virtualization for adaptive service response in 6G networks. *IEEE Networking Letters*. 2023;5(2):125–129.

[118] Duran K and Canberk B. Digital twin enriched green topology discovery for next generation core networks. *IEEE Transactions on Green Communications and Networking*. 2023;7(4):1946–1956.

[119] Su X, Jia Z, Zhou Z, et al. Digital twin-empowered communication network resource management for low-carbon smart park. In: *ICC 2022-IEEE International Conference on Communications*. Piscataway, NJ: IEEE; 2022. p. 2942–2947.

[120] Dai C, Yang K, and Deng C. A service placement algorithm based on merkle tree in MEC systems assisted by digital twin networks. In: *2022 IEEE 21st International Conference on Ubiquitous Computing and Communications (IUCC/CIT/DSCI/SmartCNS)*. Piscataway, NJ: IEEE; 2022. p. 37–43.

[121] He Y, Yang M, He Z, et al. Resource allocation based on digital twin-enabled federated learning framework in heterogeneous cellular network. *IEEE Transactions on Vehicular Technology*. 2022;72(1):1149–1158.

[122] Guo Y, Zhuang Y, Li X, et al. Time–frequency correlated network sensing edge deployment for digital twin. In: *2022 4th International Academic Exchange Conference on Science and Technology Innovation (IAECST)*. Piscataway, NJ: IEEE; 2022. p. 1183–1187.

[123] Yuan S, Zhang Z, Li Q, et al. Joint optimization of DNN partition and continuous task scheduling for digital twin-aided MEC network with deep reinforcement learning. *IEEE Access*. 2023;11:27099–27110.

[124] Luo J, Zeng J, Han Y, et al. Distributed deep reinforcement learning for resource allocation in digital twin networks. In: *Proceedings of 7th International Congress on Information and Communication Technology: ICICT 2022*, London, vol. 1. Berlin: Springer; 2022. p. 771–781.

[125] Wieme J, Baert M, and Hoebeke J. Relay selection in Bluetooth Mesh networks by embedding genetic algorithms in a digital communication twin. In: *2022 IEEE 23rd International Symposium on a World of Wireless, Mobile and Multimedia Networks (WoWMoM)*. Piscataway, NJ: IEEE; 2022. p. 561–566.

[126] Naeem F, Kaddoum G, and Tariq M. Digital twin-empowered network slicing in B5G networks: Experience-driven approach. In: *2021 IEEE Globecom Workshops (GC Wkshps)*. Piscataway, NJ: IEEE; 2021. p. 1–5.

[127] Abdel-Basset M, Hawash H, Sallam KM, et al. Digital twin for optimization of slicing-enabled communication networks: A federated graph learning approach. *IEEE Communications Magazine*. 2023;61(10):100–106.

[128] Hong H, Wu Q, Dong F, et al. NetGraph: An intelligent operated digital twin platform for data center networks. In: *Proceedings of the ACM SIGCOMM 2021 Workshop on Network-Application Integration*; 2021. p. 26–32.

[129] Lombardo A, Morabito G, Quattropani S, et al. Design, implementation, and testing of a microservices-based digital twins framework for network management and control. In: *2022 IEEE 23rd International Symposium on a World of Wireless, Mobile and Multimedia Networks (WoWMoM)*. Piscataway, NJ: IEEE; 2022. p. 590–595.

[130] Gong Y, Wei Y, Feng Z, et al. Resource allocation for integrated sensing and communication in digital twin enabled internet of vehicles. *IEEE Transactions on Vehicular Technology*. 2023;72(4):4510–4524.

[131] Lu Y, Huang X, Zhang K, et al. Communication-efficient federated learning for digital twin edge networks in industrial IoT. *IEEE Transactions on Industrial Informatics*. 2020;17(8):5709–5718.

[132] Dai Y, Zhang K, Maharjan S, et al. Deep reinforcement learning for stochastic computation offloading in digital twin networks. *IEEE Transactions on Industrial Informatics*. 2020;17(7):4968–4977.

[133] Bellavista P, Giannelli C, Mamei M, et al. Application-driven network-aware digital twin management in industrial edge environments. *IEEE Transactions on Industrial Informatics*. 2021;17(11):7791–7801.

[134] Luan F, Yang J, Zhang H, et al. Optimization of load-balancing strategy by self-powered sensor and digital twins in software-defined networks. *IEEE Sensors Journal*. 2023;23(18):20782–20793.

[135] Guo Q, Tang F, and Kato N. Federated reinforcement learning-based resource allocation for D2D-aided digital twin edge networks in 6G Industrial IoT. *IEEE Transactions on Industrial Informatics*. 2023;19(5):7228–7236.

[136] Tang L, Du Y, Liu Q, et al. Digital-twin-assisted resource allocation for network slicing in Industry 4.0 and beyond using distributed deep reinforcement learning. *IEEE Internet of Things Journal*. 2023;10(19):16989–17006.

[137] Geißler S, Wamser F, Bauer W, et al. MVNOCoreSim: A digital twin for virtualized IoT-centric mobile core networks. *IEEE Internet of Things Journal*. 2023;10(15):13974–13987.

[138] Morette N, Hafermann H, Frignac Y, et al. Machine learning enhancement of a digital twin for wavelength division multiplexing network performance prediction leveraging quality of transmission parameter refinement. *Journal of Optical Communications and Networking*. 2023;15(6):333–343.

[139] Curri V. Digital-twin of physical-layer as enabler for open and disaggregated optical networks. In: *2023 International Conference on Optical Network Design and Modeling (ONDM)*. Piscataway, NJ: IEEE; 2023. p. 1–6.

[140] Borraccini G, Straullu S, Giorgetti A, et al. Experimental demonstration of partially disaggregated optical network control using the physical layer digital twin. *IEEE Transactions on Network and Service Management*. 2023;20(3):2343–2355.

[141] Hao J, Bai J, Zhao G, et al. Intelligent scheduling method of optical transmission network based on digital twin. In: *2023 IEEE 6th Information*

[142] Wu Y, Zhang M, Zhang L, et al. Dynamic network topology portrait for digital twin optical network. *Journal of Lightwave Technology.* 2023;41(10):2953–2968.

[143] Fu Q, Chen D, Sun H, et al. AlsoDTN: An air logistics service-oriented digital twin network based on collaborative decision model. In: *International Conference on Service-Oriented Computing.* Berlin: Springer; 2022. p. 398–402.

[144] Guo TN. Connectivity-aware fast network forming aided by digital twin for emergency use. In: *IEEE INFOCOM 2022-IEEE Conference on Computer Communications Workshops (INFOCOM WKSHPS).* Piscataway, NJ: IEEE; 2022. p. 1–6.

[145] Gong Y, Liu HYX, Bennis M, et al. Computation and privacy protection for satellite-ground digital twin networks. arXiv preprint arXiv:230208525. 2023.

[146] Zhang J, Zong M, Vasilakos AV, et al. UAV base station network transmission-based reverse auction mechanism for digital twin utility maximization. *IEEE Transactions on Network and Service Management.* 2024;21(1):324–340.

[147] Dai Y and Zhang Y. Adaptive digital twin for vehicular edge computing and networks. *Journal of Communications and Information Networks.* 2022;7(1):48–59.

[148] Cazzella L, Linsalata F, Magarini M, et al. A multi-modal simulation framework to enable digital twin-based V2X communications in dynamic environments. arXiv preprint arXiv:230306947. 2023.

[149] Wang P, Gao Y, Wang J, et al. Digital-intelligent twin for UAV swarm-based 5G emergency networks. *URSI GASS'21.* 2021.

[150] Jian W, Chuang Y, and Ningning Y. Study on digital twin channel for the B5G and 6G communication. *Journal of Radio Science.* 2021;36(3):340–348.

[151] Xiang Z, Wang Z, Fu K, et al. 5G wireless network digital twin system based on high precision simulation. In: *Asian Simulation Conference.* Berlin: Springer; 2022. p. 187–199.

[152] Zhao Y, Li L, Liu Y, et al. Communication-efficient federated learning for digital twin systems of Industrial Internet of Things. *IFAC-PapersOnLine.* 2022;55(2):433–438.

[153] Liang Y, Zhou Y, Liu X, et al. Digital twin assisted automatic routing selection for electric elastic optical networks. In: *International Symposium on Robotics, Artificial Intelligence, and Information Engineering (RAIIE 2022),* vol. 12454. Bellingham, WA: SPIE; 2022. p. 26–32.

[154] Li B, Liu W, Xie W, et al. Adaptive digital twin for UAV-assisted integrated sensing, communication, and computation networks. *IEEE Transactions on Green Communications and Networking.* 2023;7(4):1996–2009.

[155] Zelenbaba S, Rainer B, Hofer M, *et al.* Wireless digital twin for assessing the reliability of vehicular communication links. In: *2022 IEEE Globecom Workshops (GC Wkshps)*. Piscataway, NJ: IEEE; 2022. p. 1034–1039.

[156] Lv Z, Dang S, Qiao L, *et al.* Deep-learning-based security of optical wireless communications for intelligent transportation digital twins systems. *IEEE Internet of Things Magazine*. 2022;5(2):154–159.

[157] Demir U, Pradhan S, Kumahia R, *et al.* Digital twins for maintaining QoS in programmable vehicular networks. *IEEE Network*. 2023;37(4):208–214.

[158] Liu Z, Sun H, Marine G, *et al.* 6G IoV networks driven by RF digital twin modeling. *IEEE Transactions on Intelligent Transportation Systems*. 2023;25:1–11.

[159] Wang H, Wu Y, Min G, *et al.* A graph neural network-based digital twin for network slicing management. *IEEE Transactions on Industrial Informatics*. 2020;18(2):1367–1376.

[160] Schippers H, Böcker S, and Wietfeld C. Data-driven digital mobile network twin enabling mission-critical vehicular applications. In: *2023 IEEE 97th Vehicular Technology Conference (VTC2023-Spring)*. Piscataway, NJ: IEEE; 2023. p. 1–7.

[161] Baert M, De Poorter E, and Hoebeke J. A digital communication twin for performance prediction and management of bluetooth mesh networks. In: *Proceedings of the 17th ACM Symposium on QoS and Security for Wireless and Mobile Networks*; 2021. p. 1–10.

[162] Ferriol-Galmés M, Cheng X, Shi X, *et al.* FlowDT: A flow-aware digital twin for computer networks. In: *ICASSP 2022–2022 IEEE International Conference on Acoustics, Speech and Signal Processing (ICASSP)*. Piscataway, NJ: IEEE; 2022. p. 8907–8911.

[163] Saravanan M, Kumar PS, and Kumar AR. Enabling network digital twin to improve QoS performance in communication networks. In: *2022 IEEE Smartworld, Ubiquitous Intelligence & Computing, Scalable Computing & Communications, Digital Twin, Privacy Computing, Metaverse, Autonomous & Trusted Vehicles (SmartWorld/UIC/ScalCom/DigitalTwin/PriComp/Meta)*. Piscataway, NJ: IEEE; 2022. p. 2151–2160.

[164] Padmapriya V and Srivenkatesh M. Digital twins for smart home gadget threat prediction using deep convolution neural network. *International Journal of Advanced Computer Science and Applications*. 2023;14(2).

[165] Fu Y, Guo D, Li Q, *et al.* Digital twin based network latency prediction in vehicular networks. *Electronics*. 2022;11(14):2217.

[166] Li B, Efimov T, Kumar A, *et al.* Learnable digital twin for efficient wireless network evaluation. In: *MILCOM 2023–2023 IEEE Military Communications Conference (MILCOM)*. Piscataway, NJ: IEEE; 2023. p. 661–666.

[167] He W, Zhang C, Deng J, *et al.* Conditional generative adversarial network aided digital twin network modeling for massive MIMO optimization. In: *2023 IEEE Wireless Communications and Networking Conference (WCNC)*. Piscataway, NJ: IEEE; 2023. p. 1–5.

[168] Lai J, Chen Z, Zhu J, *et al.* Deep learning based traffic prediction method for digital twin network. *Cognitive Computation.* 2023;15:1748–1766.

[169] Nie L, Wang X, Zhao Q, *et al.* Digital twin for transportation big data: A reinforcement learning-based network traffic prediction approach. *IEEE Transactions on Intelligent Transportation Systems.* 2024;25(1):896–906.

[170] Ji X, Yue W, Li C, *et al.* Digital twin empowered model free prediction of accident-induced congestion in urban road networks. In: *2022 IEEE 95th Vehicular Technology Conference:(VTC2022-Spring).* Piscataway, NJ: IEEE; 2022. p. 1–6.

[171] Xu M, Ma Y, Li R, *et al.* TraffNet: Learning causality of traffic generation for road network digital twins. arXiv preprint arXiv:230315954. 2023.

[172] Dangana M, Ansari S, Asad SM, *et al.* Towards the digital twin (DT) of Narrow-Band Internet of Things (NBIoT) wireless communication in industrial indoor environment. *Sensors.* 2022;22(23):9039.

[173] Zhu X, Zhao L, Cao J, *et al.* Fault diagnosis of 5G networks based on digital twin model. *China Communications.* 2023;20(7):175–191.

[174] Mayer KS, Pinto RP, Soares JA, *et al.* Demonstration of ML-assisted soft-failure localization based on network digital twins. *Journal of Lightwave Technology.* 2022;40(14):4514–4520.

[175] Wang X and Chen S. Design and application of digital twin model of power communication transmission network. In: *2022 4th International Conference on Smart Power & Internet Energy Systems (SPIES).* Piscataway, NJ: IEEE; 2022. p. 2213–2217.

[176] Calvo-Bascones P, Voisin A, Do P, *et al.* A collaborative network of digital twins for anomaly detection applications of complex systems. Snitch Digital Twin concept. *Computers in Industry.* 2023;144:103767.

[177] Li L, Chen X, Xie Y, *et al.* Anomaly detection of internet service quality degradation in digital twin for fixed access network. In: *2022 10th International Conference on Information Systems and Computing Technology (ISCTech).* Piscataway, NJ: IEEE; 2022. p. 402–408.

[178] Zhu K, Hua N, Li Y, *et al.* The impact of data acquisition inconsistency and time sensitivity on digital twin for AI-driven optical networks. In: *2021 IEEE 6th Optoelectronics Global Conference (OGC).* Piscataway, NJ: IEEE; 2021. p. 225–226.

[179] Liu J, Zhang S, Liu H, *et al.* Distributed collaborative anomaly detection for trusted digital twin vehicular edge networks. In: *Wireless Algorithms, Systems, and Applications: 16th International Conference, WASA 2021, Nanjing, China, June 25–27, 2021, Proceedings, Part II 16.* Berlin: Springer; 2021. p. 378–389.

[180] Kaytaz U, Ahmadian S, Sivrikaya F, *et al.* Graph neural network for digital twin-enabled intelligent transportation system reliability. In: *2023 IEEE International Conference on Omni-layer Intelligent Systems (COINS).* Piscataway, NJ: IEEE; 2023. p. 1–7.

[181] Dong W, Yang B, Wang K, *et al.* A dual blockchain framework to enhance data trustworthiness in digital twin network. In: *2021 IEEE 1st International*

[182] Kumar K and Khari M. Architecture of digital twin for network forensic analysis using Nmap and Wireshark. In: *Digital Twin Technology*. Boca Raton, FL: CRC Press; 2021. p. 83–104.

[183] Son S, Kwon D, Lee J, *et al.* On the design of a privacy-preserving communication scheme for cloud-based digital twin environments using blockchain. *IEEE Access*. 2022;10:75365–75375.

[184] Qu Y, Gao L, Xiang Y, *et al.* Fedtwin: Blockchain-enabled adaptive asynchronous federated learning for digital twin networks. *IEEE Network*. 2022;36(6):183–190.

[185] Zhan Y, Tan X, Wang M, *et al.* Implementation and deployment of digital twin in cloud-native network. In: *International Conference on Emerging Networking Architecture and Technologies*. Berlin: Springer; 2022. p. 13–25.

[186] Su D and Qu Z. Detection DDoS of attacks based on federated learning with digital twin network. In: *International Conference on Knowledge Science, Engineering and Management*. Berlin: Springer; 2022. p. 153–164.

[187] Yigit Y, Bal B, Karameseoglu A, *et al.* Digital twin-enabled intelligent DDoS detection mechanism for autonomous core networks. *IEEE Communications Standards Magazine*. 2022;6(3):38–44.

[188] Bagrodia R. Using network digital twins to improve cyber resilience of missions. *The Journal of Defense Modeling and Simulation*. 2023;20(1):97–106.

[189] Lu Y, Huang X, Zhang K, *et al.* Low-latency federated learning and blockchain for edge association in digital twin empowered 6G networks. *IEEE Transactions on Industrial Informatics*. 2020;17(7):5098–5107.

[190] Vakaruk S, Mozo A, Pastor A, *et al.* A digital twin network for security training in 5G industrial environments. In: *2021 IEEE 1st International Conference on Digital Twins and Parallel Intelligence (DTPI)*. Piscataway, NJ: IEEE; 2021. p. 395–398.

[191] Chen M, Shao J, Guo S, *et al.* Convoy_DTN: A security interaction engine design for digital twin network. In: *2021 IEEE Globecom Workshops (GC Wkshps)*. Piscataway, NJ: IEEE; 2021. p. 1–5.

[192] Wang K, Du H, and Su L. Digital twin network based network slice security provision. In: *2022 IEEE 2nd International Conference on Digital Twins and Parallel Intelligence (DTPI)*. Piscataway, NJ: IEEE; 2022. p. 1–6.

[193] Ozdogan MO, Carkacioglu L, and Canberk B. Digital twin driven blockchain based reliable and efficient 6G edge network. In: *2022 18th International Conference on Distributed Computing in Sensor Systems (DCOSS)*. Piscataway, NJ: IEEE; 2022. p. 342–348.

[194] Kumar P, Kumar R, Kumar A, *et al.* Blockchain and deep learning for secure communication in digital twin empowered Industrial IoT network. *IEEE Transactions on Network Science and Engineering*. 2023;10(5):2802–2813.

[195] Lv Z, Cheng C, and Song H. Digital twins based on quantum networking. *IEEE Network*. 2022;36(5):88–93.

[196] Grasselli C, Melis A, Rinieri L, *et al.* An industrial network digital twin for enhanced security of cyber-physical systems. In: *2022 International Symposium on Networks, Computers and Communications (ISNCC)*. Piscataway, NJ: IEEE; 2022. p. 1–7.

[197] Feng H, Chen D, Lv H, *et al.* Game theory in network security for digital twins in industry. *Digital Communications and Networks*. 2023. https://doi.org/10.1016/j.dcan.2022.09.014.

[198] Danilczyk W, Sun YL, and He H. Blockchain checksum for establishing secure communications for digital twin technology. In: *2021 North American Power Symposium (NAPS)*. Piscataway, NJ: IEEE; 2021. p. 1–6.

[199] Feng H, Chen D, and Lv H. Sensible and secure IoT communication for digital twins, cyber twins, web twins. *Internet of Things and Cyber-Physical Systems*. 2021;1:34–44.

[200] Jiang L, Zheng H, Tian H, *et al.* Cooperative federated learning and model update verification in blockchain-empowered digital twin edge networks. *IEEE Internet of Things Journal*. 2021;9(13):11154–11167.

[201] Bozkaya E, Erel-Özçevik M, Bilen T, *et al.* Proof of evaluation-based energy and delay aware computation offloading for digital twin edge network. *Ad Hoc Networks*. 2023;149:103254.

[202] Qian LP, Li M, Ye P, *et al.* Secrecy-driven energy minimization in federated-learning-assisted marine digital twin networks. *IEEE Internet of Things Journal*. 2024;11(3):5155–5168.

[203] Kherbache M, Maimour M, and Rondeau E. IoT network digital twins modeling using Petri nets. In: *1er Congrès Annuel de la Société d'Automatique de Génie Industriel et de Productique, SAGIP 2023*, Marseille, France; 2023. p. x1–x4.

[204] Lv Z, Qiao L. Context-Aware Cognitive Communication for Sustainable Digital Twins. In: Pathan ASK, editor. *Towards a Wireless Connected World: Achievements and New Technologies*. Cham: Springer International Publishing; 2022. p. 179–201.

[205] Wu J, Guo J, and Lv Z. Deep learning driven security in digital twins of drone network. In: *ICC 2022-IEEE International Conference on Communications*. Piscataway, NJ: IEEE; 2022. p. 1–6.

[206] Dai M, Wang T, Li Y, *et al.* Digital twin envisioned secure air-ground integrated networks: A blockchain-based approach. *IEEE Internet of Things Magazine*. 2022;5(1):96–103.

[207] He X, Chen Q, Tang L, *et al.* Federated continuous learning based on stacked broad learning system assisted by digital twin networks: An incremental learning approach for intrusion detection in UAV networks. *IEEE Internet of Things Journal*. 2023;10(22):19825–19838.

[208] Xu J, He C, and Luan TH. Efficient authentication for vehicular digital twin communications. In: *2021 IEEE 94th Vehicular Technology Conference (VTC2021-Fall)*. Piscataway, NJ: IEEE; 2021. p. 1–5.

[209] Al-Shareeda S. Data authentication algorithms. In: Alginahi YM, Kabir MN, editors. *Authentication Technologies for Cloud Technology, IoT, and Big Data*. London: IET; 2019. p. 37.

[210] Liu J, Zhang L, Li C, *et al.* Blockchain-based secure communication of intelligent transportation digital twins system. *IEEE Transactions on Intelligent Transportation Systems.* 2022;23(11):22630–22640.

[211] Zhang C, Zhu L, and Xu C. BSDP: Blockchain-based smart parking for digital-twin empowered vehicular sensing networks with privacy protection. *IEEE Transactions on Industrial Informatics.* 2023;19(5):7237–7246.

[212] He C, Luan TH, Lu R, *et al.* Security and privacy in vehicular digital twin networks: Challenges and solutions. *IEEE Wireless Communications.* 2023;30(4):154–160.

[213] Yang Y, Ma W, Sun W, *et al.* Privacy-preserving digital twin for vehicular edge computing networks. In: *2022 IEEE Smartworld, Ubiquitous Intelligence & Computing, Scalable Computing & Communications, Digital Twin, Privacy Computing, Metaverse, Autonomous & Trusted Vehicles (SmartWorld/UIC/ScalCom/DigitalTwin/PriComp/Meta).* Piscataway, NJ: IEEE; 2022. p. 2238–2243.

[214] Almasan P, Ferriol-Galmés M, Paillisse J, *et al.* Digital twin network: Opportunities and challenges. arXiv preprint arXiv:220101144. 2022.

[215] Abdel Hakeem SA, Hussein HH, and Kim H. Security requirements and challenges of 6G technologies and applications. *Sensors.* 2022;22(5):1969.

[216] Shahraki A, Abbasi M, Piran MJ, *et al.* A comprehensive survey on 6G networks: Applications, core services, enabling technologies, and future challenges. arXiv preprint arXiv:210112475. 2021.

[217] Ahmad I, Shahabuddin S, Malik H, *et al.* Machine learning meets communication networks: Current trends and future challenges. *IEEE Access.* 2020;8:223418–223460.

[218] Wang S, Qureshi MA, Miralles-Pechuán L, *et al.* Applications of explainable AI for 6G: Technical aspects, use cases, and research challenges. arXiv preprint arXiv:211204698. 2021.

[219] Zhang T, Hemmatpour M, Mishra S, *et al.* Operationalizing AI in future networks: A bird's eye view from the system perspective. arXiv preprint arXiv:230304073. 2023.

Chapter 5

Digital twin empowered Open RAN of 6G networks

Antonino Masaracchia[1], Vishal Sharma[1], Muhammad Fahim[1], Octavia A. Dobre[2] and Trung Q. Duong[1,2]

The open radio access network (O-RAN) Alliance's main mission is to lead to the evolution of the next-generation network RAN by incorporating principles of openness and intelligence. Simultaneously, digital twin (DT) technology is emerging as a cornerstone for developing services in the context of sixth-generation (6G) networks. This chapter provides a comprehensive perspective on how DT and O-RAN constitute two synergistic concepts. In particular, it illustrates how their mutual integration holds the potential to facilitate the deployment of a smart and resilient 6G RAN. Notably, DT concept will play a pivotal role in enhancing the core principles of intelligence, autonomy, and openness that underlie O-RAN. The chapter begins with a concise overview of both O-RAN and DT concepts. It then proceeds to illustrate and discuss potential use cases and services achievable through a DT-based O-RAN architecture. The chapter concludes by outlining current challenges and discussing future research direction toward the implementation of such innovative network architecture.*

5.1 Introduction

A significant proliferation of smart Internet-of-Things (IoT) connected devices, accompanied by the emergence of disruptive use cases, has been observed during the last two decades. According to data analysis and forecasts from the International Telecommunication Union Radiocommunication Sector (ITU-R), this trend is expected to maintain its rapid growth trend, which will inevitably cause a collapse of the current 5G networks in the near future. For instance, it is envisioned that mobile traffic could reach 5016 exabytes per month by 2030 and beyond [2]. In addition to these network capacity-related issues, in order to support the deployment of immersive services such as virtual/augmented/extended reality (VR/AR/XR), as

---
[1]School of Electronics, Electrical Engineering and Computer Science, Queen's University Belfast, UK
[2]Faculty of Engineering and Applied Science, Memorial University, St. John's, Canada
*This chapter has been published partly in [1].

well as Industrial automation and intelligent transportation systems (ITS) service, will result of paramount importance to guarantee communication links with very stringent requirements in terms of both communication latency – latency less than 1 millisecond and 99,99999% of reliability – i.e., error-free transmission links with near-zero communication latency. This, as already happened for the previous network generations, i.e., from 1G to 4G, is calling for the definition of a new network generation of network standard referred to as 6G [3].

This is primarily exerting pressure on mobile operators, which are constantly looking to find possible ways to increase their network capacity. Typically, this network infrastructure upgrade process involves either adding new network equipment or replacing the existing ones with more efficient, flexible and scalable ones [4]. Nevertheless, the vast majority of mobile operators are nowadays persistently seeking more cost-effective solutions since, to date, the total cost amount for a radio access network (RAN) represents approximately 65%–70% of the total network capital expenditures [5]. Moreover, there is a need for new types of resource allocation policies necessary to meet the reliability and latency requirements of the communication links. Currently, these are based on the adoption of conventional convex optimization techniques, which due to the complex nature of the communication scenario often fall short in meeting near-zero latency requirements. In addition, they only work well for small- and medium-scale optimization problems. To this end, current research direction is migrating toward the adoption of artificial intelligence (AI) techniques, which are expected to play a pivotal role in the design of optimal decision-making policies [6].

A similar challenge was already faced during the deployment of 4G/5G networks. During that period, the introduction of software-defined network (SDN) and network function virtualization (NFV) concepts represented a very revolutionary way to drastically reduce the cost of upgrading the core of the network. Following the same principle, possibilities for RAN virtualization have been also investigated and implemented during the last few years. For instance, potential solutions for the cost savings through the centralization of expensive baseband computing resources virtualized and run as software on generic hardware through a centralized baseband pool have been proposed in the centralized RAN (C-RAN) architecture. However, their reliance on proprietary interfaces represents the main issue of the entire RAN virtualization approaches, which consequently makes them less cost-effective. To eliminate vendor lock-in and further reduce the cost of potential RAN virtualization solutions, global mobile operators recently founded the Open RAN (O-RAN) Alliance [5]. Indeed, the main aim of the Alliance is to provide a RAN evolutionary path entirely based on the concepts of openness for software and interfaces, as well as on the principle of intelligence through the incorporation of AI-based logic, which will make networks self-driving and autonomous. However, since these AI models are designed and trained on particular snapshots of the network, there is the risk that they could result inefficient. In other words, they might struggle to cope, in terms of optimal decision policy, with all the possible network scenarios, leading to a deterioration in the network's quality of service and user experience.

As previously mentioned, a multitude of immersive services such as VR, AR, and XR, are expected to be delivered with the deployment of 6G networks. These, compared to 5G mobile network services, will provide a complete digitization of the surrounding environment as well as of all the network components. In this perspective, the introduction of a revolutionary concept as the DT is expected to be a keystone for future wireless communication networks. In fact, the introduction of the DT concept will mean the creation of digital representations of every real-world product and system, all hosted on either cloud or edge servers [7]. Furthermore, DT will embed AI and big data analytics features taking real-time data from real systems as input. These will allow to obtain clear and comprehensive models of each physical system, as well as a comprehensive understanding of their behavior, which in return provides the possibility to perform real-time decision-making policies aimed at improving the entire development process and efficiency of the considered product/system [8,9]. Thanks to its potentialities, DT technology has already yielded substantial breakthroughs across diverse domains such as manufacturing, healthcare, urban planning, and logistics. Notably, in the context of manufacturing, the application of digital twins is experiencing a rapid and expansive growth, with its utility encompassing multiple aspects of the production chain, such as design, production streamlining, quality assurance, and predictive maintenance. Furthermore, manufacturers now also leverage digital twins to simulate and evaluate product performance under a spectrum of scenarios, leading to cost reductions and enhanced operational efficiency. In the context of urban planning, digital twins offer the invaluable capability to simulate city development, optimize infrastructure, transportation systems, and resource allocation, thereby expediting the creation of smarter, more sustainable urban environments. Then we are assisting in a transformation process where the connection between digital twins and artificial intelligence and machine learning is closer than ever to providing smarter decision support and automation capabilities [10].

5.1.1 Motivation and contribution

Considering the earlier discussion, one can clearly notice the evidence that the DT concept seamlessly aligns with the principles of openness, autonomous operation, and intelligence embodied by O-RAN. This means that the integration of DT paradigm with O-RAN architecture will result in a very powerful synergy to further enhance the intelligence and self-driving capabilities of O-RAN itself. Indeed, since DT will permit to have a precise model that comprehends the behavior of every physical system, its integration will play a pivotal role in the ongoing optimization and training of O-RAN AI/ML models in real-time and with fresh data from the actual underlying network. From this perspective, this chapter provides the following contributions:

- A clear vision of the O-RAN ecosystem with its principles and features, as well as on DT and its development process.
- The potentialities that a DT-enabled O-RAN holds as a keystone enabling architecture for 6G-oriented services are illustrated through use cases.
- Open challenges and future research directions on this groundbreaking.

136 Digital twins for 6G

The remainder of this chapter is organized as follows. A brief overview of the O-RAN ecosystem and DT concept are provided in Sections 5.2 and 5.3, respectively. Relevant use cases of DT in O-RAN are presented in Section 5.4, while challenges and future directions are discussed in Section 5.5. Finally, conclusions and final thoughts are provided in Section 5.6.

5.2 Background on O-RAN

This section starts by providing an initial background on what is the network RAN, its disaggregation process, and how it led toward the establishment of the O-RAN alliance. Subsequently, it provides a clear illustration of the O-RAN network architecture and its significance as a critical enabling technology for 6G.

5.2.1 RAN functionalities, building blocks, and disaggregation

In wireless telecommunication systems, the RAN represents the major component that provides access to the network for individual devices/users, enabling them to exchange information or access various multimedia services. Essentially, as illustrated in Figure 5.1, the RAN provides a radio link to users, such as a cellphone, computer or any wireless connected devices, in the form of electromagnetic waves. These links provide them with remote connectivity through the core of the network. Generally, a typical wireless base station (BS) consists of a baseband processing unit (BBU) and a radio frequency (RF) processing unit, also known as remote radio unit (RRU). The BBU, usually placed in a dedicated equipment room and connected with RRU via optical fiber, provides signal processing functions on a baseband signal, i.e., before it is modulated for transmission. As regards the RRU, it is a small outdoor transceiver typically mounted near the antenna. During the transmission phase, the RRU takes the data signal provided by the BBU through optical fiber, converts it to analog, upconverts it to the required radio frequency, amplifies it, and sends it out via the antenna. On the other hand, during the receiving side, it filters and amplifies

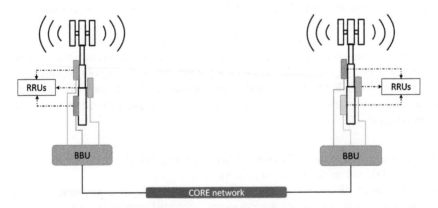

Figure 5.1 RAN components

the signal received from the antenna, converts it to a digital signal, and sends it via the fiber to the BBU and into the cellular network.

In terms of large-scale implementation, two different approaches can be identified. The first approach, representing the traditional cellular network deployment, is referred to as distributed RAN (D-RAN). In this approach (see Figure 5.2), both the BBU and the RUs are implemented on-site, with the BBU eventually connected to multiple local RRUs. Although this approach notably minimizes the fronthaul length connection, its main drawback consists in the fact that the coordination between BBUs necessary to implement carrier aggregation, multipoint, and handover services might result in difficulty. As a result, the centralized RAN (C-RAN) has been proposed, which is implemented by operators in specific deployment scenarios. In contrast to the D-RAN, the C-RAN centralizes BBU resources (see Figure 5.3), achieving then maximum inter-BBU coordination benefits. However, it has the main drawback of increasing the latency of fronthaul connections.

In addition to the physical implementation and deployment issues, the legacy version of RAN also has limitations concerning its logical functionalities. Specifically, the vast majority of BBUs used today in 4G network architecture are implemented using an all-in-one monolithic paradigm. In other words, all layers of the RAN protocol stack, i.e., physical (PHY), medium access control (MAC), radio link control (RLC), and packet data convergence protocol (PDCP) connected to the radio resource control (RRC), are implemented using proprietary hardware and software (see Figure 5.4). This vendor lock-in paradigm makes it difficult, if not impossible, to introduce new equipment from other vendors. Then, this type of approach does

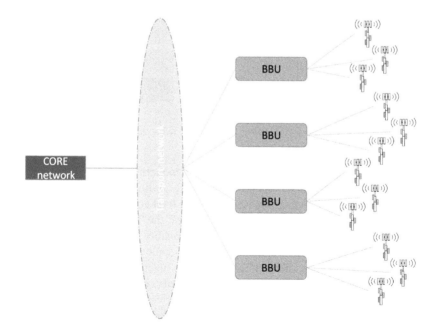

Figure 5.2 D-RAN architecture

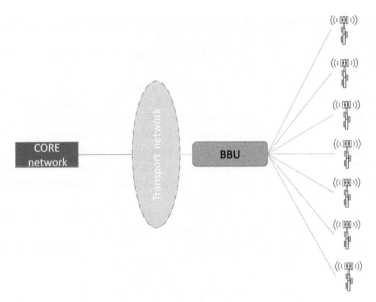

Figure 5.3   C-RAN architecture

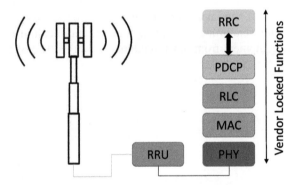

Figure 5.4   All-in-one monolithic RAN paradigm

not result definitively suitable to meet the needs of modern networks, i.e., adaptability in deployment, intelligent automation, extensive coordination among RAN nodes, quick implementation of cutting-edge features, and total cost reduction. In response to these functionality and deployment-related issues, the Third Generation Partnership Project (3GPP) studied and proposed a functionality split illustrated in Figure 5.5, with the main aim of guaranteeing enhanced deployment flexibility and efficiency in 5G [11]. This type of functionality split involves the allocation of network functions between central units (CU), distributed units (DU), and radio units (RU). In addition to provide deployment flexibility, it also offers implementation flexibility since CU and DU functions are intended to run as virtual software functions on standard commercial off-the-shelf (COTS) hardware. This allows them to be deployed in any RAN datacenter. More in detail:

Digital twin empowered Open RAN of 6G networks 139

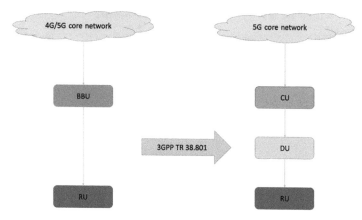

Figure 5.5   3GPP TR 38.801 functionality split

- The **RU** is the unit housing the radio hardware necessary to convert a digital signal into radio signal and received radio signals into digital signals. Generally, it implements part of PHY functionality, also known as lower PHY, such as analog-to-digital conversion, analog/digital beamforming, and fast Fourier transform.
- The **DU** software is normally deployed close to the RU. It generally runs the RLC, MAC, and parts of the PHY layer (upper-PHY).
- The **CU** software is employed to run the radio resource control (RRC) and packet data convergence protocol (PDCP) layers of the protocol stack.

According to this principle, the 3GPP defined a total of eight splitting options. Some of them are summarized in Figure 5.6. The choice of the type of solution that needs to be adopted depends on radio network deployment scenarios, constraints, and services that the mobile operator wants to provide. For example, higher functional splits are in support of network capacity enhancement in dense urban areas while lower functional splits would represent the optimal solutions for coverage use cases. Option 2 and Option 7, referred to as higher layer split (HLS) and lower layer split (LLS), have received a lot of attention within the last few years. In particular the LLS is the one on which the O-RAN architecture has been based. On the other hand, the HLS is considered as the first step toward the virtualization of the RAN. In particular, the CU is virtualized in this splitting option. In addition, in order to provide flexible network scalability and efficient RAN slicing, in this case the CU is further divided into user-plan (CU-UP) and control-plane (CU-CP).

5.2.2   Toward the concept of Open RAN

The last few years have witnessed the establishment of several forums and independent alliances whose missions are to revolutionize and transform RAN. The first initiatives can be traced back to 2016 when two independent groups were formed. More specifically, first the extensible RAN (xRAN) forum was established with the main aim of promoting a software-based and extensible architecture standardizing

140 Digital twins for 6G

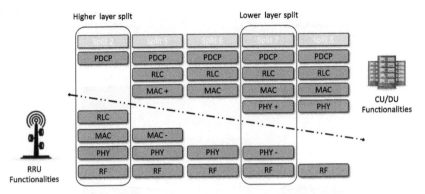

Figure 5.6 *Possible functionality split options*

all its critical elements. At the same time, the C-RAN Alliance was established with the main interests of building virtualized and open radio interfaces. In 2018, these two groups merged into the O-RAN Alliance with the intention of enriching the RANs of the next-generation wireless system with improved intelligence, cloud-scale economies, and sharpness. In order to achieve these goals, all the activities have been based on the following concepts:

- **Open standard interfaces**: These interfaces allow for interoperability between equipment from different vendors. By using open standards, network operators can mix and match components from various manufacturers, reducing their dependence on a single vendor and enhancing competition.
- **RAN functions virtualization**: This means that functions traditionally implemented in dedicated hardware can be virtualized and run on general-purpose servers or cloud infrastructure, making then the network more flexible and scalable.
- **Interoperability of multi-vendor equipment**: In this way, it will be possible to create an environment where network operators can build and manage multi-vendor RANs with greater ease. This is significant because traditional RANs often rely on proprietary interfaces and protocols, making it challenging to introduce equipment from different vendors.
- **Intelligence**: This refers to the capability of the network to dynamically adapt and optimize its operations based on real-time information processed by AI/ML intelligence mechanisms. This is a fundamental aspect to achieve improved network efficiency, performance optimization, and enhanced user experiences of modern telecommunications networks.

These will allow a reduction in capital expenditure as well as build more versatile and scalable best-of-breed RAN. Furthermore, intelligence will be helpful in handling service requirements in more complex networks such as the 6G [12–14].

In addition to O-RAN Alliance, the Telecom Infra Project (TIP) represents the second major group working toward the creation of an O-RAN. Both TIP and O-RAN are mainly focused on O-RAN interfaces, aiming at creating a flexible modular RAN for a 6G network mainly based on general-purpose hardware and open-source software. Some other open-source networking communities such as the Linux Foundation, Open Networking User Group, and Open Networking Foundation are also working in the same direction, bringing forward the concept of *open networking*.

5.2.3 Definition of O-RAN architecture

As mentioned earlier, the reference architecture for the O-RAN has been conceived in a way such that all its interfaces will be open and interoperable with other complimentary standards promoted by either 3GPP or other complimentary industry standards organizations. This level of disaggregation will enable operators to select RAN components from different vendors individually. On the other side, the virtualization will support more efficient slicing splits over the protocol stack, which thanks to the usage of AI will result in a self-organizing network RAN architecture with notably reduced costs for network deployment and scaling.

The O-RAN architecture is based on the LLS split defined by 3GPP (Figure 5.7). In particular, it consists of the following functional components described below and summarized in Table 5.1 [5]:

- Non-real-time (non-RT) RAN Intelligent Controller (RIC) embedded into the orchestration and automation layer. It is intended to provide functionalities with time requirements higher than 1 s.
- Near-real-time (near-RT) RAN Intelligent Controller (RIC) layer, which provides time-sensitive functionalities with a response time of less than 1 s.

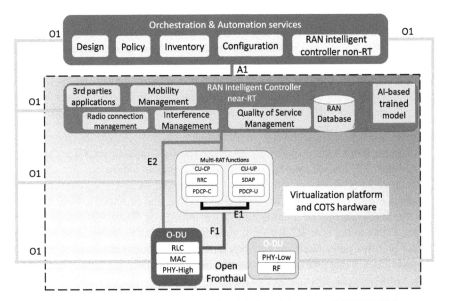

Figure 5.7 Graphical representation of O-RAN architecture. ©IEEE 2023. Reprinted with permission from [1].

Table 5.1 A comparative summary of using DT with O-RAN

O-RAN component	Functionality
Non-real-time (non-RT) RAN intelligent controller (RIC)	Part of the service management & orchestration. It supports intelligent RAN optimization in non-real time (control loops > 1 s)
Near-real-time (near-RT) RAN intelligent controller (RIC)	Enables near real-time (10 ms^{-1} s) control of the RAN
Multi-radio access technology (multi-RAT) control unit (CU) protocol stack	It is split into control plan (CP) and user plan (UP). This unit hosts radio resource control (RRC) functions, service data adaptation protocol (SDAP) functions, and packet data convergence protocol (PDCP)
The O-RAN distributed unit (O-DU)	This represents a logical node hosting radio link control (RLC) and medium access control (MAC) protocols as well as high-level functionalities of the physical layer
The O-RAN radio unit (O-RU)	This contains low-PHY layer and RF processing functions, i.e., fast Fourier transform (FFT), inverse FFT (iFFT), and digital beamforming

- A multi-radio access technology (multi-RAT) control unit (CU) protocol stack hosting data adaptation protocol (SDAP) and radio resource control (RRC) and service functions, as well as packet data convergence protocol (PDCP) for both control plane and user plane.
- The O-RAN distributed unit (O-DU) is a logical node hosting radio link control (RLC) and medium access control (MAC) protocols. High-level functionalities of the physical layer (high-PHY) such as big fat positive (BFP) compression/decompression for transmission bandwidth reduction.
- The O-RAN radio unit (O-RU), a logical node hosting low-PHY layer and RF processing functions like fast Fourier transform (FFT), inverse FFT (iFFT) and physical random access channel (PRACH) extraction.

All these modules are connected together through specific interfaces summarized in Table 5.2. The O1 interface provides operation and maintenance to CU, DU, RU, and near-RT RIC through a service management and orchestration (SMO) entity. A1 enables non-RT RIC to interact with near-RT RIC and provides it with AI models for the management and optimization of the underlying RAN. E2 is the standard interface between the near-RT RIC and underlying CU and DU units. The E1 interface is a point-to-point interface between a next-generation node B (gNB) CU-CP and a gNB CU-UP that guarantees the separation between the transport network layer and the radio network layer, as well as enables the exchange of user equipment (UE) associated information and non-UE associated information. Finally, F1 is a logical point-to-point interface between different endpoints which facilitates the interconnection of a gNB-CU and a gNB-DU supplied by different manufacturers. PHY-high

Table 5.2 O-RAN interfaces description

O-RAN interface	Description
O1	Used to allow the service SMO module to provide operation and maintenance services to CU, DU, RU, and near-RT RIC
O2	Reference interface between O-Cloud and SMO
A1	It provides direct communication between non-RT RIC and near-RT RIC in order to transport AI models for the management and optimization of the underlying RAN through the near-RT RIC
E1	A point-to-point interface between a next-generation node B (gNB) CU-CP and a gNB CU-UP. It guarantees the separation between transport network layer and radio network layer
E2	Guarantee connection between the near-RT RIC and underlying CU and DU units
F1	F1-c and F1-u are logical point-to-point interface between which facilitate inter-connection of a gNB-CU and a gNB-DU supplied by different manufacturers
Open Fronthaul	Communication between PHY-high (O-DU) functionalities and PHY-low (O-RU) functionalities

functionalities and PHY-low functionalities are connected through the open fronthaul interface.

5.2.4 O-RAN as an enabler for 6G deployment

During the last decade, AI-based solutions have been representing a common practice in addressing problems across all industry sectors. Nevertheless, the use of AI will continue to be a keystone part of 6G systems [15]. For this reason, the O-RAN architecture can be fully labeled as an enabler for 6G deployment. Indeed, both non-RT RIC and near-RT RIC are to embed AI-based functionalities to automatically and iteratively improve mobile network functionalities with minimal human intervention. In this way, the O-RAN architecture can perform complex tasks to maximize network performance such as device power consumption, load balancing, quality of service (QoS) optimization, interference management, network planning, and network configuration. These will contribute to the creation of a smart, agile, and self-adaptable 6G network able to learn and adapt itself according to the changing requirements, allowing the achievement of 6G's key performance indicators (KPIs) (see Figure 5.8). Last but not least, more efficient and flexible upgrades to the network will be also allowed. This is because the new architecture enables programmability, controllability, and intelligence in the RAN, where vendor-specific functionality traditionally incorporated in the DU, CU and the network management platform will be now handled by the RAN intelligent controllers.

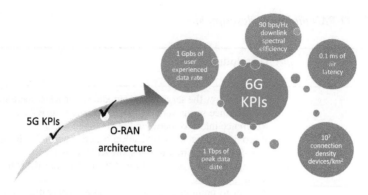

Figure 5.8 *From 5G to 6G KPIs through O-RAN. ©IEEE 2023. Reprinted with permission from [1].*

5.3 The concept of digital twin

This section provides a brief illustration of the concept of DT and highlights how it will enable the deployment of 6G-related services.

5.3.1 Definition of digital twin

The concept of DT is envisioned to play a key role in delivering 6G-oriented services. It made its appearance in 1970 during the Apollo 13 mission, where it was exploited to carry out some simulations aimed at safely bringing the Apollo 13 crew back to earth after an oxygen tank accident [16]. The first attempt for a proper DT definition occurred 40 years later when NASA gave the definition of a DT as *an integrated multi-physics, multi-scale, probabilistic simulation of an as-built vehicle or system that uses the best available physical models, sensor updates, fleet history, etc., to mirror the life of its corresponding flying twin* [17]. Then, a DT can be simply viewed as a set of computer-oriented models constantly simulating/emulating the life of a physical counterpart in the real world within its surrounding environment [18]. The DT and its physical twin (PT) are constantly connected, with a continuous data flow exchange allowing the DT to have a clear view of what is happening in the physical world, as well as to follow and optimize in a closed loop way the life cycle processes of its PT through the use of AI and big data analytics tools.

5.3.2 General architecture of a DT system

As illustrated in Figure 5.9, the general architecture of a DT system can be summarized through the following three layers:

- **User layer**: This represents the physical world where users, machines, and robots are deployed, and all the possible services are delivered in the real world, i.e., online gaming, industrial IoT, smart traffic monitoring, autonomous transportation, telemedicine, and drone-based services.

- **Networking system layer**: This layer contains all the networking devices necessary for guaranteeing the data flow exchange between DT and PT and between different DTs located in different areas of the surrounding environment.
- **DT virtualization layer**: This top layer has a complete status description of the real system. This layer is mainly equipped with a storage system used to store both historical static data and dynamic data that reflect the PT characteristics and evolution. The dynamic data is used as input for the AI-based algorithms, which will enable the DT to obtain a consistent and accurate understanding of the real system and to derive useful insights for system maintenance and decision policy optimization [19].

5.3.3 DT as an accelerator toward digitalization

In line with the discussion provided in the previous subsection, since it provides enhanced the modeling, simulation, analysis, and decision-making capabilities, on can note how AI/ML technology represents the essence of the DT paradigm. Indeed, the usage of AI/ML provides [20–22]:

- Possibility to analyze large amounts of real-time data, identify patterns and trends, and apply this information to DT models.
- Intelligent decision support systems for DTs through optimization suggestions based on real-time data and AI algorithms.
- Virtual agents and simulations representing optimal operations for autonomic and intelligent decision policy.

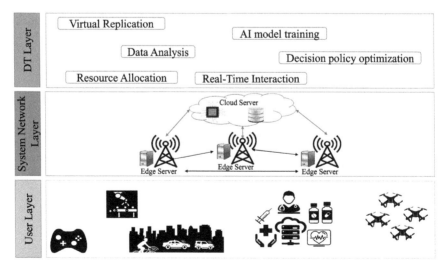

Figure 5.9 A general description of DT. ©IEEE 2023. Reprinted with permission from [1].

146 Digital twins for 6G

In other words, AI technology provides powerful analysis, decision-making, and optimization capabilities for DT, enabling them to better simulate and manage entities and processes in the physical world [23]. In turn, this means that DT technology will provide the possibility of realizing optimal decision-policies for large-scale physical systems and processes digitally represented in a cloud-based platform. This view is in parallel with the concepts of SDN and NFV already envisioned for the 5G architecture. In particular, in line with the SDN concept, a DT separates the data plane (gathered measurements) from the control plane of a real physical system. Indeed, DT is used to explore possible scenarios and to train AI/ML models which can cover all the possible corner scenarios in a real environment. Once these models are trained, they are used as control mechanisms to implement a wide range of programmable functions in the same physical infrastructure. On the other hand, according to the NFV paradigm, DT enables the possibility to design scalable virtual models for real-time monitoring/control of the correspondent PT.

As a result, thanks also to its capabilities of realizing the concepts of AI-aided learning and planning in a closed loop with its physical counterpart, DT technology generalizes the concepts of SDN and NFV to meet all the necessary KPIs for delivering 6G-oriented services. Furthermore, it will contribute to the realization of the metaverse, a completely new way to interact with real-world collected data and processes in order to perform advanced analysis of various situations and make accurate daily choices to enhance users' quality of experience (QoE) [24].

5.4 DT on O-RAN architecture: use cases

In light of the above discussion, it is clear how the DT concept perfectly fits into the principles of openness and intelligence of O-RAN. This section illustrates and discusses some of the most relevant use cases that can be deployed in the next-generation 6G network through a DT-based O-RAN architecture. A summary of the advantages of using DT in O-RAN is provided in Table 5.3.

Table 5.3 A comparative summary of using DT with O-RAN

Use case	O-RAN	O-RAN with digital twin
On-demand simulation of network scenarios	x	✓
Obtaining up-to-date data from users	✓	✓
Offline training of xApp/rApp	✓	✓
Channel modeling and prediction	x	✓
Optimal resource allocation policy	✓	✓
Real-time security and threat detection	x	✓
Real-time multiple vendor xApp/rApp testing	x	✓

5.4.1 Channel modeling for RAN optimization

Within the rapidly evolving landscape of wireless communications, the advent of 6G technology promises a transformational process, not only in terms of immersive services but also in the establishment of a seamlessly interconnected network that spans all the possible dimensions, i.e., space, air, ground, and sea. This unified network, characterized by stringent requirements, is ushering in a new era of connectivity and communication. As we look ahead to this 6G era, it becomes abundantly clear that a profound understanding of the 6G propagation channel is paramount to achieving the predefined KPIs set forth for this advanced network. In this context, traditional methodologies, including channel parameter estimation and clustering techniques, may prove inadequate in capturing the intricate relationships embedded within the data transmitted by mobile users to their respective base stations. Hence, the spotlight turns to the indispensable role that AI techniques are poised to play in this context. Indeed, the application of AI will provide the possibility to extract the full spectrum of relevant channel features, which might elude detection when relying solely on conventional approaches. Within this AI domain, deep learning (DL) and machine learning (ML) are emerging as exceptionally potent tools, offering the capacity to extract intricate details from the complex propagation environments that 6G encompasses. These AI techniques have a pivotal role in extracting critical channel parameters such as channel gain, Doppler power spectrum, power angular spectrum, and multi-path component, all information which have an important role in the optimization of transmitting and receiving antenna performance within the dynamic, multi-path terrain that characterizes 6G networks. As a result, these AI-based approaches will pave the way for unprecedented levels of connectivity, innovation, and performance optimization in the world of wireless communication. A complete discussion of these techniques can be found in [25].

Although the current O-RAN architecture provides E2 interfaces used for KPI telemetry reporting, which in turn are used to train offline rAPP/xApp models located at non-RT/near-RT RIC level, these models provided by third parties are usually trained on specific network KPI snapshots data. Hence, they may not be able to provide optimal solutions in different channel conditions/configurations. In contrast, embedding a DT-based O-RAN architecture will be perfectly suitable for performing channel modeling and optimization for 6G RAN. More specifically, the DT will be used to obtain digital replicas of physical live RAN environments by constantly collecting real channel measurements from the physical systems. These will be used by the DT to generate training and testing scenarios based on an up-to-date channel condition. As a result, different rApps/xApps can be trained offline, giving the opportunity to decide which are the most suitable based on the current network condition.

5.4.2 Network traffic forecasting and mobility management

The next-generation networks are envisioned to support a huge amount of IoT-based devices and related services which can be either static or dynamic. Due to their nature, such IoT-based services will contribute to the establishment of very complex and dynamic environments. Within these evolving ecosystems, the data traffic flow assumes a role of central importance. This flow, often characterized by its non-linearity and aperiodicity, is directed toward cloud and edge servers which undergo meticulous processing. But more importantly, the possibility of extrapolating future traffic behavior from the rich set of historical data represents a potential solution to enhance Quality of Service (QoS) and Quality of Experience (QoE). This approach is based on the principle that patterns and trends observed in past traffic data can serve as valuable points for network management decisions and resource allocation.

Even in this context, the same problem about rApp/xApp trained on outdated network data samples persists. Addressing this challenge necessitates the implementation of a data-driven framework based on AI mechanisms. Within this framework, the capability to predict data flows with remarkable accuracy becomes a reality. In particular, the virtual representation of network devices in a DT will be constantly updated by collecting historical data flows generated from sensors as well as their mobility patterns. Through the use of specific ML/DL-based structures applied to this historical data, it will be possible to extract features such as particular conditions under which a huge amount of data traffic would be generated again in the future. The correspondent-trained AI model will be then used by the RIC to perform the necessary management operations and resource allocation at the RAN level in order to maximize the QoS/QoE in the communication system as a whole. The adoption of a DT-based framework embedded with AI mechanisms will strongly support the possibility of accurately predicting data flows.

5.4.3 Security and threat detection

The realm of security threats and the detection of attacks emerge as a critical and demanding domain, deserving particular attention to ensure the seamless implementation of the forthcoming 6G networks. This urgency arises from the envisaged proliferation of IoT-based services, where an extensive array of devices converges to transmit real-time data to edge and cloud servers. These, serving as the analytical powerhouses, employ ML models to unravel essential insights and information, a pursuit that underpins the advancement of system performance. Within this context, the spectrum of potential attacks on the privacy of users' data is quite large. The implications of such attacks are primarily raising concerns over the privacy of collected data. Moreover, manipulation and poisoning of data during its journey to the respective edge or cloud server assume the guise of a direct assault on the ML models themselves. These attacks can manifest in insidious forms, including logic corruption and model evasion, exacting a significant toll on the overarching system's performance. It is worth noting that as the 6G era dawns, the anticipated shift toward blockchain as a replacement for traditional cryptography, such as elliptic curve cryptography (ECC) in 5G, holds promise. This transition is driven by blockchain's inherent attributes, including immutability and the assurance of data integrity. However, the resilience of this system is not impervious. The advent of quantum computers ushers in new possibilities for attacks, highlighting the dynamic and ever-evolving nature of the security landscape in the era of 6G.

Having a digital representation of the network and its users/devices holds the key to the development of ML algorithms adept at identifying and flagging malicious traffic and ongoing cyberattacks Furthermore, a DT representation of a device will allow the use of AI classification algorithms to evaluate the behavior of the device itself. In this way, the RIC will enact strategic measures aimed at fortifying the network security. For instance, it enables the deployment of sophisticated techniques like intelligent beamforming, meticulously designed to thwart eavesdropper attacks and safeguard the integrity of the communication environment. However, it is worth to mention that the implementation of all these security mechanisms demands the backing of quantum computers, equipped to combat and neutralize the threat posed by quantum computer-based attacks.

5.4.4 Network fault detection

Enabling the establishment of predictive mechanisms for anticipating potential network disruptions and swiftly detecting network faults stands as a pivotal cornerstone in bolstering the reliability of 6G networks. The key to realizing these enhanced functionalities lies in the adoption of a DT-empowered O-RAN architecture. In such visionary approach, every base station situated within the designated area continuously transmits performance management data to the DT, presenting a consolidated set of metrics that encompass critical parameters like service accessibility, service availability, and service quality. These data streams serve as the training sets for AI-driven fault recovery systems. Once the fault recovery system's engine has been refined through this training process, it can be seamlessly integrated into the RIC level. At this point, it takes charge of processing real-time network data, meticulously sifting through the information to unearth anomalies and trace their origins to the root causes. This proactive approach goes beyond mere detection. It also enhances the network's ability to predict potential disruptions that become a tangible reality. Consequently, remedial actions can be swiftly deployed to thwart these disruptions before they materialize, thus fortifying network resilience. Additionally, this innovative paradigm paves the way for substantial benefits for mobile network operators. The need for deploying redundant and costly copies of network components is reduced and network recovery times are expedited, translating also into improved Quality of Experience (QoE) for the end consumers.

5.5 Challenges and future directions

Although DT technology on O-RAN is a very successful combination toward the realization of a resilient, intelligent, and zero-touch 6G RAN architecture, some important issues and challenges that at the time of writing still need particular attention are discussed within this section. It is worth mentioning that most of these challenges are related to the DT concept, which is still in its infancy stage.

5.5.1 Real-time synchronization

Ensuring seamless real-time synchronization between physical entities and their DT counterparts emerges as the main challenge demanding particular attention, especially with the ever-expanding deployment of sensors that constitute an integral component of the envisioned 6G network use cases. Indeed, in these IoT scenarios, there will be sensors generating a huge amount of data traffic toward edge/cloud servers providing computing services. But, on the other hand, the same network infrastructure might need to be used for transmitting data toward the DT either for AI/ML model training or to have an updated version of the current RAN status. In light of this, it becomes evident that establishing robust connectivity and formulating optimization strategies for the strategic placement of DT cloud and edge servers represent an essential task. Indeed, such measures are instrumental in safeguarding an optimal level of synchronization within the DT representation process. Not guaranteeing optimal synchronization will exert a detrimental impact on the precision and performance of the operational DT-based O-RAN system, particularly in terms of its ability to deliver specific services [26]. Another tangible example can be found in the context of industrial IoT where the achievement of high-fidelity and real-time digital representation of all machinery involved in the industrial process assumes a central role in elevating the efficiency of the entire production line.

5.5.2 Data flow security and privacy

The entire DT-based O-RAN framework will predominantly rely on AI logic, a paradigm that introduces profound concerns surrounding security and privacy. Among these concerns, data privacy represents the foremost. This is because the establishment of a DT-based O-RAN network system necessitates the acquisition of an extensive amount of user-specific private data gathered from a myriad of underlying users and sensors. Such users, in this scenario, operate in a state of relative unawareness regarding the handling and management of their data by external cloud and edge systems. This inadvertently opens the door to potential threats, where third parties or malicious actors could potentially gain direct access to this highly sensitive and confidential data.

In addition to the privacy aspect, data transmitted from users and sensors to the DT remains perpetually vulnerable to a set of potential attacks, each carrying the potential to jeopardize the system's performance. A salient illustration of such threats is the poisoning data attack, wherein the injection and manipulation of data lead to inaccuracies that permeate the learning process of ML models at the DT level. Further, data can be poisoned in order to understand the ML models running at the RIC layer.

The path to addressing these challenges is paved with innovative solutions. Notably, the usage of federated learning, coupled with advanced techniques like homomorphic encryption and differential privacy, emerges as an effective solution. These approaches aim to fortify the security and privacy of the DT-based O-RAN framework. Additionally, the deployment of adversarial ML techniques against poisoning and logic evasion attacks assumes paramount significance. However, it is imperative to acknowledge that while these techniques hold immense potential, the existing body of literature in these domains remains relatively limited. Consequently, there is a clarion call for sustained research efforts to further enrich and fortify these methods in order to comprehensively address such security issues.

5.5.3 Data annotation

In supervised learning algorithms, the process of data annotation plays a pivotal role by establishing associations between the network attributes and their corresponding class labels. In the context of network analysis, the use of real-time data streams is essential for constructing DTs that encapsulate the dynamics of the O-RAN system. A significant hurdle arises in the form of annotating the vast and continuous data stream to effectively train ML/DL models. The accuracy of these annotations is pivotal, as it directly influences the precision of predictions and the dependability of visualization services provided by the network.

The quest for dependable and innovative architectures for data annotation becomes imperative. These architectures not only lay the foundation for sound data annotation practices but also extend their utility to validation functions at the network's edge within the O-RAN ecosystem. In essence, these architectures are instrumental in ensuring the integrity and quality of data collected at the network's periphery, reinforcing the network's overall efficacy and performance.

5.5.4 Compliance

O-RAN necessitates the creation of a digital replica of the network, encompassing workflows and automated components that streamline the shift between the network's data plane and its control plane. In this way, the DT takes on a dual role. It simultaneously oversees planning and management aspects of the network, all facilitated by AI and ML models designed for performance assurance. On the other hand, it enables the execution of operations within a virtual environment, marking a significant leap toward the improvement of network management operations. However, the effectiveness of the digitally twinning O-RAN hinges on two critical factors. First, there is the imperative need to accurately predict network mobility patterns, a fundamental task that ensures seamless network operation. Equally crucial is the preservation of the integrity and compliance of security algorithms, guarding against corruption. These prerequisites are indispensable for the success of O-RAN's digital twinning strategy. Last but not least, since the data it generates will assist in network auditing and verify that the businesses conform with one another, meticulous focus on compliance and auditing operations, is necessary to contribute toward the network's resilience, increases level of security and trust between entities. In other words, DT must be able to utilize data from the real world to analyze changes to the setup and provide interactive maps for locating the optimal settings and network performance without impacting the security of the underlying network.

5.6 Conclusions

This chapter has presented a comprehensive and lucid perspective on the advantages of deploying a DT-enhanced O-RAN architecture. It commenced with succinct descriptions and introductions to O-RAN and DT, unveiling their individual significance. Subsequently, it sheds light on the intricate synergy between DT and O-RAN, highlighting the potential of amalgamating them into a unified network architecture, which will pave the way toward the realization of an intelligent and robust 6G RAN. This was eloquently underscored by delving into pertinent use cases. Ultimately, the chapter concluded by highlighting current challenges and charting the course for future research endeavors in this nascent field of study.

Acknowledgments

This work is supported by the UK Department for Science, Innovation and Technology under the Future Open Networks Research Challenge project TUDOR (Towards Ubiquitous 3D Open Resilient Network). The views expressed are those of the authors and do not necessarily represent the project.

References

[1] A. Masaracchia, V. Sharma, M. Fahim, O. A. Dobre, and T. Q. Duong, "Digital twin for open ran: towards intelligent and resilient 6G radio access networks," *IEEE Commun. Mag.*, vol. 12, no. 6, pp. 47–55, 2023, doi: 10.1109/MCE.2022.3212570.

[2] ITU-R, "IMT Traffic Estimates for the Years 2020 to 2030," July 2015.

[3] M. Z. Chowdhury, M. Shahjalal, S. Ahmed, and Y. M. Jang, "6G wireless communication systems: applications, requirements, technologies, challenges, and research directions," *IEEE Open J. Commun. Soc.*, vol. 1, pp. 957–975, 2020.

[4] W. Saad, M. Bennis, and M. Chen, "A vision of 6G wireless systems: applications, trends, technologies, and open research problems," *IEEE Netw.*, vol. 34, no. 3, pp. 134–142, 2020.

[5] O. Alliance, "O-RAN: Towards an open and smart ran," October 2018. [Online]. Available: https://www.o-ran.org/resources.

[6] M. K. Shehzad, L. Rose, M. M. Butt, I. Z. Kovács, M. Assaad, and M. Guizani, "Artificial intelligence for 6G networks: technology advancement and standardization," *IEEE Veh. Technol. Mag.*, vol. 17, no. 3, pp. 16–25, 2022.

[7] A. E. Saddik, "Digital twins: the convergence of multimedia technologies," *IEEE MultiMedia*, vol. 25, no. 2, pp. 87–92, 2018.

[8] F. Foo, "Digital twins catalyst reflections from digital transformation world," Jun 2019. Available: https://www.ericsson.com/en/blog/2019/6/digital-twins-catalyst-booth-reflections-from-digital-transformation-world.

[9] T. Q. Duong, D. V. Huynh, S. R. Khosravirad, V. Sharma, A. D. O, and H. Shin, "From digital twin to metaverse: the role of edge intelligence-based 6G ultra-reliable and low latency communications," *IEEE Wireless Commun. Mag.*, Feb 2023 (accepted and published online).

[10] M. Liu, S. Fang, H. Dong, and C. Xu, "Review of digital twin about concepts, technologies, and industrial applications," *J. Manuf. Syst.*, vol. 58, pp. 346–361, 2021.

[11] F. Giannone, K. Kondepu, H. Gupta, F. Civerchia, P. Castoldi, A. A. Franklin, and L. Valcarenghi, "Impact of virtualization technologies on virtualized ran midhaul latency budget: a quantitative experimental evaluation," *IEEE Commun. Lett.*, vol. 23, no. 4, pp. 604–607, 2019.

[12] C.-X. Wang, M. D. Renzo, S. Stanczak, S. Wang, and E. G. Larsson, "Artificial intelligence enabled wireless networking for 5G and beyond: recent advances and future challenges," *IEEE Wireless Commun. Mag.*, vol. 27, no. 1, pp. 16–23, 2020.

[13] U. Challita, H. Ryden, and H. Tullberg, "When machine learning meets wireless cellular networks: deployment, challenges, and applications," *IEEE Commun. Mag.*, vol. 58, no. 6, pp. 12–18, 2020.

[14] A. Masaracchia, V. Sharma, B. Canberk, O. A. Dobre, and T. Q. Duong, "Digital twin for 6G: taxonomy, research challenges, and the road ahead," *IEEE Open J. Commun. Soc.*, vol. 3, pp. 2137–2150, 2022.

[15] K. B. Letaief, W. Chen, Y. Shi, J. Zhang, and Y.-J. A. Zhang, "The roadmap to 6g: Ai empowered wireless networks," *IEEE Commun. Mag.*, vol. 57, no. 8, pp. 84–90, 2019.

[16] NASA, "The ill-fated space odyssey of Apollo 13," 2002. Available: https://nssdc.gsfc.nasa.gov/planetary/lunar/apollo13.pdf.

[17] NASA, "Draft modeling, simulation, information technology & processing roadmap," Nov 2010. Available: https://ai.googleblog.com/2020/05/federated-analyticscollaborative-data.html.

[18] B. Barbara, R. Casiraghi, Elena, and F. Daniela, "A survey on digital twin: definitions, characteristics, applications, and design implications," *IEEE Access*, vol. 7, pp. 167653–167671, 2019.

[19] Q. Qinglin and T. Fei, "Digital twin and big data towards smart manufacturing and industry 4.0: 360 degree comparison," *IEEE Access*, vol. 6, pp. 3585–3593, 2018.

[20] Z. M. Fadlullah, F. Tang, B. Mao, *et al.*, "State-of-the-art deep learning: evolving machine intelligence toward tomorrow's intelligent network traffic control systems," *IEEE Commun. Surveys Tuts.*, vol. 19, no. 4, pp. 2432–2455, 2017.

[21] T. Wang, J. Li, W. Wei, W. Wang, and K. Fang, "Deep-learning-based weak electromagnetic intrusion detection method for zero touch networks on industrial IoT," *IEEE Network*, vol. 36, no. 6, pp. 236–242, 2022.

[22] B. Mao, F. Tang, Y. Kawamoto, and N. Kato, "AI models for green communications towards 6G," *IEEE Commun. Surveys Tuts.*, vol. 24, no. 1, pp. 210–247, 2022.

[23] E. Baccour, N. Mhaisen, A. A. Abdellatif, A. Erbad, A. Mohamed, M. Hamdi, and M. Guizani, "Pervasive AI for IoT applications: a survey on resource-efficient distributed artificial intelligence," *IEEE Commun. Surveys Tuts.*, vol. 24, no. 4, pp. 2366–2418, 2022.

[24] A. Moayad, B. Ouns, K. Fakhri, R. I. Al, and S. A. El, "Integrating digital twin and advanced intelligent technologies to realize the metaverse," *IEEE Consum. Electron. Mag.*, vol. 12, no. 6, pp. 47–55, doi: 10.1109/MCE.2022.3212570.

[25] C. Huang, R. He, B. Ai, *et al.*, "Artificial intelligence enabled radio propagation for communications—part I: channel characterization and antenna-channel optimization," *IEEE Trans. Antennas Propag.*, vol. 70, no. 6, pp. 3939–3954, 2022.

[26] D. V. Huynh, S. R. Khosravirad, A. Masaracchia, O. A. Dobre, and T. Q. Duong, "Edge intelligence-based ultra-reliable and low-latency communications for digital twin-enabled metaverse," *IEEE Wireless Comm. Lett.*, vol. 11, no. 8, pp. 1733–1737, 2022.

Chapter 6
Potentials of the digital twin in 6G communication systems

Bin Han[1], Mohammad Asif Habibi[1], Nandish Kuruvatti[1], Sanket Partani[1], Amina Fellan[1] and Hans D. Schotten[1]

6.1 Introduction

The digital twin (DT) technology has garnered considerable attention for its potential to revolutionize various sectors, including industrial Internet of Things (IIoT) and intelligent wireless systems. As a virtual representation of physical assets, systems, or processes, a DT offers unprecedented capabilities for monitoring, analysis, and control. Within the context of sixth-generation (6G) communication systems, DT stands as a pivotal technology with far-reaching implications. This chapter aims to elucidate the multifaceted impact of DT on 6G systems, covering its role in the development, deployment, and maintenance of 6G infrastructure, its ability to optimize network performance across different layers, and its potential to enable or enhance emerging 6G use cases.

DT technology serves as a cornerstone in the planning, service testing, and rapid development of 6G systems. It simplifies and accelerates site deployment configurations, making it easier to test the impact of configuration and functional changes in 6G networks. This technology is instrumental in reducing both the time and the cost associated with the development, deployment, and maintenance of 6G infrastructure. These aspects are discussed in detail in Sections 6.2, 6.3, and 6.4, respectively.

Meanwhile, DTs also offer significant advantages in enhancing the performance of 6G communication systems across different network layers. They serve as platforms for building artificial intelligence (AI) models for real-time analyses, tracking issues related to security and resiliency, and managing network slices efficiently. Furthermore, they play a crucial role in radio access network (RAN) optimization and effective traffic management, extending their utility to the physical (PHY) and medium access control (MAC) layers of the network. These benefits are elaborated upon in Sections 6.5–6.10.

Moreover, the integration of DT technology has the potential to either enhance or enable new use cases in 6G, such as the operation of mobile edge cloud (MEC) and

[1]Division of Wireless Communications and Radio Positioning, Department of Electrical and Computer Engineering, University of Kaiserslautern (RPTU), Germany

6G-based IIoT. By providing a real-time virtual representation of physical systems, DTs allow for more efficient and effective operation in these emerging fields. This technology is not just an add-on but a fundamental enabler that can significantly impact the range and quality of services offered by 6G systems. These emerging use cases are explored in Sections 6.11 and 6.12.

Given the comprehensive capabilities outlined above, it is evident that DT technology stands as a pivotal enabler for the future of 6G communication systems. Its applications range from the foundational aspects of network development and deployment to real-time performance optimization and the facilitation of emerging use cases. As such, the development and implementation of DT technology in 6G systems will undoubtedly be a focal point of research and innovation in the years to come.

6.2 Optimized planning, service testing, and rapid development of the 6G network

The introduction of a new generation of communication networks has almost always been accompanied with the launch of an array of new technologies and services that are meant to address the requirements set by vertical industries and network operators. For the forthcoming 6G networks, a range of revolutionary services and disruptive technologies are spanning. These are including, but not limited to: 1) Metaverse-related services such as reality (VR), augmented reality (AR), and extended reality (XR); 2) 3D holographic telepresence; 3) industrial innovations, such as collaborative robots (cobots) and self-assembling robots; and 4) highly demanding and sensitive services, such as telemedicine and healthcare [1,2].

As a result, a proactive network handling of operations is indispensable for both the industry and network operators to ensure that not only the stringent performance requirements of the aforementioned revolutionary services are met but also their extremely challenging key performance indicators (KPIs) defined in their service level agreements (SLAs) are adequately fulfilled [3]. These challenges can be characterized by managing the unparalleled number of new subscriptions, delivery of network operation support to verticals and businesses, as well as provisioning of network upgrades, maintenance, and expansions [4]. This proactive network operation has to be coupled with the proper timing in order to be able to accurately predict the future service demands. Consequently, the prerequisite capabilities and capacity of the network are established by performing a comprehensive examination of the current network's performance patterns. The established findings have to be evaluated and contrasted against different possible operation scenarios and conditions.

Based on the former reasoning, it becomes evident how the provisioning of DT technology can immensely be of benefit for the advancement of 6G services, networks' deployment and implementation. The newly derived network capabilities and functions exemplified in a DT can serve as a sandbox for testing various realistic scenarios prior to the actual implementation and deployment of the 6G network [5]. To account for the complexity associated with the futuristic 6G services, the digital twin network (DTN) should accurately model all the pertinent network layers, entities, domains, nodes, etc. with the ultimate objective of capturing the precise state of the current network. With the help of a DT, it is possible to assess and thoroughly examine different strategies, instantiate and test new 6G services on the network prior to their deployment, and evaluate and investigate the implications of rolling out new services using various analytical models [5]. Ultimately, leveraging the power of DTNs could result in an enhanced network coverage, support the expansion of the underlying infrastructures, and evaluate the resulting gains in the overall network performance. A DTN could also assist in revealing possible points of failure and weakness spots in the network to the operator while providing a platform for analyzing the effects of diverse failure scenarios on the services. It could also aid in planning to face the consequences of such scenarios [4]. The DT technology provides crucial insights into the network operation necessary to obtain a comprehensive understanding of the customer's expectations. It can also be of particular benefit in the early network planning stages by evaluating possible SLA infractions and their resulting financial risks [3], thus ensuring an optimized use of investment toward areas with the greatest return.

6.3 Simplifying and accelerating the site deployment configuration

Radio planning is an important phase that has to be carried out prior to the network deployment by examining the physical environment. The optimal coverage and performance are guaranteed through the strategic positioning of radio units and the use of ideal configurations. A DT that incorporates the quintessential characteristics of the cell sites for network deployment will be beneficial for this task, as shown in Figure 6.1. It is possible to build models that precisely capture the physical attributes and are scalable (up to a city-level), taking into account foliage, building locations, construction materials, etc. Furthermore, the network components and their respective features, antenna pattern, height, elevation, etc. can be added. This facilitates the integration of radio propagation data. The integration of radio propagation data can be applied by employing ray-tracing, enabling the computation of signal quality at any location in the modeled scenario and promoting visualization possibilities. A DT encapsulating virtual replicas of the physical objects in a radio environment and their behavior (e.g., construction material) allows an efficient projection of signal reflections and calculation of their intensities. Furthermore, signal paths and antenna beamforming can be simulated and visualized. For example, features such as transmitter beamforming and signal paths can be denoted by lobes and lines, respectively. Whereas coloring schemes can depict signal strengths (e.g., blue being the weakest to red being the strongest). Similarly, other performance indicators such as latency, link rate, coverage, etc. can be illustrated and visualized.

Combining this visualization potential with the concepts of (VR) or (XR) will open the doors for the system architects to analytically explore the model in parts or as a whole and remotely inspect network sites. It also supports system architects to visualize the effects of tuning or adaptations in real-time in the form of signal paths, strength indicators, lobes, etc., resulting in

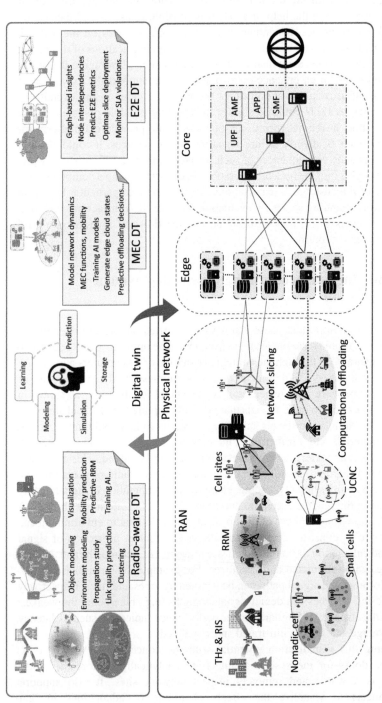

Figure 6.1 Utilizing DT technology across diverse domains and solution domains within the proposed 6G communication system to improve the system performance and streamline operational processes. The terms used in this figure are not defined elsewhere in the chapter: user plane function (UPF), access and mobility management function (AMF), session management function (SMF), and application (APP).

detailed simulation mirroring real-world conditions. Thus, network operators will be capable of designing exceptionally efficient and dependable networks, conduct remote field trials, and intensify the deployment efforts. Efforts are already underway to accelerate fifth-generation (5G) deployment in this direction [6]. Such steps are anticipated to become instrumental in the testing and deployment phases of the future beyond 5G (B5G) and 6G systems.

6.4 Testing the impact of configuration and function changes

The introduction of any new technology or protocol in the network warrants an evaluation of its functionality and performance, including new network setups and software updates. This is crucial to ensure that these changes can be implemented across the entire network without affecting performance. In light of the emergence of recent mobile networks such as 5G, B5G, and 6G, the telecommunications sector views the adoption of the continuous integration and continuous deployment (CI/CD) methodology as essential for handling the impact of the digital value chain on their operations. CI/CD is a structured framework that facilitates the continual progression of software through development, testing, deployment, and validation stages. Telecommunication companies are employing this method to augment feature flexibility and elevate software quality [7].

Canary Testing, a common feature integrated into the CI/CD pipeline, serves as a technique to verify the proper functioning of services [5]. Before a wider roll-out to all users, this testing methodology involves initially introducing new configurations or modifications to a limited subset of users accessing the system functionality and performance. This feature can be easily implemented using a DTN with the added advantage of using a safe virtual environment for testing. By simulating a network in real-time, a DTN effectively supports the validation of new changes, upgrades, and AI models. For example, it has the capability to evaluate the resources and processes needed for the cloud deployment of a new software, thereby guaranteeing a successful trial prior to transitioning to the updated version [5].

Within autonomous networks, the network management entity determines the necessary actions to sustain the desired performance and achieve network operation objectives. It is imperative to comprehend the implications of these actions, and a DTN can facilitate testing by assessing their repercussions and predicting any adverse effects on relevant KPIs. Thus, a DTN assists in making informed decisions about which actions to implement before applying them to live services [5].

6.5 Building platforms to train AI models for the 6G system

The recent advancements and rapid progress in AI are seen as a boon to many sectors, causing exponential development in the areas of B5G and 6G mobile wireless networks. Machine learning (ML) models have the potential to strengthen the end-to-end (E2E) network functionality across different layers, including but not limited to modulation and coding scheme (MCS), waveform selection, radio resource management (RRM) based on the signal-to-interference-plus-noise ratio (SINR) predictions, MEC caching, slicing of the network, power control, etc. [8]. Furthermore, the ML models can assist the network in adapting to the dynamic quality of service (QoS) needs of different services.

However, an AI model not only needs versatile and realistic data in large quantities (i.e., big data) but also needs quality in the dataset (i.e., good data). Measuring and storing all possible statistical outcomes, in any environment, can result in well-trained models with high prediction accuracy and high resilience to changes in data over time. Collecting real-time network data has high device and time cost, and it is often not feasible to capture the extremities or irregularities in a live network [5]. These shortcomings can be significantly reduced by using a DTN. The DTN has the capability to produce realistic data by emulating/simulating different devices in a wide range of scenarios repeatedly. This rich and diverse data can help the models to infer real-world conditions and make better predictions.

Many ML models can be implemented to realize this. Supervised learning, a widely recognized ML approach, involves training the model using labeled data. Reinforcement learning (RL) is another potential technique which can be used to react effectively when unforeseen patterns are available in the data. An RL agent can learn in an unsupervised fashion by using different and/or dynamic policies in different scenarios to maximize the reward in a given environment. However, implementing and converging an RL policy can potentially be costly in terms of time and may negatively impact a live network. Nevertheless, DTN can address this shortcoming by providing a secure virtual platform to train and efficiently deploy different models, sometimes even proactively. Moreover, it enables the repeated simulation of uncommon scenarios, enabling learning agents to discover innovative strategies for optimizing rewards [9].

6.6 Tackling the security and resiliency issues in 6G

B5G/6G networks are seen as a key enabler to support multiple vertical industries including agriculture, energy, retail, etc. to maintain and effectively perform their desired operations. For any vertical industry, 6G networks can enable both reliable connection and low-latency data collection from multiple low-power, low-latency Internet of Things (IoT) devices [10]. For instance, 6G has the capability to facilitate safety-critical use cases like emergency braking or intersection management in the intelligent transportation system (ITS) sector, offering low latency, traffic efficiency, and high reliability [11]. Similarly, many other vertical industries can leverage 6G networks to maximize support and services.

However, security always remains a concern when there are a plethora of devices connected to the same network. Unauthorized access to the network, data breaches, identity theft, from network glitches to outright network collapse, and various other undesirable intrusions serve as unwelcome distractions from the numerous advantages offered by 6G networks. Even the network-integrated AI models can be prone to different security attacks [12,13]. Thus, the security of the connected IoT devices and network components is of paramount importance. A majority of these concerns can be displaced during the design phase of the network and devices with robust and secure protocols. However, certain real-time threats will always be overlooked. A DTN can help overcome this shortcoming in the network design. Primarily, a DTN can help in understanding the features and various effects and repercussions of any security protocol implemented for the network. This is done by simulating scenarios under different conditions and observing the network. With these observations, a feedback loop can be established to engineer networks that are fortified against not only security breaches, degradation, and system failures but also unexpected variables. Furthermore, a DTN can also assist in creating protocols to ensure that the network remains operational even under adverse conditions.

6.7 Efficient network slice management and orchestration

Legacy mobile networks are tasked with offering connectivity for a myriad of use cases, each with its own unique and often conflicting set of QoS criteria, such as data rate, latency, reliability, and delay, to name a few. As a result, a one-size-fits-all network architecture proves to be sub-optimal in various aspects, including but not limited to resource allocation, energy conservation, and the costs associated with network deployment and upkeep [14,15]. Network slicing emerges as an innovative architectural paradigm, enabling the tailored modification of the foundational communication infrastructure to cater to specific use cases and deliver a broad spectrum of services through a software-driven, virtualized network layout [16,17].

Network slicing allows for the partitioning of the physical network and its resources among various tenants, creating logically segregated E2E virtual networks that are managed independently to enable the delivery of resilient and adaptable services [14,16]. Software defined networking (SDN) and network function virtualization (NFV) serve as the primary technological catalysts for such agile and resource-efficient network slicing approaches in communication systems [16,17]. Nonetheless, the challenge lies in maintaining E2E performance for network slices, given the plethora of services with varying and demanding requirements that utilize the same infrastructure and cross multiple network domains [14]. Effective network surveillance and the generation of accurate E2E metrics are vital for executing dynamic, autonomous network orchestration that meets the associated QoS demands [18,19].

A DTN can amplify the efficacy of managing and orchestrating network slicing in multiple dimensions [20]. It has the capability to create a digital replica of the physical slicing network, allowing for the exploration of various hypothetical scenarios and resource distribution strategies without substantial impact on the physical network. Additionally, a DTN can autonomously generate and analyze data by interfacing with the physical network. It can also forecast QoS outcomes in response to modifications in its configuration settings. Utilizing DT technology in the context of network slicing is pivotal for achieving management that is both performance-efficient and cost-effective. Moreover, it facilitates ongoing performance assessments under diverse operational conditions without compromising the integrity of the active physical network. A DTN must account for both the physical components and nodes in networks enabled for slicing, as well as their corresponding virtual counterparts. These virtual elements within a 6G network are dynamically created or terminated in real-time, adding a layer of complexity to the development of such a DTN.

Conventionally, communication networks can be represented by graphs. Due to the interconnected nature of various network nodes and the irregular graph topologies, the underlying data is organized in a non-Euclidean space. This organization makes several conventional ML frameworks, such as convolutional neural networks (CNNs), recurrent neural networks (RNNs), and auto encoders, among others, less effective in these scenarios [20]. In line with this research focus, the authors of [20] introduce a DT designed for the management and orchestration of network slicing, leveraging graph neural network (GNN) technology to probe the complex dynamics and interrelationships among network slices, resource allocation, and the foundational network infrastructure. This GNN-empowered DT utilizes an inductive graph framework to generate feature embeddings for network slices, which are represented as graphs. It then forecasts the E2E metrics for individual slices across diverse scenarios. The DT serves as an effective instrument for monitoring E2E performance by accurately predicting slice latency. The DT serves as a cost-effective tool to ensure SLA compliance when resource usage increases. It also offers advantages by mitigating the effects of link failures. It does this by identifying optimal pathways after predicting latency issues and then migrating affected network slices as needed. The architecture of this E2E DT is illustrated in Figure 6.1. Additionally, the DT employs optimal deployment strategies for network slices in instances of SLA breaches. Similarly, the authors of [21] outline a DT for network slicing that is graph-represented and employs GNNs to comprehend the complex dependencies within a network slice. A deep distributional system equipped with a DT Q-network agent learns the best network slicing policies based on graph-centric network states derived from the DT.

6.8 Enabling 6G RAN optimization and effective traffic management

The envisioned services in 6G communication systems come with diverse and challenging requirements. Furthermore, the demand for these services may change over time and space. Therefore, the planning of radio topology for peak service requirements at every point is not efficient. The network must be capable of dynamically adapting to satisfy the fluctuating demands [22]. An intelligent RAN moderation is required to guarantee optimal, adaptable, and energy-efficient utilization of the radio infrastructure and resources [23].

One of the viable ways to offload the congested macro cells is to employ small cells to serve the high density of end-users in small areas (e.g., offices, shopping malls, stadiums, and many others). However, allowing such small cells to be in an active state at all times, even in non-congested situations is energy inefficient [24]. The DTs of small cells encompassing mobile devices in its service area, their mobility attributes, service requirements, etc. can assist in anticipating high user densities and data demands in the future of the considered small cells (refer to Figure 6.1).

This enables the system to proactively activate or deactivate the small cells based on user densities and reduce power and spectral wastage.

On the other hand, Nomadic nodes (NNs) intend to equip vehicles (e.g., private cars and taxis) with low-power access nodes and wireless backhaul. Such mobile NNs may be used to increase the network capacity or extend coverage based on service and data requirements [25]. Typically, NNs are designed to be stationary, such as in a parked car, with their availability being random. However, depending on the capacity, coverage, or energy demands, these NNs can be activated or deactivated. A DT of an NN and its service area is beneficial in selecting the optimal nodes for activation. For such a task, aspects such as the qualities of the backhaul link of the available NNs and mobile base stations, co-channel interference, shadowing and multi-path fading factors, mobility factors, etc., have to be considered. Furthermore, it allows for on-the-fly network planning by considering the real-time forecast of service demands from a DT and steering the movement of an NN proactively.

In addition, advanced traffic steering methods can help improve the energy efficiency of networks by optimizing the active operational period of access points. One such promising feature of B5G and 6G communication systems is the user-centric cell-free massive multiple-input and multiple-output (MIMO) network architecture [26]. Here, the distributed units (DUs) are densely deployed and they serve fewer users. Each user in such a system is assigned a unique serving cluster. This gets rid of the concept of cell edges and provides uniform coverage and performance to all users in the network.

However, such systems are prone to the problem of channel aging [27], where the users with high mobility experience a deviation in the channel quality at the time of its estimation and during the application of this estimated channel quality for data transmission.

A DT replicating this physically distributed antenna system and its dynamics can facilitate dynamic clustering of DUs and allocation of resources as shown in Figure 6.1. The user mobility and its history must be considered to integrate channel aging effects. For this task, time-varying modeling of the channel that associates the temporal auto-correlation function of the channel with antenna features, propagation geometry, user velocity, transmission frequency, etc. has to be employed [26]. Thus, a DT of the considered solution domain helps in investigating numerous methods for RAN traffic steering and moderation. It enables real-time testing and comparison of the aforementioned solutions before applying these measures in a live network.

6.9 Optimizing the 6G radio resource management

The radio-aware DT (shown in Figure 6.1) can be devised by considering the detailed characteristics of a radio network, including the PHY and MAC layer operations of end-user devices and access points. In addition, the features of the aforementioned receiving and transmitting nodes, such as physical coordinates, trajectories, node capabilities, speed, RRM procedures, beam patterns, and link quality, are also considered. For example, the radio environment map (REM) stores the reference signal receive power (RSRP) or SINR on a grid basis in a database that may be updated in a periodic manner. Such a DT is valuable in performing a pro-active and predictive RRM [28].

A network empowered by such a DT is capable of anticipating potential errors or failures and implementing preemptive solutions. For instance, near-future prediction of a user's poor link quality, who is being served an ultra-reliable and low latency communication (URLLC) service, will enable the network to execute proactive countermeasures such as frequency band switching, new beam preparation, and adaptive resource allocation. Such proactive schemes often perform better than traditional reactive measures (e.g., re-transmissions).

Additionally, a radio-aware DT contains constantly updated and precise REM, constructed with the help of efficient propagation prediction schemes, including extended measurements and ray tracing. The radio-aware DT in controlled environments such as private factory networks has the knowledge of a device's position, its mobility, and traffic patterns (e.g., fixed-size packet transmission by sensors at regular intervals). Thus, the prediction of interference patterns and SINR at specific locations is feasible, which enables the design of a pro-active and efficient RRM scheme [28].

A radio-aware DT is also capable of predicting the link quality between access nodes and the devices even in the absence of full-scale measurements. Efficient estimation of the channel conditions is possible with the help of improved ray-tracing models and data-driven algorithms made possible thanks to the digital replication of the network's PHY layer. Thus, channel estimation overhead can be reduced and spectral efficiency can be improved [28]. Furthermore, it can improve energy efficiency as well, provided that the data-driven algorithms or ray-tracing model are less power-consuming than the comprehensive measurement campaigns.

The precise knowledge of future link conditions allows for the execution of RRM on the respective pieces of equipment for an extended period of time (composed of a fixed amount of transmit time intervals (TTIs)), and this information is delivered to the equipment in one go. These persistent short-term RRM choices enable the controller to transmit a multitude of radio control signal parameters (e.g., bandwidth-part (BWP) selection, transmit power control, modulation scheme, and coding scheme, etc.) ahead of time [28]. However, in order to make these short-term RRM decisions effective and efficient, the DT needs to have information regarding device mobility and estimation of transmit buffers over the considered number of TTIs. This approach is more practical in quasi-deterministic controlled environments (e.g., factory automation) and the aforementioned short-term RRM steps could lower the overhead of periodic measurements from devices and access nodes, which are otherwise mandatory and excessive in the case of a reactive RRM.

6.10 Terahertz wave analysis in support of reconfigurable intelligent surfaces for enhanced 6G performance

There are several distinctly challenging use cases anticipated in 6G communication systems that require extreme radio performance in terms of metrics. These metrics include but are not limited to energy efficiency, link throughput, localization accuracy, latency, network coverage, and many others. For example, immersive telepresence with AR/VR applications would need roughly 20 Gbps of data rate, whereas immersive holographic communications are only possible with data rates around 1 Tbps [1]. In addition, haptic and visual feedback present stringent requirements, such as localization accuracy below 1 cm while latency being less than 1 ms [1]. Similar strict requirements are foreseen for the cobots and massive digital twinning use case families.

The most practical means to satisfy these requirements is to make use of the larger bandwidths, ranging from 2 to 20 GHz. Such large bandwidths are available in abundance merely at higher frequencies in the upper millimeter wave (mmWave) band (100–300 GHz) and the terahertz (THz) band (300 GHz–1 THz) [1]. When the 6G system requirements are examined from the viewpoints of both network coverage and system performance, it is evident that such systems cannot operate only at higher frequencies; instead, they utilize a blend of several frequency bands in the spectrum. Therefore, it is important to study radio propagation and operations in the spectral range of 100 GHz–1 THz. At such high frequencies, the received power is dominated by the line-of-sight (LOS) signals and reflections from metallic objects. The conventional concepts of small-scale fading and shadowing alone are not sufficient to design and evaluate the 6G systems. Furthermore, the wireless channel is susceptible to molecular absorption at such high frequencies, which deteriorates the communication link [29]. Thus, it is critical to perform thorough studies of propagation aspects and carry out appropriate channel modeling at these frequency ranges.

Following a similar logic to that in Section 6.3, a DT imitating the physical domain can be helpful in performing a rigorous study of THz propagation aspects, as illustrated in Figure 6.1. DT may also help with the testing of expensive or non-trivial radio network situations in the physical domain. This enables the unbiased study and scrutiny of THz propagation aspects, thereby creating near-precise propagation models for a number of situations.

As mentioned earlier, LOS propagation is crucial at the THz bands to obtain a high enough signal-to-noise ratio (SNR) that the communication link is reliable. To this end, the re-configurable intelligent surfaces (RIS) are introduced as a feasible solution. The RIS are appropriate for adaptive reconfiguration to reflect or redirect the beams it receives from the transmitter [30]. Such a capability enables the operation of a communication link between a receiver and a transmitter steering clear of the obstacles that would lead to signal attenuation otherwise.

A DT capturing the physical propagation space, its objects, and dynamics will be advantageous in enhancing the performance of RIS. The information regarding obstacle attributes (e.g., building size, position, material composition, etc.), position and mobility of the mobile device and access nodes, environmental influences (e.g., humidity), etc., will facilitate efficient prediction of the received THz signal quality.

Thus, it is possible to use a DT to precisely model the THz propagation and study the impacts of the physical environment on signal transmission via realistic simulation. Furthermore, a DT can assist in finding the most suitable signal path that would result in maximum SNR at the receiver, among the various possible paths in the radio environment.

In addition, pro-active reconfiguration of the RIS can be carried out in the live network to alter the beam paths as necessary. The work in [31] demonstrates a DT mimicking the physical indoor space with the help of a top-view camera. Advanced image processing techniques are used by the DT to capture the features of obstacles, the receiver, and the transmitter, which are subsequently used for adapting the beam paths. Such models could be scaled up to city-level and other non-trivial outdoor environments and can assist a DT in managing the optimal functioning of RIS in the THz frequency bands.

6.11 Enhancing the operation of mobile edge clouds in 6G

MEC serves as a pivotal technology that brings computational capabilities, storage, and essential features to the proximity of end-user devices, situated at the edge of the cellular network alongside access service nodes or base stations [32,33]. It addresses key service criteria such as ultra-low latency, elevated reliability, and minimal energy usage. Moreover, it aspires to offer new functionalities and intelligent services in B5G and 6G networks by closely incorporating AI into the network infrastructure [34].

Mobile devices, functioning on cellular networks, delegate computational activities to MEC. This is a key factor in maintaining low energy usage, particularly for IoT devices. The increasing demand for services supported by MEC, each with its own prerequisites, has led to a proliferation of diverse edge servers [33]. Consequently, forecasting network behavior has become significantly more complex. Moreover, a substantial volume of mobile IoT devices necessitates offloading computational duties to MEC servers, complicating the task of establishing an efficient network management offloading strategy [35].

The network management framework must base its offloading choices on both the temporal dynamics of the user environment and sustained end-user mobility patterns [33]. Given that 6G is expected to support a variety of high-mobility scenarios with strict latency and energy requirements, the use of MEC becomes indispensable in high-mobility contexts. The complexity arises in orchestrating a sequence of offloading actions, as each current decision has implications for subsequent offloading choices.

A DT tailored for the MEC system, proficient in real-time emulation of key edge server functionalities and network dynamics, can serve as an energy-efficient base for the network management module to refine its decision-making prowess. DTs are effective instruments for monitoring the conditions of edge cloud facilities and for providing AI algorithms, such as RL agents, with essential training data (refer to Figure 6.1). As an example, the authors of [36] have pioneered the use of DTs in this arena, focusing on both the DTs of individual edge servers and the overarching MEC system. This approach elucidates the multifaceted interconnections within the MEC landscape and confronts the challenge of computational offloading from mobile devices, with the objective of curtailing offloading latency while adhering to the limitations of aggregate migration costs.

6.12 Enabling 6G-based IIoT and industrial 6G use cases

The aspirations for 6G networks transcend beyond mere incremental enhancements in established KPIs like throughput and latency. The network is poised to usher in transformative shifts in technological foundations, network configurations, and service models. A hallmark of 6G is its capacity to facilitate a wide array of nascent industrial applications, including but not limited to sustainable growth, extensive digital twinning, immersive telepresence, cobots, and localized trust domains. These innovative applications are set to catalyze the forthcoming advancements in cyber-physical systems, forging integrative links between the physical, digital, and human realms.

Additionally, the application of DT technology in such industrial scenarios can enhance the accuracy and timeliness in the monitoring, analysis, and controlling of industrial processes [9], leading to higher productivity and lower costs. Typical applications include but are not limited to predictive maintenance, process optimization, remote monitoring, and remote control. Moreover, leveraging the computing power that is brought by MEC to the edge of 6G RAN, DT delivers a high computing capacity with faster response, lower latency, and improved security, which enables not only autonomous control in time-sensitive applications such as manufacturing and driving but also the application of novel human-machine interface (HMI) technologies such as VR and AR in emerging industrial use cases, e.g., collaborative telepresence and remote operation.

6.13 Conclusion

In conclusion, the DT technology serves as a transformative force in the evolution of 6G communication systems. Its impact is broad and revolutionizes the way 6G communication systems are designed, deployed, operated, and exploited. The technology not only enhances existing functionalities but also opens avenues for innovative solutions that are integral to the 6G ecosystem [18,23]. As 6G continues to evolve, the role of DT technology will undoubtedly remain at the forefront, shaping the future landscape of communication systems.

Acknowledgments

The study in this chapter is partly supported by the EU Horizon-2020 project Hexa-X (GA No. 101015956).

References

[1] Uusitalo MA, Rugeland P, Boldi MR, *et al.* 6G vision, value, use cases and technologies from European 6G Flagship Project Hexa-X. *IEEE Access.* 2021;9:160004–160020.

[2] Jiang W, Han B, Habibi MA, et al. The road towards 6G: a comprehensive survey. *IEEE Open Journal of the Communications Society*. 2021;2:334–366.

[3] Habibi MA, Nasimi M, Han B, et al. The structure of service level agreement of slice-based 5G network. In *IEEE International Symposium on Personal, Indoor and Mobile Radio Communications*, 9–12 September 2018, Bologna, Italy.

[4] Kuruvatti NP, Habibi MA, Partani S, et al. Empowering 6G communication systems with digital twin technology: a comprehensive survey. *IEEE Access*. 2022;10:112158–112186.

[5] Öhlén P. The future of digital twins: what will they mean for mobile networks? *The Ericsson Blog*; 2021. Available from: https://www.ericsson.com/en/blog/2021/7/future-digital-twins-in-mobilenetworks.

[6] Kerris R. Ericsson builds digital twins for 5G networks in NVIDIA omniverse. *NVDIA Blogs*; 2021. Available from: https://blogs.nvidia.com/blog/2021/11/09/ericsson-digital-twins-omniverse/.

[7] Patil H and Price G. CI/CD and CD/D: continuous software delivery explained. *The Ericsson Blog*; 2020. Available from: https://www.ericsson.com/en/blog/2020/9/cicd-and-cdd-continuous-software-delivery-explained.

[8] Sun Y, Peng M, Zhou Y, et al. Application of machine learning in wireless networks: key techniques and open issues. *IEEE Communications Surveys Tutorials*. 2019;21(4):3072–3108.

[9] Habibi MA, Yousaf FZ, and Schotten HD. Mapping the VNFs and VLs of a RAN slice onto intelligent PoPs in beyond 5G mobile networks. *IEEE Open Journal of the Communications Society*. 2022;3:670–704.

[10] Mahmood NH, Alves H, López OA, et al. Six key features of machine type communication in 6G. In: *2020 2nd 6G Wireless Summit (6G SUMMIT)*; 2020. p. 1–5.

[11] Naimi M, Habibi MA, and Schotten HD. *Platoon-Assisted Vehicular Cloud in VANET: Vision and Challenges*. ESCC, Paris, France, 2–4 September 2019.

[12] Hu C and Hu YHF. Data poisoning on deep learning models. In: *2020 International Conference on Computational Science and Computational Intelligence (CSCI)*; 2020. p. 628–632.

[13] Sharma P, Austin D, and Liu H. Attacks on machine learning: adversarial examples in connected and autonomous vehicles. In: *2019 IEEE International Symposium on Technologies for Homeland Security (HST)*; 2019. p. 1–7.

[14] Habibi MA, Han B, Yousaf FZ, et al. How should network slice instances be provided to multiple use cases of a single vertical industry? *IEEE Communications Standards Magazine*. 2020;4(3):53–61.

[15] Habibi MA, Han B, Fellan A, et al. Toward an open, intelligent, and end-to-end architectural framework for network slicing in 6G communication systems. *IEEE Open Journal of the Communications Society*. 2023;4:1615–1658.

[16] Habibi MA, Han B, and Schotten HD. Network slicing in 5G mobile communication: architecture, profit modeling, and challenges. In *Proceedings*

of the 14th International Symposium on Wireless Communication Systems, September 28–October 1, 2017, Bologna, Italy, accessed: 30 November 2021.

[17] Zhang S. An overview of network slicing for 5G. *IEEE Wireless Communications*. 2019;26(3):111–117.

[18] Habibi MA, Han B, Nasimi M, *et al*. Towards a fully virtualized, cloudified, and slicing-aware RAN for 6G mobile networks. In Wu Y, *et al*. (eds), *6G Mobile Wireless Networks. Computer Communications and Networks*, Springer, Cham, 2021. p. 327–358.

[19] Habibi MA, Sanchez AG, Pavo'n IL, *et al*. Enabling network and service programmability in 6G mobile communication systems. In *IEEE FNWF*; 2022. p. 320–327.

[20] Wang H, Wu Y, Min G, *et al*. A graph neural network-based digital twin for network slicing management. *IEEE Transactions on Industrial Informatics*. 2022;18(2):1367–1376.

[21] Naeem F, Kaddoum G, and Tariq M. Digital twin-empowered network slicing in B5G networks: experience-driven approach. In *2021 IEEE Globecom Workshops (GC Wkshps)*; 2021. p. 1–5.

[22] Han B, Habibi MA, and Schotten HD. Optimal resource dedication in grouped random access for massive machine-type communications. In *2017 IEEE Conference on Standards for Communications and Networking (CSCN)*; 2017. p. 72–77.

[23] Habibi MA, Nasimi M, Han B, *et al*. A comprehensive survey of RAN architectures toward 5G mobile communication system. *IEEE Access*. 2019;7:70371–70421.

[24] Kuruvatti NP, Klein A, and Schotten HD. Prediction of dynamic crowd formation in cellular networks for activating small cells. In *2015 IEEE 81st Vehicular Technology Conference (VTC Spring)*; 2015. p. 1–5.

[25] Bulakci O, Kaloxylos A, Eichinger J, *et al*. RAN moderation in 5G dynamic radio topology. In *2017 IEEE 85th Vehicular Technology Conference (VTC Spring)*; 2017. p. 1–4.

[26] Ammar HA, Adve R, Shahbazpanahi S, *et al*. User-centric cell-free massive MIMO networks: a survey of opportunities, challenges and solutions. *IEEE Communications Surveys Tutorials*. 2022;24(1):611–652.

[27] Truong KT and Heath RW. Effects of channel aging in massive MIMO systems. *Journal of Communications and Networks*. 2013;15(4): 338–351.

[28] Han B, Richerzhagen B, Scheuvens L, *et al*. Deliverable D7.2 Special-purpose functionalities: intermediate solutions. *Deliverables—Hexa-X Project*; 2022. Available from: https://hexa-x.eu/wp-content/uploads/2022/05/Hexa-X_D7.2_v1.0.pdf.

[29] Petrov V, Komarov M, Moltchanov D, *et al*. Interference and SINR in millimeter wave and terahertz communication systems with blocking and directional antennas. *IEEE Transactions on Wireless Communications*. 2017;16(3):1791–1808.

[30] Liaskos C, Tsioliaridou A, Pitsillides A, *et al*. Using any surface to realize a new paradigm for wireless communications. *Communication ACM*. 2018;61(11):30–33.

[31] Pengnoo M, Barros MT, Wuttisittikulkij L, *et al*. Digital twin for metasurface reflector management in 6G terahertz communications. *IEEE Access*. 2020;8:114580–114596.

[32] Mao Y, You C, Zhang J, *et al*. A survey on mobile edge computing: the communication perspective. *IEEE Communications Surveys Tutorials*. 2017;19(4):2322–2358.

[33] Nasimi M, Habibi MA, Han B, *et al*. Edge-assisted congestion control mechanism for 5G network using software-defined networking. In *2018 15th International Symposium on Wireless Communication Systems (ISWCS)*; 2018. p. 1–5.

[34] Letaief KB, Chen W, Shi Y, *et al*. The roadmap to 6G: AI empowered wireless networks. *IEEE Communications Magazine*. 2019;57(8):84–90.

[35] Abbas N, Zhang Y, Taherkordi A, *et al*. Mobile edge computing: a survey. *IEEE Internet of Things Journal*. 2018;5(1):450–465.

[36] Sun W, Zhang H, Wang R, *et al*. Reducing offloading latency for digital twin edge networks in 6G. *IEEE Transactions on Vehicular Technology*. 2020;69(10):12240–12251.

Chapter 7
Digital twins for optical networks
Agastya Raj[1], Dan Kilper[2] and Marco Ruffini[1]

7.1 Introduction

Digital twins (DTs) are becoming a key technology to promote digital transformation and technical evolution in various areas. As responsive digital models of physical environments, DTs could potentially serve as advanced control and management tools, enabling real-time simulation, analysis, and optimization of physical systems. Unlike traditional methods, DTs allow for more precise and dynamic control of network elements, informed by real-time data and advanced analytics. In doing so, they provide an answer to the gap between the static and manual control methods of the past and the increasing complexity and dynamism of modern optical networks. The development of DTs is further accelerated by the continual advancements in computing power and intelligent algorithms, thereby positioning them as a promising solution to address outstanding challenges in a multitude of fields, ranging from manufacturing industries [1], to wireless [2] and optical networking [3].

This chapter will focus on the application of DTs to optical networks, where it has the potential to revolutionize the way that networks are designed, operated, and maintained. Optical networks are essential for the delivery of high-speed Internet, telecommunications, and other services, but they are also complex and costly to design, operate, and maintain. Moreover, with the continued growth in demand for lower latency and high bandwidth networks, optical networks need to support higher capacity as they are already hitting technological limits and at the same time improve their reliability, efficiency, and ability to adapt to changing network conditions.

A DT in optical networks should have the ability to simulate and test network operations before they are applied to a live network (see Figure 7.1). This can help to optimize network configurations and identify potential issues before they occur, reducing the risk of costly and time-consuming mistakes. In addition, they can increase network efficiency by enabling reliable dynamic operations. A high level of reliability could enable autonomous network configuration at short-time scales, which would allow the network control plane to adapt lightpath configurations to the actual capacity demand, rather than relying on less efficient static provisioning.

---
[1]CONNECT Center, School of Computer Science and Statistics, Trinity College Dublin, Ireland
[2]CONNECT Center, School of Engineering, Trinity College Dublin, Ireland

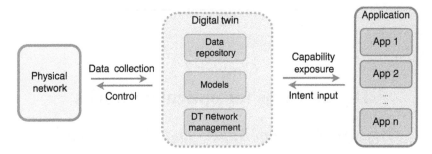

Figure 7.1   Broad structure of a DT. DTs model the whole life-cycle of a system, establishing a bidirectional and real-time interaction with the physical instance.

Additionally, DTs might be used to monitor and analyze network performance in real-time, providing valuable insights into network behavior and helping to identify and diagnose problems more quickly.

One prospective benefit of implementing digital twins (DTs) in optical networks could be their potential for enabling predictive maintenance, which can help to prevent network failures and minimize downtime. The theoretical premise is that by analyzing data collated from sensors and various monitoring systems, DTs may have the capacity to provide early warnings of impending issues, thereby assisting in scheduling maintenance prior to the manifestation of problems. This could considerably reduce the necessity for reactive maintenance and enhance overall network availability.

While the development of DTs in optical networks is still under exploration, there is substantial potential suggested by existing research. The potential of DTs in this context is especially underscored by advancements in machine learning (ML) techniques. ML methodologies have demonstrated success in optical network management tasks such as resource allocation [4], failure prediction [5], and service provisioning [6]. The integration of these ML algorithms into a DT framework could further optimize these tasks, facilitating more efficient and reliable network management [3].

Moreover, the concept of a DT is not entirely new to the field of optical networks. The practice of quality of transmission (QoT) estimation [7], a form of offline simulation used to predict signal quality based on network and traffic conditions, can be viewed as a specific implementation of a DT. Even though these QoT estimations do not encompass all aspects of a comprehensive DT, they share the core principle of utilizing a virtual model for predictive purposes.

However, it is crucial to acknowledge the inherent complexities and potential limitations of this approach. A DT might produce false positives, indicating potential issues which, upon further investigation, prove to be non-existent. This can lead to unnecessary, expensive, and time-consuming maintenance procedures. Given the reliance of modern optical systems on the accuracy and precision of the information they utilize, such inaccuracies could significantly impact the cost efficiency and productivity of network management.

Additionally, the development of a comprehensive DT of an entire optical network represents a significant undertaking, involving numerous challenges. These include managing the complexity and heterogeneity of optical network data, developing reliable and scalable DT models, and integrating these models with network control and management systems. Despite these challenges, the potential benefits of a comprehensive DT, such as improved network reliability and fault management, make it a promising direction for future research and development. This reaffirms the necessity of further research and development to mitigate potential failings and fully realize the promising benefits of DTs in optical networks.

The rest of the chapter is organized as follows: Section 7.2 explores some of the current issues plaguing optical networks, in particular, current suboptimal performance and limited scope of automation. We also discuss how DTs could pave the way towards resolving these issues. In Section 7.3, we explore the structure of a DT for optical networks. This section also highlights recent work being done in this field. Section 7.4 discusses technologies which are instrumental in the development of DTs. Optical testbeds which simulate large-scale optical networks, particularly in regards to development of DTs, are discussed. This section also highlights the importance of open source emulation tools focusing on the Mininet Optical and GnPy model, as well as open data and networking protocols. The final section discusses the existing challenges and future work in DTs.

7.2 Current issues in optical networks

While many of the Internet and mobile applications that we enjoy today are based on highly automated processes, the underlying control of optical networks is still heavily based on human operation and proprietary customized controls. A key reason is that while the physics of optical communications is well understood, an optical network is an analog system with highly complex controls and dynamics. An optical network is formed from a set of analog components that transmit and condition optical carriers as analog signals, even though they carry digitally modulated data. These components are sensitive to changes in configuration parameters, such as signal power, and exhibit a dynamic response to a modification of network load (i.e., adding, removing, or rerouting wavelength channels or tuning network components) that can influence signals throughout the network, causing long-range transmission issues, which can cascade throughout the whole network [8].

Such network dynamics and other analog characteristics of optical systems are managed through proprietary system controls that are not exposed to the network operators, particularly at the network element level. This means that network operators have limited control and information on their networks which leads to uncertainty on how the network will react in response to changes in load or configuration.

This leads to two major issues: (1) suboptimal network operation and (2) limited automation. We discuss these issues in detail below.

7.2.1 Issue of suboptimal network operation

Optical networks are operated in regimes with significant optical noise and impairments, and changes in network configuration have the potential to induce significant and complex changes in transmission behavior related to optical noise, channel powers and QoT. These changes also have the capacity to cascade throughout the network, and are difficult to predict. As a result, network operators tend to approach optical network design and operation with caution.

One possible solution to reduce impairments and avoid unpredictable scenarios in network operation is to reduce uncertainties by incorporating single-vendor solutions for the entire network. One advantage of this approach is that these solutions reduce uncertainty of operation because the vendor can thoroughly test a system and constrain its control within deterministic bounds. QoT metrics are used to estimate signal performance based on a vendor's end-to-end characterization of the system.

This information is embedded into a planning tool that is delivered to the operator, as shown in Figure 7.2. The operator can then activate the planning tool every time a new service needs to be provisioned. The tool will first estimate the QoT to determine the best system configuration based on prior system lab testing, then apply margins to account for uncertainties in performance and behavior, after which it will reconfigure the network with the new service (e.g., a wavelength channel).

Margins are commonly used when planning network operations. Taking the generalized signal to noise ratio (GSNR) as an example, if a channel with a given modulation and baud rate requires a certain GSNR threshold of X decibels (dB) to operate properly (i.e., within the forward error correction (FEC) coding limit), the margin is an increment in the target GSNR threshold that takes into account the uncertainty in estimating various types of impairments. Thus, adding a margin M to the threshold X gives the operator enough leeway to keep the system working with a given reliability, even if the GSNR deteriorates (up to the extent covered by the margin M). These margins are applied even when the given channel is deployed below its performance limits, for example, a transceiver with a 2000 km reach being deployed on a 500 km path, resulting in underutilization of resources.

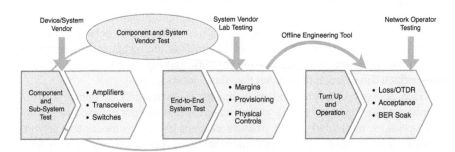

Figure 7.2 Current network architecture. Current system allows for end-to-end system testing to be performed only by the vendor, making it highly vendor dependent.

The major drawback in today's widespread use of margins is that the network is constantly under-performing, i.e. operating at lower data rates than could be achieved by having a better knowledge of the network behavior. The range of allocated beginning of life margins used in practice can be as high as 6–8 dB, with another 0–7 dB of unallocated margins [9,10]. A difference of 10 dB in optical signal to noise ratio (OSNR) threshold, for example, can make the difference between the ability to run a 400 Gb/s (DP-16QAM at 56 Gbaud) versus 100 Gb/s (DP-QPSK at 28 Gbaud) service [11].

In addition, while the single-vendor approach provides a reliable network configuration, it operates as a black box, meaning the operator does not know the operating margins during operation, and thus how far from optimal is the network utilization. Also, the planning tools are based on static models, which are created and tuned in the laboratory and typically make use of only limited network data that could be collected from the live network of the operator. This lack of information translates into the use of sub-optimal margins. Finally, another important issue is that single-vendor solutions lead to vendor lock-in, which is a major contributor to the high cost and in some aspects slow evolution of optical networks.

These issues have recently become a focus for the research community, following from recent introduction of coherent and white-box technologies(i.e., hardware designed to be controlled by external third-party software). Although open networking has been around and attempted in the past, the development of open interfaces and white box equipment has made open networking systems viable and more likely to be commercially successful. Until recently, control plane experiments were limited in scale and capability due to the lack of affordable and customizable hardware. In addition, open networking enables building a network by combining sub-system equipment from different vendors. This has created a new marketplace with several new players, including hardware vendors, device operating systems, system integrators, and a plethora of open-source solutions (e.g., ONOS, OpenDayLight, SONiC, etc.). While the multi-vendor approach has the potential to bring important economic benefits, reducing costs, and increasing innovation, it also introduces new challenges. The wider choice of available optical equipment increases the uncertainty of the end-to-end network behavior, leading to even higher margin requirements, which further constrains network performance. In addition, a disaggregated multi-vendor system cannot avail of the thorough characterization and end-to-end system engineering that a single-vendor system can provide.

Thus, there is a strong need for novel methods of performance analysis that can bridge the performance gap between closed proprietary systems (Figure 7.2) and open disaggregated optical systems (Figure 7.3). Moving to a DT approach is a logical extension of the single-vendor planning tool to address the needs of multi-vendor, open network systems that do not have the benefit of end-to-end system testing in a development lab. As shown in Figure 7.4, this enables a new, scalable, disaggregated system engineering model that breaks vendor lock-in, while preserving performance. A key innovation of the DT is that it is trained on data from the live network. This means it can adapt to the specific behavior of each individual network component (e.g., erbium doped fiber amplifier ripple gain function), actual

176 Digital twins for 6G

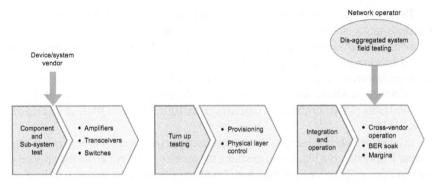

Figure 7.3 Disaggregated optical network architecture. Disaggregated systems do not allow end-to-end system testing unless performed by the network operator, which is expensive and complex.

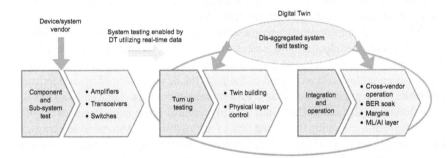

Figure 7.4 DT-based network architecture. New DT engineering model enables highly scalable disaggregated systems and makes up for lack of end-to-end system testing through the use of a DT and ML-based controls using enhanced and real-time data interaction.

network configuration (i.e., network channel load), and component aging. This has the potential to go beyond even what single-vendor planning tools can offer today.

7.2.2 Issue of limited automation

Operators are currently using slow and methodical network reconfiguration. New wavelength channels are typically only provisioned when necessary to add new revenue-generating services or capacity upgrades, which are usually set indefinitely. There is no dynamic wavelength channel optimization, for example, following the actual IP traffic load between network nodes, even for diurnal patterns. This means that the optical network is likely to be poorly utilized, which would lead to excess capital costs. It also would likely increase operational costs due to the larger footprint of additional provisioned equipment and increased maintenance requirements and energy consumption of the network. Telecom equipment racks are at or near thermal density limits for central offices. It is estimated that globally in 2020, data transmission networks, which include optical networks, consumed between 260 and 340 TWh in 2020, which is equivalent to 1.1%–1.4% of global electricity use (in comparison, energy used by data centers was 1%) [12].

Remote control of devices through standardized application programming interface (API) is available, but there is little automation in the decision-making process. These APIs are only exposed to the network operators as a network management interface, without visibility to the underlying controls and therefore a human operator is required for most operations. This increases substantially the time required to support new services and their cost. While this was generally the desired practice in the past when optical networks were only used in the large long haul, backbone networks, it is a blocking issue for many applications at the edge of the network that need optical scale capacity today, but at much tighter cost points. Cell densification and use cases that require short-lived high-capacity connectivity (e.g., augmented reality experience for large events, support for mission critical services, emergency response using ultra-high definition drone footage, etc.) require lower cost and more operationally scalable capacity management [13].

Also a DT would be a key element for network automation, as it can be used to run use cases and predict the outcome of a set of actions, before they are applied to a production network [3]. Once a reliable DT can be produced, new algorithms can be developed that autonomously identify possible actions (i.e., parameter configuration, route and wavelength selection), test them over the DT sandbox environment, and identify the optimal configuration to be deployed on the production network. This level of automation will produce more reliable and faster response to events such as service requests, mitigation against network failures and network attacks, and lightpath optimization to increase the overall network throughput.

7.3 DT development for optical networks

Based on the common overarching theme of DTs across fields [14,15], and keeping in mind the requirements of developing a DT for optical networks; the structure of a DT for optical networks can be broadly sub-divided into five main modules, which are discussed explicitly in the remainder of this chapter: (1) Data Collection, (2) Data Fusion, (3) Model Layer, (4) Simulation, and (5) Application.

In the physical layer, real-time data is collected from objects using sensors and monitoring techniques. This data is then processed and stored in a database. Since data is collected from the underlying components of the process, a data analysis and cleaning process is required at this stage. Using data mining and processing techniques, useful information is extracted, redundant features are discarded, and deeper insights are subsequently explored.

This data is utilized in the virtual space to refine existing virtual models, modifying them according to new data insights. These models produce specific results based on application needs and provide optimization strategies for the physical space. The interaction between the virtual models and the real-world process is achieved

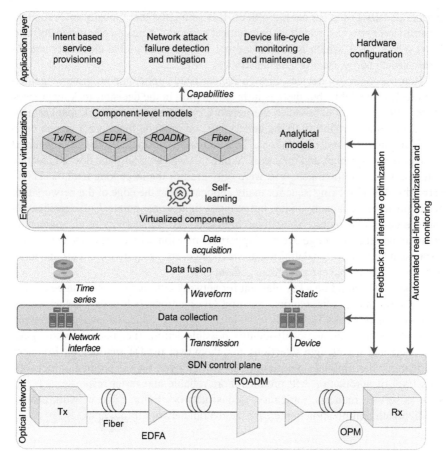

Figure 7.5 DT framework for optical networks. The DT framework for optical networks consists of Physical Layer, Data Collection Layer, Data Fusion Layer, Modeling Layer encapsulated in an Emulation Layer, and Application Layer. All the layers have feedback processes to allow iterative optimization through ML methods.

through the physical object's forward response to the virtual models and the models' backward feedback to physical objects.

In order to integrate DT and optical networks, we adopt a bottom-up approach and present a broad overview of DT structure (Figure 7.5).

7.3.1 Data collection layer

One of the key challenges in constructing DTs lies in data collection, which can be particularly complex in the context of optical networks. The physical layer of optical networks is a complex ecosystem comprising multiple components from various vendors with varying degrees of compatibility. Moreover, these components are often located in different parts of the network, making it challenging to access and monitor them effectively. Additionally, facilitating the latest low latency and high bandwidth networks necessitates the use of specialized equipment and techniques to collect data such as OSNR, GSNR, bit error rate (BER) and QoT in general.

Real-time data can be extracted from diverse sources, such as physical equipment, transmission systems, and network interfaces, through sensor detection, optical performance monitoring, quality of transmission estimation, and network message reporting. The successful implementation of DTs depends heavily on the accumulation of massive data from all elements and throughout the network's lifetime. Only when sufficient data are collected from the physical space can a virtual model be accurately established in the digital realm.

The data on the network status (e.g., resource utilization information, notification message, delay jitter, and blocking rate) can be obtained from the central network log viewer, as well as optical performance monitors. The data on the equipment operating state (e.g., environment temperature, unusable time, input/output optical power, and laser bias current) are acquired from internal device mechanisms. Modern components such as re-configurable optical add-drop multiplexers (ROADMs), in-line amplifiers (ILAs), and transceivers have robust network configuration protocol (NETCONF) processes which can be utilized to record the device status. The data on the performance of the transmission system (e.g., OSNR, modulation format, transmission speed, and system impairment) are obtained from optical performance monitor (OPM) modules and digital coherent detectors.

7.3.2 Data fusion layer

However, not only there is the challenge of collecting data from different sources but there is also the challenge of making the data time synchronized, and homogeneous. Additionally, different types of data possess varying structural characteristics, operation modes, storage mechanisms, and matching algorithms. For example, some sources such as Network status data are represented as time-series, some sources, such as gain spectrum, are represented in waveform. Table 7.1 describes different types of data that can be collected across a typical optical network. There is a need for a central mechanism to ensure homogeneous ordering of data in database, especially in instances such as fault detection, where events occur on a microsecond level. Implementing data fusion is crucial for integrating multi-source heterogeneous data at the data layer. Data fusion involves several processes, including data desensitization, cleaning, labeling, naming, normalization, sampling, augmentation, and balancing, to transform the raw data into usable information that can be processed.

7.3.3 Modeling layer

DTs create a near-real-time connection between physical and virtual networks using virtual models. The model layer is the heart of the DT system and can describe the physical state, simulate operational processes, predict trends, and optimize device performance. It is responsible for simulating and predicting the performance of

Table 7.1 Data types associated with optical components and services[1]

Data type	Optical component	Description
Time series	Optical fiber	I/O power
	Amplifier	Gain, pump power,
	Optical switch	Insertion loss, switching loss, attenuation tilt
	Optical transceiver	Transceiver noise, receiver noise, laser phase noise
	ROADM	Active channels, filtering penalty, channel attenuation, channel power spectrum
	OPM	BER, OSNR, GSNR, latency, throughput
	Network management	Alarms, network health, status
Static	Optical fiber	Length, fiber type, attenuation coefficient
	Transceiver	Bit Rate, modulation format, type, channel spacing
	Network management	Topology, hardware configuration of components
	Analytical model parameters	CM model, optical switch penalty model, WSS in-line filtering penalty model
Waveform	OSA	Power distribution, noise figure
	Transceiver	Modulated signal waveform, eye diagrams
	Amplifier	Wavelength-dependent gain spectrum

[1]This table is not exhaustive, and there might be other data types or additional data sources depending on the specific network configuration, components used, and monitoring tools employed. The data fusion layer should be designed to be able to scale to new data formats.

real-world networks, as well as adjusting dynamically to changes introduced in the network.

7.3.3.1 Analytical versus machine learning

Traditional modeling methods in optical communication face several challenges, such as reliance on expert experience and mathematical–physical models, limited efficacy in static and isolated scenarios, and an inability to conduct real-time iterative optimization.

Optical networks are largely dependent on physical phenomena. One of the central demands to ensure a low latency optical network is to have all-optical components, which means devices employed perform operations such as amplification and routing directly on the light beam [16]. The optical domain deals with complex physics involving interaction between light and other sensitive materials such as silica (in optical fibers) [17]. This makes the operation and maintenance of optical networks intricate and challenging due to various optical physical impairments that can impact the performance of provisioned lightpaths. Optical devices such as erbium doped fiber amplifiers (EDFAs) have many physical impairments that impact performance, such as amplified simulated emission [18]. Optical crosstalk and signal filtering can occur in ROADMs, leading to interference between adjacent channels and reduced signal quality [19]. polarization mode dispersion (PMD) in single-mode fibers can lead to pulse broadening and distortion, limiting the system's capacity to transmit high data rates [20]. Kerr non-linearities, such as self phase modulation (SPM), cross phase modulation (XPM), and four wave mixing (FWM), result from the intensity-dependent refractive index of the optical fiber [21]. These effects can cause signal distortion, crosstalk, and additional noise.

Let us consider the case of propagation of optical signals in optical fibers. The non-linear Schrodinger equation (NLSE) is the basic mathematical model to characterize optical signal propagation [22]. The NLSE is a partial differential equation that models the propagation of optical signals in fibers, accounting for the interplay between chromatic dispersion and Kerr non-linearities. However, the NLSE does not generally lend itself to analytical solutions. There are methods such as the Fourier method which obtain only an approximate solution to the NLSE, which, however, scale poorly with respect to complexity and run time. This limits the DT to perform real-time simulations, especially since optical signal propagation is one of the integral steps to realizing an accurate virtual model.

Another case where traditional mathematical models are used is in predicting EDFA gain spectra. This model is based on solving a set of differential equations that describe the energy levels and transitions in erbium-doped fiber. The rate equations take into account the pump power, signal power, and fiber characteristics to predict the gain and noise figure of the EDFA [23]. However, its accuracy can be limited due to simplifying assumptions made during the derivation of the rate equations. Another analytical model widely used for EDFA gain spectrum prediction is the centre of mass (CM) model, which assumes that the gain spectrum can be modeled using a center of mass function based on single-channel and fully-loaded ripple functions [24]. This center of mass equation provides an estimate of the wavelength-dependent gain based on simple EDFA measurements, but it does not capture many important features of the EDFA wavelength-dependent gain and its behavior over the full operating range. These traditional analytical models rely on expert knowledge and mathematical–physical models, which can limit their effectiveness in dynamic scenarios and hinder real-time iterative optimization.

Machine learning-based approaches, in contrast, can adapt to changing conditions and optimize performance using large amounts of data collected from optical communication systems. Machine Learning algorithms are at the heart of modern approaches to DTs. Most of the recent approaches to the DT model layer widely employ machine-learning-based approaches, with analytical models being applied in conjunction as necessary [3,25–27]. ML models, since the advent of deep learning in 2012 have become state of the art in prediction, simulation and forecasting problems [28]. One key requirement for highly accurate deep learning models is large amounts of data. Essentially, the performance of deep learning algorithms increases at an exponential rate with an increase in data [29]. Because of the collection and storage of massive real-time data in the data layer, which contain a wealth of information and various time-varying characteristics, ML-based approaches are the perfect candidate for modeling DTs.

Deep neural networks (DNNs) are a natural starting point to address the issues mentioned above. Theoretically, deep neural networks (DNNs) are universal function approximators, which means that given enough data they can map any non-linear function [30]. A DNN is capable of learning complex and high-dimensional representations of input data, and can extract intricate mapping between input and output data. Their hierarchical learning approach means that the lower layers of a DNN map broader features, while the deeper layers build upon the lower layers by mapping more complex features [31].

7.3.3.2 Modeling layer components

A model layer typically consists of several interconnected sub-modules that represent and simulate different parts of the optical network. Two general categories of models can be considered:

(1) Component-level models: This sub-module represents the virtual simulation of optical devices, such as ROADMs, wavelength selective switches (WSSs), Transceivers, EDFAs, filters, and modulators. This is perhaps the most challenging sub-module to develop, as optical devices can be challenging to model. When designing an optimal component-level model for optical DT, it is important to consider the following constraints and ensure that it can generalize reasonably well across all of the domains listed below:

- **Complex physical phenomena**: Optical devices work on principles of physical phenomena that are based on complex and non-linear interaction between channels, as mentioned in the previous section. The individual components also interact with other components in intricate ways, making it challenging to create an individual model in isolation without accounting for the effect of other components in the experiment [17]. Typically in device characterization experiments, the other devices in the network are assumed to be ideal. This is useful to characterize the device in isolation; however, the bias introduced in the data from other devices should be accounted for when integrating a model in DT.
- **Heterogeneity**: Even optical devices of the same make exhibit variability in their characteristics due to sensitive manufacturing tolerances and defects. Most optical devices are sensitive to material fabrication and impurities which can be difficult to create within a narrow tolerance. For example, the core of an EDFA is manufactured by doping Erbium ions into silica fibers. Consequently, the gain spectrum of EDFA is dependent on the Erbium ion concentration, which can be difficult to standardize across batches of commercially produced EDFAs. While manufacturers strive to maintain high levels of consistency and narrow tolerances, a degree of variability is inevitable. This is evident in Figure 7.6, which shows the gain spectrum of 8 EDFAs of the same vendor Lumentum, with the same channel loading conditions and input power. The gain spectrum across these EDFAs varies even with similar operating conditions [32].

Devices across different vendors are even more difficult to model. Different vendors have their own distinct design and manufacturing philosophies, with varying levels of quality control. Some vendors also have their proprietary elements, and specific calibration strategy which lead to a large variability between

Digital twins for optical networks   183

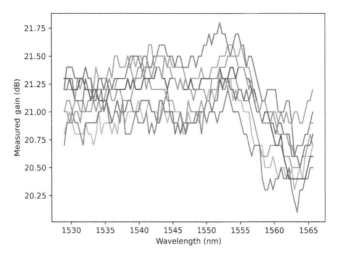

Figure 7.6   EDFA gain spectrum of different devices of the same make. The gain spectrum of different EDFA devices of the same make shows varying characteristics. In the above figure, all the EDFAs are of the same make (Lumentum), with the same input power and channel loading. However, a degree of variability still exists. Data sourced from COSMOS EDFA Dataset.

devices with other vendors. And in a large optical network spanning countries, it is inevitable to deal with devices of multiple vendors.

To account for intra and inter-vendor variability, a model needs to be *generalizable* and *transferable*. Generalizability of a model means that it can accurately reproduce the general principle of the device across all vendors and makes, and transferable refers to a single model which can be transferred to different devices with minimal additional data.

- **Limited data availability**: Most machine-learning models require large amounts of data to be able to make prediction with operational accuracy. The data layer of the DT provides a good abstraction for model layer to receive data. However, in already operational optical networks, large amounts of diverse data may be difficult to obtain. Since network configurations cannot change significantly during live operation, ML models need to train and predict accurately while working with a restricted data set. This is a cross-domain problem in machine learning, where a large part of current research is diverted towards developing machine learning models which utilize ever-shrinking datasets, such as transfer learning [33] and semi-supervised learning [34]. Especially unsupervised and semi-supervised learning algorithms can be utilized to work with smaller datasets. Particularly, authors in [35] utilize unsupervised algorithms to analyze EDFA automatic gain control (AGC) in an amplifier cascaded system. Recently work such as [36] uses transfer learning based ML models to characterize the gain spectrum of EDFA using minimized data collection.

- **Scalability**: The large scale of modern optical networks spanning different geographical locations makes it difficult to create scalable models. Introducing additional components from different manufacturers and increases the dimensionality of the parameter space of machine-learning algorithms exponentially, which leads to the curse of dimensionality problem [37]. With an increase in parameters, the search-space becomes so high dimensional that the data itself becomes sparse. Although the capacity of GPU distributed data processing is increasing exponentially in accordance with Moore's law, implementing these techniques introduce additional complexity in terms of implementation and infrastructure. Simple, but more efficient learning algorithms will be helpful in this regard.

(2) Integration with end-to-end models: End-to-end models are generally trained in the process of experimentation and include all the data associated with a given experiment or end-to-end system configuration. Once component models enable virtual components to be created and then composed to form different experiments, as one would in a physical testbed, the end-to-end models can provide an important source of validation of the DT models' accuracy at reproducing the physical testbed's behavior.

One key necessity to incorporate end-to-end models in the modeling layer, on top of component-level models, is the fact that in a complex topology, the interactions between different components can significantly impact the overall system performance. These interactions depend on network topology, lightpath routes, insertion losses, and the specific characteristics of the components involved. Especially with the increase in the number of EDFAs in a system, the potential for non-linear effects, such as FWM and SPM, also increases [17]. FWM is a significant non-linear effect in wavelength division multiplexing (WDM) systems, especially when the channel spacing is tight. FWM occurs when three or more wavelengths propagate simultaneously in the fiber and interact with each other, generating new signals at frequencies that are the sum or difference of the original frequencies. The severity of FWM can depend on several factors such as channel spacing, fiber length, modulation factor, and input power. Having an end-to-end model for the network topology can account for these interactions, while maintaining component-level accuracy.

End-to-end models can also facilitate the simulations and analysis of the network routing, and resource allocation strategies. Once the component based models are accurate up to a certain threshold, this sub-module can be used for QoT estimation. QoT estimation is one of the key outcomes towards building a DT. Fast QoT estimation is a useful tool for network planning, however, QoT degradation is induced by various link propagation impairments. Table 7.2 describes the various noise impairments added by optical components.

The overall QoT of the network changes dynamically with topology changes. The component-level models alone cannot collectively account for overall QoT estimation. An end-to-end network topology model can be used to estimate QoT by modeling the optical signal propagation through the network and considering the impairments introduced by components. This approach is applied successfully to

Table 7.2 Impairments introduced by optical components

Optical component	Noise impairments
Optical fiber	Attenuation, polarization mode dispersion, non-linear effects
Optical amplifier	ASE noise, gain saturation, non-linear effects
Optical switch	Insertion loss, switching loss
Optical transceiver	Transmitter noise, receiver noise, phase and thermal noise
WSS	Optical filtering penalty
Splitter	Uneven splitting factor

optical networks for QoT estimation [38,39]; however, there has not been a comprehensive comparison against methods that use individual device models. While such end-to-end models provide a comprehensive view, they face challenges with respect to scalability, particularly in larger metro-level networks. These models, in striving to capture a network's entire complexity, can become unwieldy and computationally heavy, slowing down response times. On the other hand, component-level models, despite their intricate nature, scale better due to their modularity. Each component can be separately modeled, tested, and updated, allowing for parallel development and enhancing overall scalability. Furthermore, component-based models enable better fault isolation and can be more efficiently customized to fit specific requirements. A balanced approach utilizing both component and end-to-end models may offer the best trade-off between precision and scalability.

With numerous sub-modules handling different tasks, synchronization and data sharing need to be established between these individual modules that will make them work in conjunction. This is one of the key challenges to developing the model layer for a DT. There exist many state-of-the-art methods which characterize different components of an optical network at individual device levels, such as EDFAs [36], fibers [40], Raman Amplifiers [41], WSSs [42], and transceivers [43]. However, integrating them to represent optical networks as a whole remains a challenge.

To mitigate the issue of limited automation as described in Section 7.2, the model layer should be resilient, possessing the capability to accommodate network modifications, anomalies, and enhancements. With real-time telemetry, the modeling layer needs to continuously adapt in response to the dynamic network environment and topology changes. This can be achieved by techniques such as transfer learning and self-learning. Transfer learning, a training paradigm, enables the adaptation of machine learning models from a pre-established source domain to a specified target domain [33]. This approach is particularly valuable for open disaggregated networks that incorporate components from diverse vendors, facilitating the creation of component-level models for incremental components introduced in the network. For instance, when expanding a network with additional EDFAs, transfer learning can utilize data from pre-existing EDFA models to characterize new amplifiers. This technique has been explored in recent EDFA gain characterization papers [36,44], where a target EDFA model is characterized from pre-trained source EDFA models using minimal additional data collection.

Furthermore, the models must be measured by the system, or supplied by the vendor [45], where the actual values of physical layer parameters may not align with the values set by the controller. Given such discrepancies, it is crucial for the DT model to minimize the difference between its output and actual network behavior. An iterative refinement technique, utilizing gradient descent has been proposed to reduce the uncertainties of the output power and Noise Figure for EDFAs [46]. To enable the DT to provide accurate outputs amid seasonal fluctuations, time series analysis can be employed to integrate significant information regarding performance fluctuations and component aging. Extreme value statistics of a QoT dataset from an operational network is investigated in [47]. By applying a generalized extreme value (GEV) distribution, the frequency of minimum QoT values is predicted, which may be extrapolated to estimate the likelihood of unforeseen network changes.

7.3.4 Simulation and virtualization

The scope of a DT is to make use of a set of individual physical models, to build a virtual platform which emulates real network facilities.

Emulation platforms provide thus a controlled environment that allows researchers and engineers to test and validate various network configurations, optimization algorithms, and control strategies without affecting the live network it represents. This enables the identification of potential issues and optimization of network performance before deploying changes to the real-world network.

A real-time simulation operated over a DT emulation platforms can provide accurate prediction and analysis of optical network performance under varying conditions. Integration of real-time data from physical networks into DTs enables dynamic and precise simulation and prediction of network performance. Emulation platforms also facilitate cost and time efficiency in developing and testing new technologies, techniques, and strategies for optical networks, accelerating innovation. Risk mitigation is also achieved through safe testing and refining of new approaches on emulation platforms before actual implementation in operational networks. Additionally, emulation platforms offer scalability and flexibility for modeling various network sizes, topologies, and configurations, making the DT adaptable to different optical network scenarios, thereby enabling the testing and evaluation of a wide range of potential solutions.

A key element of an optical DT emulator is the availability of software defined networking (SDN) interfaces for monitoring and collecting data, and then producing suitable changes that are tested first on the DT platform and then automatically applied to the live network. In doing so, the network operators will get a consistent, accurate, responsive, and comprehensive view of the real-world network, which will drive automated decision-making algorithms. We discuss some of the best emulation platforms in Section 7.4.2 "Open Emulation Tools."

7.3.5 Application layer

Once the target accuracy is achieved in the DT modeling of physical network devices, intelligent applications can be built on top of the virtual simulation that can autonomously and reliably perform various functions which would otherwise require human supervision. We have outlined below some of the ways in which a DT can be used.

7.3.5.1 Intent-based service provisioning

A DT can be used to evaluate different network design options, such as wavelength routing, modulation format, and launch power, to optimize end-to-end throughput, GSNR, and other key performance metrics, without disrupting existing network traffic and services. The key goal here would be to provide an optimal solution given a service request with only high-level parameters, such as desired BER, OSNR, and latency.

Once an optimal network configuration has been identified using the DT, it can be deployed in the live network with confidence that it will perform as expected. With the backward feedback from the real network, the DT can adjust weights as necessary and continuously monitor the performance of the live network, including key performance indicators such as path BER, and prompt the network operator to make adjustments if required. The success of the operations can be measured in terms of the ability to carry out the service request without disrupting existing signals.

7.3.5.2 Network attack failure detection and mitigation

With the advent of open networks giving rise to transparent optical switching gives flexibility to dynamically allocate light-paths and bandwidth. However, this also leaves the network more vulnerable to denial of service (DoS) attacks (whether malicious or device misconfiguration), malware and botnets [48]. Key technologies such as optical spectrum as a service (OSaaS) allow multiple consumers over a single spectrum which makes attack mitigation more complex [49]. Traditional fault management systems monitor only typical operating parameters and lack precision and effective fault diagnosis functions. A DT can be useful in this case in multiple ways.

Since the data layer of the DT is configured to collect fused data from multiple sources, data related to network status and real-time information can be used in a self-learning model to learn normal network behavior, and detect anomalous events such as unexpected traffic patterns or unusual message flow. Once a potential attack is detected, the model layer of the DT can help mitigate the attack by proposing and simulating different countermeasures [50]. For example, if an EDFA pump fails, the DT can simulate the effects and identify the configuration leading to minimal disruption or maximum throughput. This allows network operators to evaluate the effectiveness of different mitigation strategies before applying them in the live network.

By continuously monitoring the network and simulating different attack scenarios, optical DTs can also help improve network security over time. They can be used to identify vulnerabilities in the network's existing security protocols.

7.3.5.3 Device life-cycle monitoring and maintenance

Active optical devices such as EDFAs and transceivers have a life cycle which depends on various factors such as quality of manufacturing, operating conditions, and maintenance practices. Typically, the life cycle of EDFA is estimated to be between 10 and 15 years, with some devices lasting even longer under ideal conditions. However, as EDFA ages, its performance may degrade due to various factors such as the loss of pumping power, the degradation of optical fibers, and the increase of noise and distortion. This important problem has not been fully considered yet; however, it has a significant impact as optical networks become larger and older. Examples of work aiming at detecting equipment faults are available in [51,52]. However, more comprehensive datasets are required in order to trains models that can provide some accuracy in fault prediction.

A DT can monitor the trend of equipment state's parameters and predict potential failure in advance if the parameters deviate from the trend. Since a network uses many devices of the same make, identifying failure on a single device will help prevent potential failures on all other devices. Additionally, DTs can be used to simulate the performance of the equipment under different operating conditions and configurations, enabling network operators to optimize the device settings to maximize its lifespan and minimize the risk of failures.

7.3.5.4 Hardware configuration

In open optical networks, programmable components serve an integral role in optimizing the QoT and need dynamic monitoring to adapt to changing network requirements. For example, a programmable optical transceiver has configurable parameters such as the modulation format (MF), symbol rate (SR), FEC coding, and probabilistic constellation shaping. These could be changed by a control plane depending on the real-time link state, optical signal quality, and capacity demand, in order to optimize the network with respect to overall capacity and wavelength/bandwidth allocation plan. One of the key issues in configuring such parameters for optimal performance is the fact that they can affect multiple other components in the optical line, making performance optimization more complex to achieve, especially as networks grow larger.

Since a DT contains a virtual simulation of all the network components, it learns the complex non-linear dependence between network performance and underlying hardware parameters. The search-space of possible hardware configurations can be explored to select the optimal configuration to provide maximum QoT.

[3] propose a DT-enabled programmable optical transceiver (POT) adaptive configuration scheme that uses deep reinforcement learning (DRL) to provide a smart control strategy for the POT configuration. In the physical space, POT provides transmission bandwidth resources flexibly; in the data layer, multidimensional monitoring data on POT are collected, desensitized, cleaned, and normalized. The data fusion platform combines multidimensional monitoring data and information from various layers to input the monitoring data into the digital space. In the DRL agent, a double-deep Q-learning network is selected, consisting of two deep neural networks: modeling and evolving networks. The modeling network determines the relationship

between the network state and transmission performance utilization under various POT control actions, while the evolving network learns real-time POT control experience and adjusts the modeling network parameters to adapt to the dynamic network environment. The proposed DRL-based scheme reduces spectrum consumption and improves spectral utilization while meeting transmission requirements and average network delay performance.

7.3.6 Recent work and case studies on optical DTs

The authors in [3] presented a high-level concept of digital twin optical network (DTON). The proposed DTON framework employs a four-layer architecture—physical, data, model, and application layers. Through the use of a cognitive optical network, real-time data from physical equipment, transmission systems, and network interfaces can be collected, processed and stored in the data layer. The model layer then focuses on building intelligent fault management, flexible hardware configuration, and dynamic transmission process simulation models in the model layer. These models are expected to perform various functions in the application layer, including fault prediction, hardware resource optimization, and network efficiency improvement. Finally, optimization strategies were formulated based on the comprehensive analysis results, and response actions were taken in the application layer. The use of deep learning algorithms was employed throughout the entire process to implement data acquisition, model establishment, numerical analysis, and strategy execution. It should be noted that as of today we are still at the early, conceptual stage of DT for optical networks. Up until now, the main focus has been on using machine learning to create data-driven models of optical devices and systems. Early DT architectures have only recently started to come together. One such example is the work in [25], where the authors build upon DTON in [3], and propose and demonstrate an architecture to deploy and operate a DT. The study focuses on the life-cycle of a DT, showcasing measurements of setup and tear-down times. A central module, digital twin entity manager (DTEM), is responsible for deploying and operating the DT. The life-cycle is divided into four phases:

- Deployment: A DT is requested by the operator, and the entity manager will deploy a DT and virtual SDN agents using virtualization.
- Calibration: This step involves updating the DT with real-time data so it reflects current physical state
- Operation: Allows the operator with tools to run simulations and what–if scenarios.
- Removal: The DTON is removed upon request, freeing up the virtualization resources.

The feasibility of this architecture has been demonstrated on a physical optical network consisting of four whitebox ROADM/optical cross connect (OXC) nodes, sliceable bandwidth variable transceiver (S-BVT), and five bidirectional amplified optical links. The DTEM is able to virtualize and deploy SDN agents and controllers, on top of a cloud infrastructure. The set-up and tear-down times of the DT is studied, and the variance is found to be very low (±0.4 s). It shows that DT can be deployed practically in a physical testbed. However, further development is needed to demonstrate operations and further capabilities.

The use of DT technology has also started to appear in some of the research carried out by telecom vendors. However, also in this case, the use cases are still limited to laboratory experiments. For example, authors in [26] demonstrate AI-Light, a SDN framework which combines telemetry with stored static information to create a reliable DT. The controller combines telemetry data from the network equipment with static data (like device configuration, etc.) to create a reliable DT. The applications are demonstrated by computing the optimal configuration of signal to noise ratio (SNR)-based optimization algorithms over the DT using a QoT estimator. The best power per channel is computed for each optical multiplex section (OMS) by iteratively modifying the attenuation of the WSSs and assessing the corresponding SNR using the QoT estimator. Once the optimization algorithm converges, the final configuration is applied to the physical network equipment. The resulting spectrum is shown to be flatter, with the BER improving for the critical services, although the BER of the non-critical services has slightly increased. However, further details are needed on DT creation processes.

A similar work on DT-enabled SNR optimization was demonstrated in [27]. A DNN-based model of OMS and EDFA was trained on a large dataset containing 10,000 power profiles and SNR measurements. The trained models are transferred to an autoencoder (AE)-based decoder for output of SNR optimized power profiles. The DNN-based encoder combined with transferable AE decoder is a promising architecture for generative outputs in DTs. However, the training of these models still requires large amounts of data, which needs to be further studied.

Finally, in the last year, we have seen the emergence of some field trials. Although the example we refer to in [53] does not mention DT specifically, it is one of the first demonstrations of the use of an AI engine to make decisions on routing, across more than one domain. A flexi-grid multi-channel Qot estimator, operating over a fiber network connecting two testbed that span multiple domains across 493 km, is trained according to BER measurements collected for different transmission formats. The AI controller is then capable of rerouting signals, using the information from the estimator, to reroute signals, when their BER drops below a pre-defined threshold.

7.4 Open testbeds, open emulation tools, and open data

The research towards the development of DTs in optical networks is just at its dawn. With recent advancements in ML methodologies, high-performance computing and increased accessibility to graphics processing units (GPUs), and other acceleration engines, we anticipate a steep increase in the ability to make use of DTs in production networks over the coming years. The development of large-scale testbeds is a crucial factor in accelerating the progress of DTs, especially if they are integrated with live production networks and are based on open programmable hardware. This can also facilitate the creation of publicly available datasets for the research community, which are essential for progressing research on DTs. The adoption of

SDN-based testbeds can also facilitate third-party access to researchers and enable running multiple experiments in parallel. For this purpose, open source solutions like open network operating system (ONOS) [54] become vital for development of DTs. However, a key concern remains as to whether such testbeds can be used to address the needs of control and management experiments for optical networks at scale.

In the next section, we will discuss examples of testbed infrastructures that are fueling the development of DTs in optical networks. Section 7.4.1 details how large-scale testbeds are enabling essential network research. We will discuss some of the most popular and efficient network testbeds, from an optical networks point of view. We will also explore design objectives, challenges, and future work of these large-scale testbeds.

Section 7.4.2 discusses some of the popular emulation platforms for optical networks, and how they can be used to develop and design network experiments. Finally, Section 7.4.3 discusses the importance of open data and open network frameworks.

7.4.1 Large-scale testbeds

The realization of DTs is majorly based on machine-learning models, which require large amounts of data to achieve operational accuracy. These models also require the data to be collected across all elements of the network, as well as across the entire lifetime. Accurate estimations and out-of-box scenarios also need to be simulated to allow the DT to anticipate and predict network operations. However, these simulations and data collection cannot be performed in legacy optical networks. Commercial optical networks have stringent business practices and operating conditions, such as customer privacy regulations. Any application of DT needs to be tested thoroughly before deploying in a real network.

Optical networking experiments in the laboratory typically consist of a small number of nodes. These networking experiments are effective at studying network operation and management, but do not reproduce the operating conditions of real networks with hundreds of nodes. Noise or other impairments can be artificially introduced, but this does not always capture the associated dynamics. Modeling a DT on such limited experiments might lead to the DT performing unexpectedly during real-world operation as the network size scales up, and will ultimately require a large amount of human-assisted operation. As optical networks today can scale to hundreds or thousands of nodes, new network control systems need to be evaluated under conditions where the associated transmission impairments are active.

This creates the need for programmable testbeds which allow the academic research community as well as the commercial industry to explore the most challenging and real-world environments, to collect data, simulate, and design the goal of low-latency and high bandwidth network. These testbeds should be designed in such a way that allows researchers to study the physical effects at scale and how they interact with machine-learning automated controls as well as simulate and collect data to train the machine-learning algorithms. Multi-user operation is particularly important at this stage, which enables a much larger group of researchers to make use of the infrastructure, often within collaborative projects.

The key idea of a DT is to create frameworks that can turn these network emulators into reliable twins of a target testbed or even live networks. These can be used in a number of different ways. For example, training over a real network can create a more precise simulation environment for researchers, who can then scale the number of nodes to develop and test new control plane algorithms over large-scale networks. Alternatively, a DT can be permanently connected to a target network, with models that evolve with the network (i.e., as the network behavior changes due to faults, environmental conditions, etc.). This enables accurate prediction of control plane actions, which can increase use of fully automated decision-making.

7.4.1.1 Example of optical testbeds

In this section, we will discuss four examples of large-scale testbeds, and how they are addressing the needs for research of optical network control and development of optical DTs at scale.

(1) GENI

the global environment for networking innovation (GENI) [55] is a distributed large-scale testbed in the United States. Inspired by PlanetLab, GENI utilizes slicing to incorporate resources from various certified labs, providing more programmatic control and network isolation through OpenFlow-enabled network components. One of the major attractions of GENI is its ability to include real users in experiments, as many studies rely on end-user interaction. To support end-user opt-in while maintaining network slicing, GENI employs a two-phase opt-in process that safeguards end-user privacy and requires consent.

To enable a wide variety of experiments, instead of relying on a single experiment platform, GENI employs Open APIs to encourage multiple experiments of different scales. Though this approach does provide a deeper programmability to operators, it also introduces an extra layer of complexity.

In order to enable large-scale experimentation, GENI is built as a federated model, i.e. GENI can incorporate resources from different labs. The labs also called GENI-enabled campuses must adhere to specific requirements, including the deployment of GENI technologies such as GENI racks, OpenFlow switches, and WiMax. However, to enable a larger network, commercial optical systems have been used, which excluded most optical systems research. Extensions have been enabled at specific nodes, which allow local optical system experiments to communicate across the larger national network. Integrating multiple optical network testbeds with a common control plane allows for larger-scale control experiments, although the physical layer effects are still missing. This federation of testbeds has been widely used to investigate control system complexity and scaling [56]. GENI's single control plane approach is more suitable for experiments involving control planes. Moreover, optical experiments on GENI have been significantly less than its usual popularity in wireless research.

(2) FABRIC

FABRIC [57] is a more recent national scale programmable testbed in the United States that is based similarly on GENI, using a federation of optical testbeds, which allows it to have connection with university campuses and labs, as well as commercial cloud systems. One of the key advantages of FABRIC lies in the fact that it is not isolated from the Internet. Usually testbeds like GENI are isolated from the Internet leads to improvement in reproducibility of experimental results, it does place a limit on what experimenters can do with the testbed. This integration allows for a combination with existing popular public clouds, such as amazon web services (AWS), Azure, and Google Cloud. Since a large community of researchers already use these public clouds for experiments and simulations, FABRIC provides a way to test experimental topologies in public clouds directly to production.

There have been numerous studies in FABRIC testbed, particularly focusing on programmable data switches. However, experiments focusing on optical networks still remain elusive on FABRIC, owing to lack of physical layer controls.

(3) COSMOS

Smart city testbeds utilizing commercial networks provide an opportunity to experiment with commercially available technologies, but future technologies will require networks with higher performance capabilities than what is currently available. To enable experimentation with such future applications and technologies, research networks with capabilities beyond commercially available networks are necessary. Such a city-scale research optical network would enable experimentation on both applications and the underlying networks on a large scale.

The COSMOS testbed [58] in New York City utilizes this approach by combining a city-scale advanced wireless network with a fully disaggregated and programmable optical network with hub labs at Columbia University, Rutgers University, and City College of New York. This multilayered computing architecture includes millimeter-wave (mmWave) and sub-6 GHz software defined radios (SDRs), a disaggregated ROADM optical network with user-programmable SDN control, and a core and edge cloud. The availability of open and programmable optical components such as whitebox ROADM units and Ethernet switches supporting coherent transceivers has opened the door to building large-scale research testbeds. Further capabilities of COSMOS have been demonstrated in [59–61].

(4) Open Ireland

Open Ireland [62] is a reconfigurable testbed headquartered at Trinity College Dublin. It focuses on interplay between future optical networks and next-generation radio technologies. The testbed includes both indoor and outdoor 5G new radio, cloud, and optical transmission equipment deployed across Trinity College Dublin, around the Dublin Docklands area, and out to the Dublin City University (DCU) Campus in North Dublin. The testbed has dark fiber access to HEAnet's Dublin metro ring, giving the ability to carry out experimentation at metro-scale level. OpenIreland also has data links to COSMOS and is a node of the European P4 RARE network.

The OpenIreland testbed includes over 1,700 km of fiber spools, disaggregated ROADMs, amplifiers, coherent transponders, optical signal and noise monitors, and standard optical laboratory equipment. The testbed's full reconfigurability is achieved through the use of a large port count optical fiber switch that manages the topology for any experiment. This switch is controlled by an SDN control plane, primarily using the NETCONF protocol, enables to reconfigure the network, and allows to collect data such as QoT and OSNR.

The architecture of Open Ireland testbed enables to build and deploy multiple isolated testbed environments, which allows multiple researchers to work on using the same hardware simultaneously. The testbed cloud is controlled through OpenStack, while JFed allocates resources according to experimenters' requests. Recent work has been demonstrated in [63,64].

7.4.2 Open source emulation tools

Emulation platforms are a great tool to help scaling up existing testbed and are the foundation stone for the development of DTs.

In packet network emulators like Mininet [65], it is relatively easy to emulate networking nodes by virtualizing the network interface of a device, as the operation emulated are of digital nature. However, optical network operations are composed both by a control plane, which is controlled digitally and a physical transmission layer, which is analog and requires a level of physical layer simulation. Based on this philosophy, Mininet-Optical [66] integrates physical simulations of optical networks with Mininet's existing emulation of packet networks. Mininet-Optical simulates the physical behavior and impairments of the analog optical network (using open source models such as gaussian noise simulation in python (GNPy) [67]), emulates the data plane of both packet and optical networks, and exposes SDN control API to open SDN controllers. Discrete emulation of optical components such as transceivers, amplifiers, fiber spans, and ROADMs can allow researchers to create topologies and simulate optical transmissions. Perhaps, one of the biggest advantages of Mininet-Optical is providing an end-to-end environment for dynamic cross-layer SDN control.

Capabilities of Mininet-Optical have been demonstrated in numerous papers recently. The authors of [68] evaluate the ability of an SDN control plane to use OPM measurements to improve QoT estimation in real-time, using Mininet-Optical. The authors of [59] implement EDFA gain spectrum models in Mininet-Optical as early prototypes of component-specific DTs of COSMOS testbed. Additional work was also carried out in the same paper investigating the power dependence of the stimulated Raman scattering (SRS) effect. Although these works are good demonstrations of some of required functions of an optical emulation capabilities, further development is needed to include a wider range of functions. One of the key goals of Mininet-Optical is to support the development of DTs, a work that is currently being carried out with respect to the COSMOS and OpenIreland research testbeds.

7.4.3 Open software and open data

Open data and open software have a significant role in the modeling layer of optical DT systems and the broader adoption of SDN in optical networks.

Open data refers to data that is freely available for anyone to access and use. This facilitates the development of standardized benchmarks across different DT that can help making meaningful performance comparisons across proposed solutions. One key area where open datasets are proving instrumental is for implementing modeling techniques that use transfer learning. Transfer learning is an ML technique that aims at improving the performance of models on target domains by transferring the knowledge contained in different but related source domain. This increases the convergence speed of models, as well as provide higher accuracy and generalizability, while using lower amount of data. The publicly available dataset are thus essential to boost the development of DT models across entire research communities. Recently, the authors of [69] produced an open-source dataset on gain spectrum measurements from 16 EDFAs in the COSMOS testbed. A subsequent paper [36] used the same dataset, to demonstrate the use of transfer learning to model the EDFA gain spectrum using minimal data.

Since the control and feedback processes in a DT are based on SDN, the availability of open-source SDN controllers such as ONOS [54] and OpenDaylight [70] will be instrumental to accelerating their adoption, once dependable DTs become available. Open and programmable optical components, such as whitebox ROADM units and transponder white boxes supporting coherent transceivers, provide greater flexibility and control over the operation of optical networks. Open Networking standards have also opened the doors to several new vendors in optical components marketplace (e.g., Lumentum, EdgeCore, etc.), whereas earlier this space was dominated by a much smaller number of large telecom vendors.

In conclusion, the recent emergence of open networking devices, open source control planes, open source emulations and modeling tools, open datasets and accessible open networking testbeds, lays the groundwork for the future development of optical DTs.

7.5 Conclusions

The potential of DTs in optical networks is indeed promising, yet its realization demands considerable effort. We conclude this chapter by highlighting some of the current barriers that future research work should address in order to foster the advancements and developments of optical DTs.

Architectures: To date, various architectures have been proposed. For instance, an architecture for control of an open and disaggregated optical network has been presented, highlighting the independent operation of data and control planes by means of a physical layer DT [71]. Another architecture for a time domain DT suggests the sequential integration of multiple pre-trained DNN models, each corresponding to specific network elements [72]. However, the practicality of implementing these frameworks within a standard, open, and multi-vendor environment remains to be investigated.

Modeling approaches: Modeling of optical networks has progressed using both analytical and ML-based strategies, and with diverse approaches, spanning from models of individual components to holistic end-to-end network models. However, for a DT to be effective, an integrated approach that melds these diverse models within a cohesive architectural framework is required. A less explored aspect of DT modeling is the in-depth statistical analysis of the model themselves. ML models, often perceived as black boxes, offer limited transparency into their operational mechanics. Implementing rigorous statistical analyses could not only bolster the precision of DTs but also provide operators with more reliable confidence intervals. Moreover, the ML models tend to struggle with generalization across hardware from different vendors. Further development of advanced modeling techniques such as AutoML, transfer learning, and self-learning could address the modeling challenges.

Data collection mechanisms: There have been efforts to develop mechanisms to collect data from optical equipment in recent years. However, since operators still use a limited amount of data today, some devices also offer limited data. For instance, optical channel monitors (OCMs) are still not integrated in in-line amplifiers as its higher cost is not yet justified. Moreover, components from different vendors can use different APIs for data collection, which could result in a lack of homogeneity, and real-time synchronization for a DT. An automated mechanism for data collection, with consistent API across vendors, would significantly increase the accuracy of DT ML models.

Accessing network data: Another significant challenge to the development of DTs for optical networks is the research community's limited access to data from live networks. While there are testbeds (across vendors, operators and academia, as mentioned in the previous section), their size is limited. DTs development would benefit from access to data from large-scale and live infrastructure. Without this, it is challenging to develop models that not only generalize effectively but function seamlessly with real-time telemetry data, adapting to realistic network environments. Other domains within artificial intelligence (AI), like natural language processing, or human speech recognition have advanced rapidly, largely attributed to the availability of extensive and accessible datasets. A similar effort in optical networks, consolidating network data from various testbeds and live network, would considerably accelerate the development of DT.

Appendix A
Acronyms

ASE	amplified spontaneous emission
DT	digital twin
ROADM	re-configurable optical add-drop multiplexer
ILA	in-line amplifier
OSNR	optical signal-to-noise ratio
GSNR	generalized signal-to-noise ratio
QoT	quality of transmission
IoT	Internet of Things
FEC	forward error correction
BER	bit error rate
NETCONF	network configuration protocol
OPM	optical performance monitor
ML	machine learning

EDFA	erbium-doped fiber amplifier
XPM	cross-phase modulation
SPM	self-phase modulation
FWM	four-wave mixing
PMD	polarization mode dispersion
NLSE	Non-linear Schrodinger equation
CM	center of mass
DNN	deep neural network
WSS	wavelength-selective switch
DRL	deep reinforcement learning
MF	modulation format
SR	symbol rater
POT	programmable optical transceiver
DTON	digital twin optical network
OXC	optical cross connect
DTEM	digital twin entity manager
S-BVT	sliceable bandwidth variable transceiver
SDN	software-defined networking
API	application programming interface
SRS	stimulated Raman scattering
AGC	automatic gain control
GPU	graphics processing unit
DoS	denial of service
OSaaS	optical spectrum as a service
SDR	software-defined radio
dB	decibels
dBm	decibel milliwatts
MANO	management and orchestration
GEANT	Gigabit European Academic Network
DCU	Dublin City University
AWS	Amazon Web Services
GHz	Gigahertz
TCD	Trinity College Dublin
WDM	wavelength division multiplexing
OSA	optical spectrum analyzer
GENI	The Global Environment for Networking Innovation
ONOS	open network operating system
GNPy	Gaussian noise simulation in Python
OMS	optical multiplex section
SNR	signal-to-noise ratio
AE	autoencoder
AI	artificial intelligence
GEV	generalized extreme value
OCM	optical channel monitor

References

[1] F. Tao, H. Zhang, A. Liu, and A. Y. C. Nee, "Digital twin in industry: State-of-the-art," *IEEE Transactions on Industrial Informatics*, vol. 15, pp. 2405–2415, 2019.

[2] L. U. Khan, Z. Han, W. Saad, E. Hossain, M. Guizani, and C. S. Hong, "Digital twin of wireless systems: Overview, taxonomy, challenges, and opportunities," *IEEE Communications Surveys & Tutorials*, vol. 24, no. 4, pp. 2230–2254, 2022.

[3] D. Wang, Z. Zhang, M. Zhang, *et al.*, "The Role of digital twin in optical communication: Fault management, hardware configuration, and transmission simulation," *IEEE Communications Magazine*, vol. 59, pp. 133–139, 2021.

[4] Y. Zhang, J. Xin, X. Li, and S. Huang, "Overview on routing and resource allocation based machine learning in optical networks," *Optical Fiber Technology*, vol. 60, p. 102355, 2020.

[5] F. Musumeci, C. Rottondi, G. Corani, S. Shahkarami, F. Cugini, and M. Tornatore, "A tutorial on machine learning for failure management in optical networks," *Journal of Lightwave Technology*, vol. 37, pp. 4125–4139, 2019.

[6] T. Panayiotou, M. Michalopoulou, and G. Ellinas, "Survey on machine learning for traffic-driven service provisioning in optical networks," *IEEE Communications Surveys & Tutorials*, vol. 25, no. 2, pp. 1412–1443, 2023.

[7] Y. Pointurier, "Machine learning techniques for quality of transmission estimation in optical networks," *Journal of Optical Communications and Networking*, vol. 13, p. B60, 2021.

[8] D. Kilper, S. Chandrasekhar, and C. White, "Transient gain dynamics of cascaded erbium doped fiber amplifiers with re-configured channel loading," in *2006 Optical Fiber Communication Conference and the National Fiber Optic Engineers Conference*, pp. 3 pp.–, Mar. 2006.

[9] Y. Pointurier, "Design of low-margin optical networks," *Journal of Optical Communications and Networking*, vol. 9, pp. A9–A17, 2017.

[10] J.-L. Augé, "Can we use flexible transponders to reduce margins?," in *2013 Optical Fiber Communication Conference and Exposition and the National Fiber Optic Engineers Conference (OFC/NFOEC)*, pp. 1–3, Mar. 2013.

[11] F. E. Office, "Next-generation photonic transport network using digital signal processing," *Fujitsu Scientific and Technical Journal*, vol. 48, no. 2, pp. 209–217, 2012.

[12] "World Energy Outlook 2021 – Analysis." https://www.iea.org/reports/world-energy-outlook-2021.

[13] N. Andriolli, A. Giorgetti, P. Castoldi, *et al.*, "Optical networks management and control: a review and recent challenges," *Optical Switching and Networking*, vol. 44, p. 100652, 2022.

[14] T. Ruohomaki, E. Airaksinen, P. Huuska, O. Kesaniemi, M. Martikka, and J. Suomisto, "Smart city platform enabling digital twin," in *2018 International Conference on Intelligent Systems (IS)* (Funchal – Madeira, Portugal), pp. 155–161, IEEE, Sept. 2018.

[15] R. Minerva, G. M. Lee, and N. Crespi, "Digital twin in the IoT context: A survey on technical features, scenarios, and architectural models," *Proceedings of the IEEE*, vol. 108, pp. 1785–1824, 2020.

[16] A. A. M. Saleh and J. M. Simmons, "All-optical networking—evolution, benefits, challenges, and future vision," *Proceedings of the IEEE*, vol. 100, pp. 1105–1117, May 2012.

[17] R. Ramaswami, K. Sivarajan, and G. Sasaki, *Optical Networks: A Practical Perspective*. Morgan Kaufmann, 2009.

[18] "Optical Fiber Amplifiers. Scott Freese. Physics May PDF Free Download." https://hobbydocbox.com/Radio/71895914-Optical-fiber-amplifiers-scott-freese-physics-may-2008.html.

[19] Y.-T. Hsueh, A. Stark, C. Liu, *et al.*, "Passband narrowing and crosstalk impairments in ROADM-enabled 100G DWDM networks," *Journal of Lightwave Technology*, vol. 30, pp. 3980–3986, 2012.

[20] B. Hakki, "Polarization mode dispersion in a single mode fiber," *Journal of Lightwave Technology*, vol. 14, pp. 2202–2208, 1996.

[21] S. K. O. Soman, "A tutorial on fiber Kerr nonlinearity effect and its compensation in optical communication systems," *Journal of Optics*, vol. 23, p. 123502, 2021.

[22] "Nonlinear Fiber Optics | SpringerLink." https://link.springer.com/chapter/10.1007/3-540-46629-0_9.

[23] A. Saleh, R. Jopson, J. Evankow, and J. Aspell, "Modeling of gain in erbium-doped fiber amplifiers," *IEEE Photonics Technology Letters*, vol. 2, pp. 714–717, 1990.

[24] K. Ishii, J. Kurumida, and S. Namiki, "Experimental investigation of gain offset behavior of feedforward-controlled WDM AGC EDFA under various dynamic wavelength allocations," *IEEE Photonics Journal*, vol. 8, pp. 1–1, 2016.

[25] R. Vilalta, R. Casellas, Ll. Gifre, *et al.*, "Architecture to deploy and operate a digital twin optical network," in *Optical Fiber Communication Conference (OFC) 2022* (San Diego, California), p. W1F.4, Optica Publishing Group, 2022.

[26] A. Ferrari, V. V. Garbhapu, D. L. Gac, I. F. de Jauregui Ruiz, G. Charlet, and Y. Pointurier, "Demonstration of AI-light: An automation framework to optimize the channel powers leveraging a digital twin," in *2022 Optical Fiber Communications Conference and Exhibition (OFC)*, pp. 1–3, Mar. 2022.

[27] X. Pang, S. Li, Q. Fan, *et al.*, "Digital twin-assisted optical power allocation for flexible and customizable SNR optimization," in *2022 Optical Fiber Communications Conference and Exhibition (OFC)*, pp. 1–3, Mar. 2022.

[28] Y. LeCun, Y. Bengio, and G. Hinton, "Deep learning," *Nature*, vol. 521, pp. 436–444, 2015.

[29] X.-W. Chen and X. Lin, "Big Data deep learning: Challenges and perspectives," *IEEE Access*, vol. 2, pp. 514–525, 2014.

[30] K. Hornik, "Approximation capabilities of multilayer feedforward networks," *Neural Networks*, vol. 4, pp. 251–257, 1991.

[31] "Representation Learning: A Review and New Perspectives | IEEE Journals & Magazine | IEEE Xplore." https://ieeexplore.ieee.org/abstract/document/6472238.

[32] Y. Liu, X. Liu, L. Liu, et al., "Modeling EDFA gain: Approaches and challenges," *Photonics*, vol. 8, p. 417, 2021.

[33] F. Zhuang, Z. Qi, K. Duan, et al., "A comprehensive survey on transfer learning," *Proceedings of the IEEE*, vol. 109, pp. 43–76, 2021.

[34] J. E. van Engelen and H. H. Hoos, "A survey on semi-supervised learning," *Machine Learning*, vol. 109, pp. 373–440, 2020.

[35] U. C. de Moura, J. R. F. Oliveira, J. C. R. F. Oliveira, and A. C. César, "EDFA adaptive gain control effect analysis over an amplifier cascade in a DWDM optical system," in *2013 SBMO/IEEE MTT-S International Microwave & Optoelectronics Conference (IMOC)*, pp. 1–5, 2013.

[36] Z. Wang, D. Kilper, and T. Chen, "Transfer learning-based ROADM EDFA wavelength dependent gain prediction using minimized data collection," in 2023 Optical Fiber Communications Conference and Exhibition (OFC), San Diego, CA, USA, pp. 1–3, 2023.

[37] T. Poggio, H. Mhaskar, L. Rosasco, B. Miranda, and Q. Liao, "Why and when can deep-but not shallow-networks avoid the curse of dimensionality: A review," *International Journal of Automation and Computing*, vol. 14, pp. 503–519, 2017.

[38] I. Sartzetakis, K. Christodoulopoulos, C. P. Tsekrekos, D. Syvridis, and E. Varvarigos, "Quality of transmission estimation in WDM and elastic optical networks accounting for space–spectrum dependencies," *Journal of Optical Communications and Networking*, vol. 8, pp. 676–688, 2016.

[39] M. Bouda, S. Oda, O. Vassilieva, et al., "Accurate prediction of quality of transmission based on a dynamically configurable optical impairment model," *Journal of Optical Communications and Networking*, vol. 10, pp. A102–A109, 2018.

[40] Q. Zhuge, X. Zeng, H. Lun, et al., "Application of machine learning in fiber nonlinearity modeling and monitoring for elastic optical networks," *Journal of Lightwave Technology*, vol. 37, pp. 3055–3063, 2019.

[41] D. Zibar, A. Ferrari, V. Curri, and A. Carena, "Machine learning-based Raman amplifier design," in *Optical Fiber Communication Conference (OFC) 2019 (2019), Paper M1J.1*, p. M1J.1, Optica Publishing Group, Mar. 2019.

[42] A. Minakhmetov, T. Zami, B. Lavigne, and A. Ghazisaeidi, "Accurate prediction via artificial neural network of OSNR penalty induced by non-uniform WSS filtering," in *26th Optoelectronics and Communications Conference (2021), Paper M4A.1*, p. M4A.1, Optica Publishing Group, July 2021.

[43] Z. Tao, Y. Fan, X. Su, K. Zhang, et al., "Characterization, measurement and specification of device imperfections in optical coherent transceivers," *Journal of Lightwave Technology*, vol. 40, pp. 3163–3172, 2022.

[44] A. Raj, Z. Wang, F. Slyne, T. Chen, D. Kilper, and M. Ruffini, "self-normalizing neural network, enabling one shot transfer learning for modeling EDFA wavelength dependent gain," 2023. arXiv:2308.02233.

[45] Q. Zhuge, X. Liu, Y. Zhang, et al., "Building a digital twin for intelligent optical networks [Invited Tutorial]," *Journal of Optical Communications and Networking*, vol. 15, p. C242, 2023.

[46] E. Seve, J. Pesic, C. Delezoide, S. Bigo, and Y. Pointurier, "Learning process for reducing uncertainties on network parameters and design margins," *Journal of Optical Communications and Networking*, vol. 10, p. A298, 2018.

[47] J. W. Nevin and S. J. Savory, "Analysis of the extremes of SNR time series data using extreme value statistics," in *Optical Fiber Communication Conference (OFC) 2021* (Washington, DC), p. W6A.35, Optica Publishing Group, 2021.

[48] N. Skorin-Kapov, M. Furdek, S. Zsigmond, and L. Wosinska, "Physical-layer security in evolving optical networks," *IEEE Communications Magazine*, vol. 54, pp. 110–117, 2016.

[49] K. Kaeval, T. Fehenberger, J. Zou, et al., "QoT assessment of the optical spectrum as a service in disaggregated network scenarios," *Journal of Optical Communications and Networking*, vol. 13, pp. E1–E12, 2021.

[50] M. Furdek and C. Natalino, "Chapter Ten - Machine learning for network security management, attacks, and intrusions detection," in A. P. T. Lau and F. N. Khan (eds.) *Machine Learning for Future Fiber-Optic Communication Systems*, pp. 317–336, New York, NY: Academic Press, 2022.

[51] Z. Wang, M. Zhang, D. Wang, et al., "Failure prediction using machine learning and time series in optical network," *Optics Express*, vol. 25, pp. 18553–18565, 2017.

[52] C. Zhang, D. Wang, L. Wang, et al., "Temporal data-driven failure prognostics using BiGRU for optical networks," *Journal of Optical Communications and Networking*, vol. 12, p. 277, 2020.

[53] R. Yang, H. Li, Y. Teng, et al., "Field trial demonstration of AI-engine driven cross-domain rerouting and optimisation in dynamic optical networks," in *2023 European Conference on Optical Communications (ECOC)*, 2023.

[54] P. Berde, M. Gerola, J. Hart, et al., "ONOS: Towards an open, distributed SDN OS," in *Proceedings of the Third Workshop on Hot Topics in Software Defined Networking* (Chicago, Illinois, USA), pp. 1–6, ACM, 2014.

[55] M. Berman, J. S. Chase, L. Landweber, et al., "GENI: A federated testbed for innovative network experiments," *Computer Networks*, vol. 61, pp. 5–23, 2014.

[56] L. Liu, W. R. Peng, R. Casellas, et al., "Experimental demonstration of OpenFlow-based dynamic restoration in elastic optical networks on GENI testbed," *2014 European Conference on Optical Communication, ECOC 2014*, Nov. 2014.

[57] I. Baldin, A. Nikolich, J. Griffioen, et al., "FABRIC: A national-scale programmable experimental network infrastructure," *IEEE Internet Computing*, vol. 23, pp. 38–47, 2019.

[58] D. Raychaudhuri, I. Seskar, G. Zussman, et al., "Challenge: COSMOS: A city-scale programmable testbed for experimentation with advanced wireless," in *Proceedings of the 26th Annual International Conference on Mobile*

Computing and Networking (London, United Kingdom), pp. 1–13, ACM, Apr. 2020.

[59] E. Akinrintoyo, Z. Wang, B. Lantz, T. Chen, and D. Kilper, "(INVITED) reconfigurable topology testbeds: A new approach to optical system experiments," *Optical Fiber Technology*, vol. 76, p. 103243, 2023.

[60] J. Yu, T. Chen, C. Gutterman, S. Zhu, G. Zussman, I. Seskar, and D. Kilper, "COSMOS: Optical architecture and prototyping," in *Optical Fiber Communication Conference (OFC) 2019 (2019), Paper M3G.3*, p. M3G.3, Optica Publishing Group, 2019.

[61] C. Gutterman, A. Minakhmetov, J. Yu, *et al.*, "Programmable optical x-Haul network in the COSMOS testbed," in *2019 IEEE 27th International Conference on Network Protocols (ICNP)*, pp. 1–2, 2019.

[62] "OpenIreland Testbed – OpenIreland." https://open-ireland.atlassian.net/wiki/spaces/OP/overview.

[63] J. Müller, F. Slyne, K. Kaeval, *et al.*, "Experimental demonstration of ML-based DWDM system margin estimation," Feb. 2023.

[64] K. Kaeval, F. Slyne, S. Troia, *et al.*, "Exploring service margins for optical spectrum services," in *2022 European Conference on Optical Communication (ECOC)*, pp. 1–4, Sept. 2022.

[65] "Mininet: An instant virtual network on your laptop (or other PC) – Mininet." http://mininet.org/.

[66] A. A. Díaz-Montiel, B. Lantz, J. Yu, D. Kilper, and M. Ruffini, "Real-time QoT estimation through SDN control plane monitoring evaluated in Mininet-optical," *IEEE Photonics Technology Letters*, vol. 33, pp. 1050–1053, 2021.

[67] A. Ferrari, M. Filer, K. Balasubramanian, *et al.*, "GNPy: An open source application for physical layer aware open optical networks," *Journal of Optical Communications and Networking*, vol. 12, pp. C31–C40, 2020.

[68] A. A. Díaz-Montiel, B. Lantz, J. Yu, D. Kilper, and M. Ruffini, "Real-time QoT estimation through SDN control plane monitoring evaluated in Mininet-optical," *IEEE Photonics Technology Letters*, vol. 33, pp. 1050–1053, 2021.

[69] Z. Wang, D. Kilper, and T. Chen, "An open EDFA gain spectrum dataset and its applications in data-driven EDFA gain modeling". *Optica Open*, April 2023.

[70] S. Badotra and J. Singh, "Open daylight as a controller for software defined networking," *International Journal of Advanced Research in Computer Science*, 2017.

[71] G. Borraccini, S. Straullu, A. Giorgetti, *et al.*, "Experimental demonstration of partially disaggregated optical network control using the physical layer digital twin," *IEEE Transactions on Network and Service Management*, vol. 20, pp. 2343–2355, 2023.

[72] D. Sequeira, M. Ruiz, N. Costa, A. Napoli, J. Pedro, and L. Velasco, "OCATA: A deep-learning-based digital twin for the optical time domain," *Journal of Optical Communications and Networking*, vol. 15, p. 87, 2023.

Chapter 8
Dynamic decomposition of service function chain using a deep reinforcement learning approach

Swarna B. Chetty[1], Hamed Ahmadi[2], Massimo Tornatore[3] and Avishek Nag[1]

The sixth generation of mobile networks (6G), which is anticipated to provide applications and services with faster data rates, ultra-reliability, and lower latency than the fifth generation (5G) of mobile networks, will enable the Internet of Things (IoT) to expand further. These highly demanding 6G applications will burden the network and impose stringent performance requirements. Although network function virtualization (NFV) presents a promising solution to these complex requirements, it also introduces significant resource allocation (RA) challenges. Because 6G network services will be very complex and relatively short-lived, network operators will be forced to deploy them flexibly, on-demand, and agilely. Microservice techniques are being researched to address the aforementioned concerns, in which the services are decomposed and loosely connected, resulting in enhanced deployment flexibility and modularity. This research looks into a novel RA technique for microservices-based NFV that allows for the faster and more dynamic deployment and decomposition of virtual network function (VNF) onto substrate networks. VNF decomposition introduces additional overheads that have a negative impact on network resources; thus, finding the correct balance of when and how much decomposition to allow is crucial. Therefore, we propose a criterion for determining the potential/candidate VNFs for decomposition and the granularity of such decomposition. The joint problem of microservice decomposition and efficient embedding is challenging to model and address using exact mathematical models. As a result, we constructed a reinforcement learning (RL) model using double-deep Q-learning. This demonstrated that the microservice method had a nearly 50% higher normalized service acceptance rate (SAR) than the monolithic deployment of arriving services.

[1]School of Electrical and Electronic Engineering, University College Dublin, Ireland
[2]Department of Electronic Engineering, University of York, UK
[3]Department of Electronics and Information in Politecnico di Milano, Italy

8.1 Introduction

With the emergence of cutting-edge technologies and applications, the wireless communication industry is considered one of the fastest-growing industries. Despite the fact that the fifth-generation of mobile networks (5G) is already providing critical support for enhanced mobile broadband (eMBB), ultra-reliable low-latency communications (URLLC), and massive machine-type communications (mMTC). It is questionable if the initial 5G infrastructure can support more mature applications and technologies such as digital twin (DT), connected robotics, autonomous systems, augmented reality (AR)/virtual reality (VR)/mixed reality (MR), and blockchain and trust technologies [1,2], as these emerging applications will expect to demand services with rigid standards, such as lower latency, high reliability and significant data rates [1]. As a result, significant progress toward the sixth generation of mobile networks (6G) is continuing, which must be tailored to support the upcoming service types like computation-oriented communications (COC), contextually agile eMBB communications (CAeC), and event-defined uRLLC (EDuRLLC) in addition to eMBB, URLLC, and mMTC services [3]. Consequently, this modifies the experience of heterogeneous networks by dynamically altering the load, also 6G system is intended to deliver new services alongside existing ones instantly. Augmenting the complexity of the 'Network Dynamics', causing an increase in human error. Introducing advanced technologies such as digitalization (DT) and virtualization, along with machine learning, can improve operations like decision-making and upgrading and diminish human errors.

8.1.1 NFV as we know

The transition to 6G is facilitated by several initiatives on both the access and network sides. On the network side, one of the initiatives is network function virtualization (NFV), which was launched in 2012 [4]. Traditionally, the NFV framework isolates the network functions (NFs) from their dedicated and proprietary substrate appliance and allows them to run as softwarized NFs on commodity hardware. NFV-based networks enable flexible migration of virtual network functions (VNFs) from one server to another in response to dynamic variation in resource demand. These softwarized NFs provide network operators with the freedom and agility to embed and re-embed VNFs and service function chaining (SFC) to satisfy resource scaling (e.g., vertical scaling, horizontal scaling) and migration requests in response to variation in the resource demand.

An SFC is a collection of various VNFs and their corresponding virtual links (VLs) with varying resource demands (e.g., CPU, memory, bandwidth) that should be satisfied by the underlying infrastructure. The complexity grows even further for the constrained SFC (e.g., affinity and anti-affinity constraints). The VNFs in an SFC are connected in a specific chronological order; their deployment should follow the same order, satisfying the resource requirements while increasing complexity. This is the case for each SFC. The SFC's major goal is to offer users with guaranteed Service Level Agreements (SLAs), which is accomplished in three stages: VNF-chain composition (CC), VNF-Forwarding Graph (FG)*, and VNF-scheduling (SCH). Our work is restricted to stages 2 and 3 only; we consider stage 1 as a predetermined factor provided by the network operator. The NFV-based network provides considerably more flexibility for the placement of the online arriving services but exploiting this flexibility mapping can cause a significant computational problem. One of the main challenges is 'NFV-resource allocation (RA)', which is described as the provisioning of guaranteed resources by the network or the underlying infrastructure to the requested SFCs for an effective deployment. NFV-RA, due to its requirements and constraints, normally becomes an NP-hard problem [5]. To efficiently address the NFV-RA, different heuristics and machine-learning-based approaches have been investigated in the literature (discussed in the next section in detail with appropriate references).

8.1.2 Microservices decomposed NFVs

Microservices are a software architecture model that breaks down (disintegrates) large monolithic systems into smaller (micro) manageable, independent, and self-contained components. Because they are composed of loosely connected software codes, the independent micro-components provide upgrading, scaling, testing, and migration benefits [6]. The microservice concept has established itself as a successful technique in cloud-based apps like Netflix [7] and Amazon [8]. Moreover, the 'cloud-native' nature of telecom networks makes the microservices-based NFV framework an excellent choice for future networks. The virtualization method is the core of the cloud computing architecture, enabling diverse applications to coexist and simultaneously utilize the same resource infrastructure [9]. The NFV-RA challenge grows more intriguing as the solution (theoretically) gets closer to the optimum by decomposing the monolithic NFs into microservices. In comparison to monolithic systems, these loosely linked micro-VNFs (mVNF) offer the freedom of migration, scalability, maintenance, and software update without interfering with the neighboring VNFs. Additionally, it is more fault tolerant [6,9]. These advantages, nevertheless, bring additional expenses and limitations. Especially when the VNF-forwarding graph (FG) embedding issue is addressed using the microservice approach, the deployment and architectural complexity increased [9], resulting in striving for more bandwidth and latency resources. As shown in Figure 8.1, the arrived service with 5 VNFs (0–4) and 7 Virtual Links (VLs) ((0,1),(0,4),(2,0), (2,1),(3,1),(4,3), and (4,0)) is finely-decomposed, according to the proposed granularity criteria (as described in Section 8.4.7). The decomposed service is represented in Figure 8.2. In the example provided, each monolithic VNF (hereinafter referred to as VNF) is divided into 5 independent and deployable micro-functionalities, resulting 25 mVNFs and 100 micro-VL (mVLs), causing increased resource usage. Finding the ideal decomposing scenario for each SFC is crucial since decomposing all VNF is an ineffective strategy.

A significant number of standard functionalities are repeated during the deconstruction of monolithic VNFs; these are known as redundant mVNFs, which increase

---

*It is a graphical representation of VNFs that are sequentially chained and deliver an end-to-end network service.

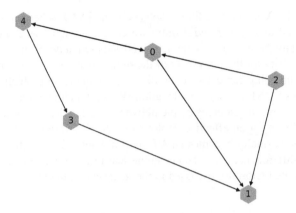

Figure 8.1   Monolithic service

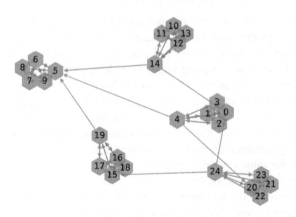

Figure 8.2   Decomposed service

processing overheads, causing high consumption of resources. The deconstructed VNF-FG needs to be re-architected appropriately to overcome this limitation. Because *m*VNFs are characterized as self-contained, i.e., lightweight and reusable, they can be scalable and quickly migratable.

8.1.3   Digital twin

One of the latest approaches in the wireless sector is system digitalization, which provides support for conceptualizing, prototyping, testing, designing, and operational phases. These digitalized phases benefit more from the availability of real-time data, which brings better possibilities for monitoring and enhancing the operational functions throughout the system's life cycle [10]. 'Digital Twin' can be defined as a synchronized virtual version (prototype) of a physical system or process that contains all historical data. This approach provides an end-to-end understanding of an entity, allowing it to efficiently operate and manage the dynamic topology and deploy monolith or decomposed services. Implementing this strategy contributes to a better understanding of the NFV-RA challenges, i.e., the digital twin of topology will provide overall information on commodities' availability and performances, which can be monitored effectively.

In this work, we solve a deep-reinforcement-learning-based NFV-RA problem that allows VNFs to be dynamically decomposed based on the substrate node's resource availability and re-architecting these decomposed VNF-FG before deploying, providing for more agile and fast-orchestrated future networks. Optimal embedding is needed as the VNFs are progressively chained; further decomposition will make the VNF-FG even more complex, i.e., the VNF-FG becomes denser in terms of interconnected nodes and links. This optimal embedding solution should satisfy all the physical resource constraints and their interdependencies of the decomposed services. Achieving this optimal solution using a mathematical optimization approach will be expensive and impractical for larger topologies. Thus, in this work, we proposed a deep reinforcement learning (DRL)-based solution tool, as it offers the ideal balance between optimality, computational efficiency, and the flexibility to decide on and respond to a change in the situational conditions (i.e., the disintegration of monoliths into microservices between episodes). The effectiveness of using DRL-based models, meta-heuristics, and integer linear programming (ILP) to solve an NP-hard problem has been studied in [11]. The authors have shown the superiority of deep reinforcement learning (DRL)-based models over other models and that integer linear program (ILP) performs better for topologies with fewer nodes, such as those with 10 nodes. But for much denser topologies, it is both costly and unfeasible. Models like ILP cannot handle the NFV-RA problem with stringent quality of service (QoS) requirements since QoS measures like latency have a non-linear relationship with traffic flow on the links [12]. In summary, the following are the important contributions of this paper:

- Develop a deep Q-learning-based framework for optimum VNF embedding on a substrate network.
- Determine granularity criteria for decomposing monolithic VNFs into microservices using network resources.
- Develop a strategy for re-architecting the deconstructed microservices so that no microservices are redundant.
- Adapting the deep Q-learning model for VNF embedding to optimally embed the microservices also (along with the microservices interdependencies) if there are decomposed microservices according to the granularity criteria.

8.2 Literature review

Traditionally, network service deployment was regarded as a Bin-Packing problem that had been thoroughly explored [12]. Various techniques, such as ILP and mixed-integer linear program (MILP), have been implemented to solve the online VNF-FG embedding (FGE) problem based on the demanded characteristics.

In [13], for example, the optimization is based on local and global fitness values obtained using the rational rule. Only a section of the substrate network is evaluated for deployment to reduce execution time, limiting the model's performance. The authors of [14] developed a queuing-based strategy that splits the problem and implements the VNF-FGE issue in sequential order. In order to provide a joint optimization for VNF mapping and scheduling for denser and more complex networks, the authors of [15] utilized the MILP model. However, because of the repetitious approach, the technique is not scalable for more extensive networks, and [15] ignored the virtual link mapping and its corresponding latency. Machine learning (ML) was commonly proposed as an effective tool for handling such complicated difficulties, in addition to heuristic or meta-heuristic techniques or linear formulations. For example, to address the VNF-placement issues in optical networks, the authors of [16] proposed a genetic method called Genetic Algorithm for Service mapping with VIrtual Topology design (GASVIT). To map the VNFs and VLs, the authors of [17] proposed an ML architecture based on the experiential networked intelligence (ENI) framework. These ML-aided solutions might be supervised, unsupervised, or Reinforcement Learning (RL)-based. However, the authors should have illustrated the advantages of one ML model over another or the criteria by which the proposed architecture will select these models. The authors of [18,19] introduced a Q-Learning (QL)-based model (QL). A few additional studies apply various variations of RL and DRL. For example, the authors of [12] used DRL to create an Enhanced Exploration Deep Deterministic Policy (E^2D^2PG) but did not investigate the microservices concept. Most researchers choose heuristic or meta-heuristic approaches to tackle the VNF-FGE issue, resulting in a suboptimal solution; this implies a clear trade-off between solution quality and computing time. Because of the problem's complexity, the RL model could not provide the best answer. As a result, effectively resolving the VNF-FGE dilemma remains a challenge. In this study, we analyze the advanced RL, double-deep Q learning (DDQL) model to enhance these suboptimal solutions, where learning from past experience by updated models might be beneficial. The implementation of decomposed VNF-FGs was examined in the research in [20–23] in the context of microservices-based NFV. The authors in [21] proposed a MILP model to select the appropriate decomposition for each SFC from the pool of existing SFCs based on traffic demands. When embedding decomposed VNFs, the authors of [22] employed the DRL model; however, they did not specify the micro-segmentation criterion for each VNF. Thus, the authors in the earlier studies had prior knowledge of the structure of the decomposed VNF or mVNFs-FG; however, this is not a feasible solution for online VNF mapping when the arriving SFC type is unknown to the network. As a result, in relation to these studies in this work, we suggest a dynamic decomposition approach based on network availability; a thorough explanation is provided in the next section.

Furthermore, most researchers in the literature used the shortest-distance approach to determine the path between the substrate nodes to deploy the VNFs and mVNFs, resulting in a significantly unbalanced load on specific links. To preserve topological equilibrium, our research [27,28] uses a practical link-selection approach and explores a more realistic connection between the CPU and RAM (nodal resource), which is also used in our study. As in our initial study [27], QL models were adopted for deploying the services, and the model's performance was studied for various degrees of nodal failures, communication delays (like 30 ms, 50 ms, and 100 ms) between the VNF and topology density. When network density and complexity increased, it was observed that the model suffered from the curse of dimensionality, resulting in poor performance. On the other hand, the [28] model is based on the flaws of the [27] paradigm. The authors of [28] use the deep Q learning (DQL) model, and its performance is assessed for (a) varying nodal capacity, (b) VNF complexity, (c) nodal outage, and (d) network density. We identified the DQL model's overestimation problem during this investigation. Solving this would improve the service's acceptability. As a result, the modified DRL model, DDQL, is used in this work to address the limitations of the DQL model in [28].

This research aims to improve model learning by introducing an improved DQL version known as DDQL, which overcomes the overestimation problem produced by QL and DQL models. Our decomposition methodology initially employs analytical criteria to identify potential/candidate VNFs for decomposition from arriving network services before determining the degree of micro-segmentation. These analytical criteria are defined by the current network availability and required resources by VNFs, making it appropriate for an independent and intelligent zero-touch system. Thus, our decomposition occurs dynamically, where the structure of the decomposition is static or pre-defined by the engineers. However, because our model does not have any prior information on the decomposition structure of the potential/candidate VNF, it is more pragmatic. Table 8.1 lists relevant works.

8.3 Problem statement

The NFV-RA optimization problem will be presented in this section. The arrival of the SFC is assured in a discrete time-step form, i.e., individual arriving SFC is deployed over the substrate network at each time step. The deployment of an arriving SFC is considered successful if the placement is accomplished within the time-step, else the SFC is discarded or rejected. Regarding the decomposition scenario, the candidate VNF is divided into multiple mVNFs depending on the granularity criteria (described later). These deconstructed mVNFs and mVLs are visually depicted as micro-VNF-FG (mVNFs-FG). The model must conform to the same limitations as the VNF-FG for a successful mVNFs-FG deployment. As a result, the primary aim is to increase service acceptability by allocating resources efficiently to SFCs.

8.3.1 Objective

Our suggested model's goal is to maximize the deployment of online SFCs or service acceptance rate (SAR) on the network so that all the VNFs and VLs of an SFC are embedded according to the promised QoS.

8.3.2 Constraints

The SFC must meet all established conditions for a successful deployment. These limitations are as follows.

Table 8.1 Summary of related work

Ref	Model	Resource type	Rel. CPU and RAM	Decomposition	Online decomposition
[20]	M/M/1	CPU	No	Yes	No
[21]	MILP	CPU, BW, traffic	No	Yes	No
[12]	DRL	CPU, RAM, BW, latency, LR	No	No	No
[24]	ILP	CPU, BW, and delay	No	No	No
[25]	ILP	CPU, BW	No	No	No
[13]	MILP	CPU, RAM, BW, and PD	No	No	No
[14]	Heuristic	CPU, RAM, and BW	No	No	No
[26]	Meta-heuristic	Storage	No	No	No
[16]	Genetic algorithm	CPU, RAM, and storage	No	No	No
[18]	QL	CPU, BW	No	No	No
[19]	QL	CPU, RAM, and storage	No	No	No
[27]	QL	CPU, RAM, BW, and latency	Yes	No	No
[28]	DQL	CPU, RAM, BW, and latency	Yes	No	No
[23]	DQL	CPU, BW, and latency	No	Yes	No
Our Model	DDQL	CPU, RAM, BW, and latency	Yes	Yes	Yes

Rel. CPU and RAM, BW, loss rate, and PD represent relationship between the CPU and RAM, bandwidth, loss rate, and propagation delay, respectively.

8.3.2.1 VN mapping

Depending on the SFC type, the criteria for each VNF varies [12]. Some VNFs, such as firewalls and deep packet inspection, may necessitate more significant CPU and RAM resources than others. The delivery of appropriate resources by the substrate nodes according to the requested resources is critical for a successful VNF deployment. Similarly, each mVNF requires a significantly lesser amount of resources, which will be incorporated into the substrate node. Again, for a successful deployment, the substrate node must provide the demanded resources.

8.3.2.2 VL mapping

It is vital to analyze the link availability during the deployment in order to give the promised QoS to the users. The inadequate network performances are caused by incorrect VL/mVLs mapping. Thus, if the substrate link satisfies the specified (requested) link requirements, such as bandwidth and latency, then the VL/mVLs are successfully deployed.

8.3.2.3 Seamless path

These VL/*m*VLs should establish a continuous pathway between the head VNF/*m*VNFs and the end VNF/*m*VNFs without any additional loops and should be according to the guaranteed QoS. For example, if the head VNF/*m*VNF and end VNF/*m*VNF are positioned on the same substrate node, an unwanted loop may be created. We applied constraint (7) from [12] to prevent similar situations. Because VNFs/*m*VNFs can be placed on the same substrate node, intra-communication between them on the same substrate node takes place via a vSwitch or bridge.

8.3.2.4 Latency

Our study focuses on latency-sensitive applications, which determine a candidate path for a VL among all feasible paths based on latency requirements. We examined an upper bound delay of 50 ms for the *m*VNF-FGs and 100 ms for monolithic VNF-FGs. Rather than focusing on one mode of communication, such as eMBB, URLLC, or mMTC, our suggested model covers services delivered by all modes of communication. These upper bound values were chosen randomly for experimental reasons. Our adaptable suggested model can be adjusted to meet future communication needs. This is a broad representative issue formulation for NFV-RA that we looked at in [28]. In this study, we modified [28]'s framework for microservices-based VNF deployment and solved it with the advanced DRL approach.

8.4 Deep RL solution for microservice decomposition

8.4.1 Reinforcement learning

RL is a self-learning process where the agent or, say, the decision-maker observes a state $s_t \in S$ from the environment at each time-step t and responds with an appropriate action $a_t \in A$ by applying a π_t policy. Based on the selected action, the environment provides numerical reward $r_t \in R \subset \mathbb{R}$ and causes a new state s_{t+1}. The agent's primary objective is to maximize the total rewards (expected discounted return R_t), which promote the model's training even during the network dynamism. The sum of received discounted rewards is represented as the projected discounted return.

$$R_t = \sum_{k=t}^{T} \gamma^{k-t} r_k \qquad (8.1)$$

where T is the terminal state. The discounted rate γ is a critical parameter as it regulates the agent's behavior, expressed as the current value of the future rewards, in ranges from $0 \leq \gamma \leq 1$. The agent operates as myopic when γ is at zero, i.e., focusing on maximizing the immediate rewards for the chosen action. On the other hand, the agent becomes more farsighted, taking into account the value of future rewards as it approaches 1 [29]. Because the agent's current behavior affects future activities, in our model, we chose a 0.99 discount rate for our research. The optimization problem in this study is expressed as a Markov Decision Process (MDP), a mathematical framework (functions) for investigating decision-making issues. This decision-making process is partially random and partially controlled by the agent.

The agent seeks to improve the decision strategy by establishing an optimal policy $\pi^* = S \rightarrow A$ to get an ideal solution based on the observations collected (i.e., state, action, rewards, and next state). By maximizing the optimal action-value function, the optimal policy is calculated.

$$\pi^*(s_t) = \arg\max_a Q^*(s_t, a_t) \tag{8.2}$$

and this action-value function is expressed using the Bellman Equation, i.e.,

$$Q^\pi(s_t, a_t) = \sum_{s_{t+1}} P(s_{t+1}, r_t | s_t, a_t)(r_t(s_t, a_t)) \\ + \gamma \sum_{a_{t+1}} \pi(a_{t+1} | s_{t+1})(Q^\pi(s_{t+1}, a_{t+1})) \tag{8.3}$$

where s_t, a_t, and r_t are the state, action, and reward achieved at time-step t, respectively. For a given current state and action, $P(s_{t+1}, r_t | s_t, a_t)$ is the probability of the next state, where s_{t+1} and a_{t+1} are the next state and action for time-step $t+1$. Following the policy π with the γ discounting factor, $\pi(a_{t+1} | s_{t+1})$ is the probability of determining the next action. It will be challenging to measure the transition probability in our study due to the environment's unexpected behavior; consequently, learning from prior and present experiences will be advantageous.

QL is an off-policy model-free technique with target and behavior policies. The agent learns the action value in the target policy, and the behavior policy is used for action selection by interacting with the environment. In other words, the target policy in QL learns from the behavior policy. Equation (8.3) is revised as (8.4) which reflects the fundamental principle of QL. Using the μ policy, the most qualified (suitable) action will have the highest estimated action-value (Q-value). The agent learns and updates its knowledge after each iteration, as the unawareness of the environmental behavior.

$$Q^\mu(s_t, a_t) = r_t(s_t, a_t) + \gamma \max_a Q^\mu(s_{t+1}, a; \theta_t) \tag{8.4}$$

Equation (8.4) consists of two parts, immediate reward function ($r_t(s_t, a_t)$) and maximizing function ($\max_a Q^\mu(s_{t+1}, a; \theta_t)$). The maximizing function is a single estimator with θ parameters (weights and biases) whose purpose is to select and evaluate actions with the same parameters. This introduces noise and substantial divergence in the Q-values, as the selected action is an over-optimistic Q-value. This is known as the over-estimation problem, indicating that the attained action is not an ideal solution but rather a non-optimal action for the observed state. The considerable noise in Q-value leads to significant positive biases, which will disrupt the updating and learning procedures.

To eliminate this drawback, we opted for double QL [30]. The fundamental difference is, instead of using a single estimator to execute two activities, the Double QL uses two different estimators; Q, Q'. Similar to basic QL, the Q estimator determines the optimal action using the maximizing function with θ_t parameter and Q' estimator (with θ'_t parameters) is used to estimate the Q-values for the achieved action, i.e., assessing the current policy or say assessing the obtained action. Equations (8.5) and (8.7) are the target estimates for the basic QL and the double QL, respectively and

(8.6) and (8.9) are the Q-value updating procedure for QL and double QL, respectively. Convergence of the Q-value to the true action-value is required to achieve the goal, which is possible by minimizing the distance between the true and estimated Q-values using the mean squared error loss function (8.10).

Basic QL:

$$y_t = R_t + \gamma \max_a Q(s_{t+1}, a; \theta_t) \tag{8.5}$$

$$Q(s_t, a_t) \leftarrow Q(s_t, a_t) + \alpha(y_t - Q(s_t, a_t)) \tag{8.6}$$

Double QL:

$$y_t = R_t + \gamma Q'(s_{t+1}, \mathbf{a}; \theta_t') \tag{8.7}$$

$$\mathbf{a} = \max_a Q(s_{t+1}, a; \theta_t) \tag{8.8}$$

$$Q(s_t, a_t) \leftarrow Q(s_t, a_t) + \alpha(y_t - Q(s_t, a_t)) \tag{8.9}$$

$$Loss = (R_t + \gamma Q'(s_{t+1}, \mathbf{a}; \theta_t') - Q(s_t, a_t; \theta_t))^2 \tag{8.10}$$

For solving this NFV-RA problem, we consider the DDQL model with two deep neural networks (DNNs); primary DNN (Q estimator) and target DNN (Q' estimator). Both are of identical architecture, but the parameters (such as weights and bias) for the target DNN are updated smoothly at every τ iterations, rather than at every iteration. This diminishes the correlation between the DNNs and reaches stable learning. The learning can be improved further by adopting the experience replay, which stores all observed transitions (state, action, rewards, and next state) in a memory buffer. These previously stored transitions are consistently sampled and fed into the model for training. The DDQL process, including decomposition and re-architecture, is depicted in Figure 8.3, and each component of the process is described in detail below.

8.4.2 Environment

The physical topology on which the VNFs/mVNFs will be deployed has been established as the environment; this topology consists of high-volume servers with distributed processing resources. We designed the topology as a directed graph $G = (N, L)$, where N represents the set of physical nodes and L is the set of physical links of the topology. The amount of nodal resources like CPU cores and RAM, etc, is indicated as n and the nodal resources type is denoted as $\Im_{node}$ which is indexed from 0, 1,..., $\Im_{node} - 1$. The available resource on a physical node y is expressed as $n_y = [n_{y,0}, \ldots, n_{y,\Im_{node}-1}]$, where each term like $n_{y,w}$ represents the amount of available resource of type w on the physical node y. Similarly, each l signifies the amount of link resources like bandwidth, latency, packet loss rate, jitters, etc. $\Im_{link}$ stands for the number of link resource types, the available resource for physical link u is $l_u = [l_{u,0}, \ldots, l_{u,\Im_{link}-1}]$, where $l_{u,w'}$ is the amount of available w' resource type on a physical link u. All the notations are described in Table 8.5.

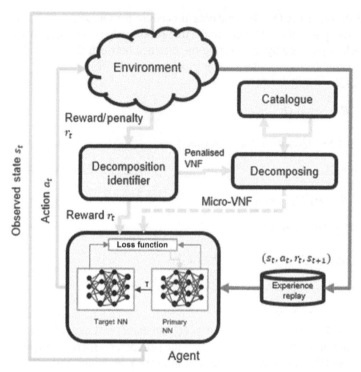

Figure 8.3 RL-based NFV-RA with microservices decomposition

8.4.3 State space

A graphical representation of an SFC is called VNF-FG, which is modeled as a directed graph $\mathbf{G}' = (V, J)$, where V and J are the set of VNFs and VLs, respectively, with their corresponding resource requirements. The demanded nodal resources (CPU core, RAM) and link resources (latency, bandwidth) are denoted as v and j, respectively. Similarly, considering microservices, decomposition of each potential VNF (ρ) candidate is represented as a directed graph $\mathbf{G}''_\emptyset = (E_\emptyset, F_\emptyset)$, where $E_\emptyset$ and $F_\emptyset$ are the set of mVNFs and mVLs, respectively. The state-space for the requested SFC (ξ) is expressed as $(\mathbf{V}_\xi, \mathbf{R}_\xi, \mathbf{E}_\xi, \mathbf{H}_\xi)$. The chronological order of the VNFs of an SFC is denoted as $\mathbf{V}_\xi = [v_{\xi,0}, \ldots, v_{\xi,|V|-1}]$ and the vector of the requested computing resources by the corresponding VNFs is indicated as $\mathbf{R}_\xi = [r_{\xi,0}, \ldots, r_{\xi,|V|-1}]$, where $|V|$ is the length of SFC ξ. The $r_{\xi,x}$ represents the details of requested resources by the VNF x of the SFC ξ, which is expressed as $r_{\xi,x} = [r_{\xi,x,0}, \ldots, r_{\xi,x,\Im_{node}-1}]$, where $r_{\xi,x,w}$ describes the amount of resource requested by the VNF x of SFC ξ of type w. The status of VNFs in terms of microservices (mVNFs) and their requested resources is described as $\mathbf{E}_\xi$ and $\mathbf{H}_\xi$. $\mathbf{E}_\xi = [e_{\xi,0}, \ldots, e_{\xi,|V|-1}]$, each $e_{\xi,\emptyset}$ determines the vector of finely decomposed VNF ($\emptyset$) sequences, that is., $e_{\xi,\emptyset} = [e_{\xi,\emptyset,0}, \ldots, e_{\xi\emptyset,|E_\emptyset|-1}]$, where $|E_\emptyset|$ is the length of decomposed mVNFs. $\mathbf{H}_\xi = [h_{\xi,0}, \ldots, h_{\xi,|V|-1}]$, again, $h_{\xi,\emptyset} = [h_{\xi,\emptyset,0}, \ldots, h_{\xi,\emptyset,|E_\emptyset|-1}]$, depicts set of the requested resource by the mVNFs of the VNF (ø). For the mVNF (z), for example, the $\mathbf{h}_{\xi,\emptyset,z} = [h_{\xi,\emptyset,z,0},...,h_{\xi,\emptyset,z,\Im_{node}-1}]$ offers precise information on the requested resource amount, where $h_{\xi,\emptyset,z,w}$ determines the amount of requested resource of type w by the mVNF z of VNF ø.

When it comes to resource initialization, most related works have allocated computational resources at random, ignoring the fact that resources like CPU cores and RAM are highly correlated. As a result, in our study, we use the relationship outlined in [31] between the CPU core and RAM. The $\Im_{node}$ in this study is 1, which reduces the state-space complexity without oversimplifying the system model while keeping the integral connection between the resources. $\Im_{link} = 2$, indicating latency and bandwidth. In future, we will investigate the performance of the developed model for various resource kinds.

8.4.4 Action space

The overall physical nodes/servers present in the topology are defined as action space (A), which is expressed as $A_\xi = [A_\xi^{vnf}, A_\xi^{mvnf}]$. $A_\xi^{vnf} = [a_{\xi,0}^{vnf},...,a_{\xi,|V|-1}^{vnf}]$ represents the action-space for monolithic VNFs, where for each VNF (x) the action-space is $a_{\zeta,\lambda}^{vnf} = [a_{\xi,x,0}^{vnf},...a_{\xi,x,|N|-1}^{vnf}]$. The introduction of microservices increases the dimensional complexity of the action-space size. Let us consider, an overall action-space vector for an SFC, represented as $A_\xi^{mvnf} = [a_{\xi,0}^{mvnf},...,a_{\xi,|V|-1}^{mvnf}]$, $|A_\xi^{mvnf}|$ is the service-chaining length. $a_{\xi,x}^{mvnf} = [a_{\xi,x,0}^{mvnf},...,a_{\xi,x,|E_x|-1}^{mvnf}]$ vector describes the sequentially decomposed mVNFs per VNF, $|a_{\xi,x}^{mvnf}|$ is the fine-granularity of mVNFs. For each mVNF (f), the action-space will be $a_{\xi,x,f}^{mvnf} = [a_{\xi,x,f,0}^{mvnf},...,a_{\xi,x,f,|N|-1}^{mvnf}]$.

8.4.5 Reward function

The agent receives the rewards from the environment upon successfully deploying the VNFs/mVNFs and VLs/mVLs by the obtained action; else it receives a strict penalty. As previously stated, the reward function is an important aspect of learning since the environment offers feedback at each movement. To elaborate, the reward function ($R(\xi)$) at a time-step is composed of two sub-reward functions: Local ($L_{reward}(\xi_{vnf/mvnf})$) and Global ($G_{reward}(\xi)$) reward function as in (8.11). The local reward function indicates the model's performance for each VNF/mVNF and the global reward function for the overall VNF-FG/ mVNF-FG.

$$R(\xi) = L_{reward}(\xi_{vnf/mvnf}) + G_{reward}(\xi_{VNF-FG/mVNF-FG}) \qquad (8.11)$$

The quality of the agent's decisions can be analyzed by the value of the local rewards which are estimated based on the constraints in Section 8.3.2. The local reward functions for VNFs/mVNFs and VLs/mVLs are mentioned in (8.12), (8.14), (8.19), and (8.24). Φ_x^y and Φ_i^w are the binary parameters which specify the deployment status of VNF/ mVNF (x) onto the substrate node (y) and VL/mVL (i) onto the substrate link (w), respectively. On successful placement (i.e., satisfying the constraints 8.3.2.1–8.3.2.4), the decision Φ parameter is 1, else 0. Likewise, the global reward, which is for the overall VNF-FG/mVNF-FG, is defined based on the definite continuous path between the head and tail VNFs and is subject to the constraints specified in Section 8.3.2. The conditions stated in (8.17), (8.18), (8.22), and (8.23) are used to verify the successful deployment of a VNF-FG/mVNF-FG. The formulated reward function is a point-based system, where the environment feeds the agent with high positive points on success action and negative points (penalty) for each failed action. This technique supports the agent in choosing an optimal solution. Local reward function for VNF:

$$L_{reward}(\xi_{vnf}) = \sum_{x=0}^{|V|-1} L_{reward}(x) \tag{8.12}$$

$$L_{reward}(x) = \begin{cases} R^{pt}_{vnf}, & \text{if } \Phi^y_x = 1 \\ P^{pt}_{vnf}, & \text{otherwise} \end{cases} \tag{8.13}$$

Similarly, local reward function for mVNF:

$$L_{reward}(\xi_{mvnf}) = \sum_{x=0}^{|E|-1} L_{reward}(x) \tag{8.14}$$

$$L_{reward}(x) = \begin{cases} R^{pt}_{mvnf}, & \text{if } \Phi^y_x = 1 \\ P^{pt}_{mvnf}, & \text{otherwise} \end{cases} \tag{8.15}$$

Global reward for VNF-FG (**G′**):

$$G_{reward}(\xi_{\mathbf{G'}}) = \begin{cases} GR^{pt}_{\mathbf{G'}}, & \text{if } C1 \& C2 \text{ are satisfied} \\ 0, & \text{otherwise} \end{cases} \tag{8.16}$$

$$C1 : |V| \times R^{pt}_{vnf} = L_{reward}(\xi_{vnf}) \tag{8.17}$$

$$C2 : |J| \times R^{pt}_{vl} = L_{reward}(\xi_{vl}) \tag{8.18}$$

$$L_{reward}(\xi_{vl}) = \sum_{i=0}^{|J|-1} L_{reward}(i) \tag{8.19}$$

$$L_{reward}(i) = \begin{cases} R^{pt}_{vl}, & \text{if } \Phi^w_i = 1 \\ P^{pt}_{vl}, & \text{otherwise} \end{cases} \tag{8.20}$$

Similarly, Global reward for mVNF-FG (**G″**):

$$G_{reward}(\xi_{\mathbf{G''}}) = \begin{cases} GR^{pt}_{\mathbf{G''}}, & \text{if } C3 \& C4 \text{ are satisfied} \\ 0, & \text{otherwise} \end{cases} \tag{8.21}$$

$$C3 : |E| \times R^{pt}_{mvnf} = L_{reward}(\xi_{mvnf}) \tag{8.22}$$

$$C4 : |F| \times R^{pt}_{mvl} = L_{reward}(\xi_{mvl}) \tag{8.23}$$

$$L_{reward}(\xi_{mvl}) = \sum_{i=0}^{|F|-1} L_{reward}(i) \tag{8.24}$$

$$L_{reward}(i) = \begin{cases} R_{mvl}^{pt}, & \text{if } \Phi_i^w = 1 \\ P_{mvl}^{pt}, & \text{otherwise} \end{cases} \quad (8.25)$$

As previously stated, most works opted for the Shortest-Distance method to identify the path between the VNFs. This approach introduces biases, resulting in substantial congestion on a few connections and prolonged link latency. This approach will not be viable, especially for time-sensitive applications. To overcome this, we investigated a method of finding the path perVNF-FG based on the upper bound of delay. For our analysis, we selected this restriction to 100 ms for VNF-FGs and 50 ms for mVNF-FGs; however, the created model may be trained for alternative latency values. When the definite end-to-end path is built under the latency constraint for VL/mVL, the appropriate reward is delivered to the agent. As a result, the reward function is formulated based on the placement of the VNFs and VLs on the most suitable substrate nodes and links.

Algorithm 2 summarizes the placement of VLs. Line 4 discovers all the feasible paths between the source VNF/mVNF and the destination (end) VNF/mVNF of an SFC. Lines 5–17 instruct the model to determine the constrained satisfying path from the pool of feasible paths. On success, positive incentives are given, else penalties are provided. All these transitions are stored in the memory buffer for better learning, as mentioned in line 18.

8.4.6 Decomposition Identifier

If every arriving VNFs were continuously decomposed, as shown in Figure 8.2, the performance of the microservice-based system would not be beneficial because it would increase architectural and computational complexity by requiring more link resources, such as bandwidth and latency, than it anticipated. Thus, the model's primary goal is to determine the potential VNF (say candidate VNF) for decomposition, which would improve the optimization problem by increasing the deployment rate rather than degrading the performance. To achieve this, each VNF is examined against a decomposition identification module to determine if it qualifies as a prospective candidate or not for dynamic decomposition. One of the considered criteria is when the agent (Ξ) is unable to find a suitable/satisfying substrate node. This can be due to low network resource availability as compared to the demanded resource $r_{\xi,x}$, as in (8.27). $\mathscr{R}\left(k_{\xi,x}^{vnf}\right) = \left[k_{\xi,x,0}^{vnf}, \ldots, k_{\xi,x,\mathfrak{I}_{node}-1}^{vnf}\right]$ is the current resource availability of the action ($k_{\xi,x}^{vnf}$), where $k_{\xi,x}^{vnf}$ represents the action (substrate node) achieved from the agent for VNF (x) for all resource types. K_ξ^{vnf} and K_ξ^{mvnf} are the achieved action vectors from the agent for the monolithic VNFs, and decomposed VNFs, respectively, for an SFC (ξ), as in (8.26). $k_\xi^{vnf} = \left[k_{\xi,0}^{vnf}, \ldots, k_{\xi,|V|-1}^{vnf}\right]$, where $k_{\xi,x}^{vnf}$ is the obtained action for VNF (x). $K_\xi^{mvnf} = \left[k_{\xi,0}^{mvnf}, \ldots, k_{\xi,|V|-1}^{mvnf}\right]$, where $k_{\xi,}^{mvnf}$ describes the VNF's ($\emptyset$) decomposition status. $k_{\xi,\emptyset}^{mvnf} = \left[k_{\xi,\emptyset,0}^{mvnf}, \ldots, k_{\xi,\emptyset,|E_\emptyset|-1}^{mvnf}\right]$, where $k_{\xi,\emptyset,f}^{mvnf}$ is the action selected by the agent for mVNF (f) of potential VNF candidate ($\emptyset$).

In this way, only essential VNFs are decomposed, optimizing the resource allocation efficiently.

$$\Xi(V_\xi, R_\xi, E_\xi, H_\xi) = [K_\xi^{vnf}, K_\xi^{mvnf}] \tag{8.26}$$

$$r_{\xi,x} > \mathscr{R}(k_{\xi,x}^{vnf}) \tag{8.27}$$

8.4.7 Granularity criteria

After the identification of potential VNFs, the next concern is how finely these VNFs should be decomposed. Instead of accepting the operator's static (pre-defined) decomposition model (as in the literature), we suggest dynamic granularity criteria. This granularity criterion governs the extent of potential VNF's micro-segmentation, which is stated in terms of the Granularity Index (GI). The GI is estimated using two parameters: (1) requested processing resource (CPU core) of the currentVNF; (2) the Network Availability Index (NAI). To perform the dynamic decomposition, we require continuously updated network conditions, i.e., the current network resource availability, NAI (Δ). The NAI is the ratio of currently available processing resources in a topology $\varphi_{topology,t}$ to the initialized resource in a network, i.e., maximum nodal capacity $\zeta_{topology}$. The model estimates NAI at each time step, as defined in (8.28). The NAI ranges between 0 and 1.

$$\Delta_t = \frac{\varphi_{topology,t}}{\zeta_{topology}} \tag{8.28}$$

With 0 being an extremely exhausted network and 1 representing a fully available network (a plethora of resources). The product of the currently requested processing resource (CPU cores) by the VNF (R_V) to the NAI (Δ_t), is represented as GI (Γ_t^V), as in 8.29.

$$\Gamma_t^V = R_V \times \Delta_t \tag{8.29}$$

Figure 8.4 portrays the GI for a VNF with 5 CPU cores resources.

Considering, when NAI is 0.8 (i.e., only 80% of the topology's processing resource is available), then the calculated GI is 4 (0.8×5 CPU cores). In this case, the micro-segmentation more or less will be coarse-grained, possibly the VNF may be split into 2, or 3 VNFs with (2,3) or (4,1), or (3,1,1) CPU core(s) as requested processing resources combinations. With the reduction of NAI, the GI drops too, instructing the model toward finer-granularity. Considering the same example, for

NAI	1.0	0.9	0.8	.....	0.2	0.1	0
GI for 4 CPU cores	4.0	3.6	3.2	.....	0.8	0.4	0
GI for 5 CPU cores	5.0	4.5	4.0	.....	1.0	0.5	0

——————————Finely-grained Micro-VNFs——————————▶

Figure 8.4   Granularity criteria

NAI as 0.2, the GI dropped to 1 (0.2 × 5 CPU cores), implying that VNF can be split into 5 *m*VNFs with maximum of 1 CPU core each. The GI is a granularity-determining function for decomposition. With the decrease of GI, the granularity of the VNFs increases, indicating that finer micro-segmentation is required when network resources are scarce.

8.4.8 Re-architecture of VNF-FG

Each decomposed *m*VNF has an independent functionality. These micro-functionalities are repeated over a decomposed SFC. From the example given in Figure 8.5, the SFC comprises five VNFs (presented vertically): WAN optimizer, Edge Firewall, Monitoring Functions, Application Firewall, and Load Balancer [32] chained in a specific order. Further, each VNF, such as WAN Optimizer, is decomposed into *m*VNFs like 'Read from NIC', 'Parse Header', 'Classify on L7 Type', 'Decompressed HTTP Payload', and 'Send to NIC', as shown (horizontally in the figure). Figure 8.5 reveals that several functionalities are repeated throughout the decomposed SFC, such as 'Read from NIC', 'Parse Header', and 'Classify on L7 type'. These repeated micro-functionalities are referred to as overlapping/redundant functions; deploying these functions will prompt the unnecessary consumption of overheads/CPU cycles. Instead, by sharing these redundant functionalities among the decomposed SFC can be eliminated the extra CPU cycles. As in the example, overlapping *m*VNFs like 'Read from NIC' are prevalent and embedding this *m*VNF frequently will consume additional resources; sharing such *m*VNFs will be beneficial. Identifying and removing such redundant functionalities is advantageous for resource usage, such models are highly desirable. Figure 8.6 shows an example of re-architecture of the above-mentioned SFC, in which overlapping functions are detected and removed, and accordingly, the decomposed SFC is re-structured without any disturbance in the flow.

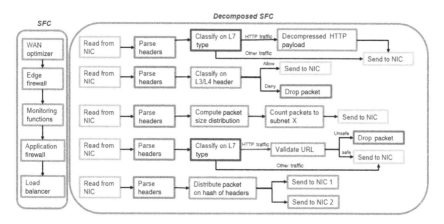

Figure 8.5   Decomposed VNF-FG

220 Digital twins for 6G

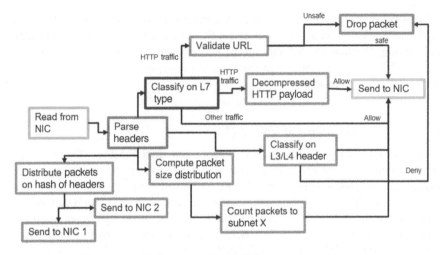

Figure 8.6 Decomposed VNF-FG after re-architecture

Therefore, in this work, we proposed an identification and re-architecture model. After the decomposition of potential VNFs, the identification model inspects for the existence of these newly created mVNFs in the network. On success, the model retrieves the desired information, such as its deployed location (substrate node) from the repository and the re-architecture model is triggered, indicating the need for re-architecture of mVNF-FG. Rather than placing the identical mVNF onto the network, the present mVNF is connected to the precedent mVNF to construct an mVNF-FG that meets the promised QoS requirements. On an unsuccessful attempt, the newly created mVNF details (resource requirements) are fed to the DDQL agent to achieve a suitable substrate node for successful deployment. These details are saved in a catalog for future reference. This optimizes the nodal and link resources, as will be evident in the simulated results.

8.4.9 Overview of the proposed model

Algorithm 1 provides an overview of the proposed model. Lines 1–4 illustrate the initialization of the fundamental aspects for establishing the primary and target neural network (NN) and replay memories for stable learning. As noted in lines 5–7, the environment is restored to its previous state with sufficient resources at the start of each episode. Once the agent determines the action using the ε-Greedy Exploration-Exploitation strategy (lines 9–18) for the observed state. The agent receives incentives or penalties based on whether the chosen action meets or fails to meet the required resources (lines 19–31). In the event of a failed attempt, the model looks for decomposition opportunities. The model commences the decomposition phase; otherwise, a penalty is applied (lines 21–29). The replay buffer stores all of these transitions (line 33). Following that, the target Q values and loss are estimated (lines 38 and 39), and the primary and target NN are updated at the specified

Algorithm 1: VNF-FG placement based on DDQL

1. Initialize Learning Rate η, Discounting Factor γ, Exploration Rate ε, Memory Size, mini-Batch size, Replace τ and Latency
2. Initialize Replay Buffer Bu
3. Initialize Primary NN $Q(s, a)$ with weights θ
4. Initialize Target NN $Q'(s, a)$ with weights $\theta' \leftarrow \theta$
5. **foreach** *episode i = 1... epi* **do**
6.   Reset the Environment
7.   Initialize substrate node resource R_N, substrate link R_L
8.   **foreach** *time-step t = 1... T* **do**
9.   Observer the state s_t
10.   **while** *all VNFs are given to the agent* **do**
11.   Using ε-Greedy Action Selection Method a_t
12.   **if** $random(0, 1) > \varepsilon$ **then**
13.   Exploitation
14. $a_t = \arg\max_a Q(s_t, a_t; \theta_t)$
15.   **else**
16.   Exploration
17. $a_t = random(A(s))$
18.   **end**
19.   **if** *Action a_t successfully embeds VNF v_t* **then**
20.   Reward r_t
21.   **else**
22.   **if** *check if Decomposition is possible* **then**
23.   Initialize Decomposition Identifier
24.   Initialize Granularity criteria
25.   Observe decomposed v_t, choose actions A
26.   Store micro-segmented v_t transitions in Bu
27.   Initialize Re-architecture model
28.   **else**
29.   Penalty r_t
30.   **end**
31.   **end**
32.   Take action a_t, then observe reward/penalty r_t and next state s_{t+1}
33.   Store transitions (s_t, a_t, r_t, s_{t+1}) in Bu
34.   **end**
35.   **foreach** *update step* **do**
36.   Uniformly Sample the mini-batch size transitions
37.   sample $b_t = (s_t, a_t, r_t, s_{t+1}) \sim Bu$
38.   Estimate the Target Value Q using Equation 8.7
39.   Estimate the Loss function Equation 8.10
40.   Update the Primary NN
41.   Update the Target NN after every τ steps
42. $\theta' \leftarrow \tau * \theta + (1 - \tau) * \theta'$
43.   **end**
44.   Execute Algorithm 2
45.   **end**
46. **end**

Algorithm 2: Virtual link placement

1 **if** *all VNFs/ mVNFs are deployed* **then**
2     $x_0 \leftarrow$ source VNFs/ mVNFs,
3     $x_j \leftarrow$ Destination VNFs/ mVNFs
4   Estimate all path P between x_0, and x_j
5   **foreach** *substrate path p = 1... P* **do**
6   **while** $b = j$ **do**
7   Estimate Latency (x_0, x_b)
8   **if** $Latency(x_0, x_b) \leq Latency$ **then**
9   Reward is given
10   **else**
11   Penalty is given
12   **end**
13   **end**
14   **end**
15 **else**
16   Penalty is given
17 **end**
18 Store transition (s_t, a_t, r_t, s_{t+1})

rate (lines 40–42). The performance of our model is not limited to the parameters provided; it may also be trained for different use-cases such as URLLC or eMBB.

8.5 DNN architecture

A suitable NN architecture is equally as important as creating the model's other properties, such as states, actions, environment, and reward functions. In order to deliver an optimal solution, the NN architecture must be constructed depending on the dynamicity and complexity of the problem. Two hyper-parameters—the depth of the NN, or the existence of hidden layers in an NN, and the width of the NN, or the number of neurons in each layer—control the amount of feature extraction. Because an analytical calculation cannot be used to determine the depth of the NNs, a careful experiment is conducted. As a result, this section describes the impact of our model's completely connected hidden layers. As illustrated in Figure 8.7, we deployed 100 SFC using the DDQL model on the BtEurope topology, with hidden layers varying from 2 to 10. In addition, each SFC consists of five VNFs with varying resource requirements. Figure 8.7 shows that the model's performance improves as the number of layers (depth) rises, but this comes with an increase in calculation time. From the standpoint of learning, learning stability happens at several stages depending on the depth of NNs. The learning process starts after 200 episodes for shallower NNs (2 and 4 hidden layers), but much earlier for deeper NNs (6, 8, and 10 hidden layers). The overall performance of the shallow layer in terms of the mean of normalized SAR is low when compared to the upper layers, showing that the lower layers failed to collect characteristics from the input to provide acceptable SFC placement. However, the number of layers equal to 6, 8, and 10 has a more significant

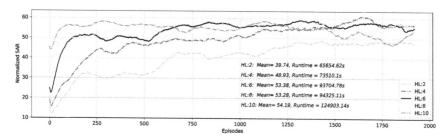

Figure 8.7 Performance of DDQL model with various hidden layers (HL)

performance quality (i.e., the model extracts the features more accurately), but at the expense of a longer runtime. As the depth increases above 6, the model produces equivalent results. As a result, our study chose a depth of 6 to find a balance between performance and computation.

8.6 Simulation results

The performance of our suggested DRL-based strategy across various network topologies is discussed in this section. We began our analysis with a smaller architecture, Netrail, which has 7 nodes and 10 full-duplex links, and then went to a denser network, BtEurope, which has 24 nodes and 37 full-duplex lines. We began our analysis with a smaller architecture, Netrail, which has 7 nodes and 10 full-duplex links, and then went to a denser network, BtEurope, which has 24 nodes and 37 full-duplex lines. Moreover, we evaluated the effectiveness of monolithic and microservice-based SFC placement under varied nodal capacities for these topologies. We anticipated that each substrate node would commence with either 12 cores and 4 CPUs (a total of 48 CPU cores) or 8 CPUs (a total of 96 CPU cores) as its nodal capacity (processing resource). The substrate link resources are configured uniformly, with link capacity (bandwidth) ranging from 1 to 100 Gbps and link delay (latency) ranging from 0 to 10 ms.

The generation of the directed graph, VNF-FGs, is performed using the Erdős-Rényi (ER) model [33]. The nodes' connections are produced probabilistically with ε as 0.3, which introduces diversity in the VNF-FGs. As a result, the connection probability p of VLs is dependent on the number of VNFs per VNF-FG (Υ) and the ε, i.e., $p = \frac{(1+\varepsilon)\log \Upsilon}{\Upsilon}$. The ER model randomly selects the directions of these VL to enhance the diversity of the VNF-FGs, resulting in a broad selection of VNF-FGs that improves NN learning. The VL resource in VNF-FGs is initialized with a uniform distribution with bandwidth ranging from 1 to 10 Gbps. Our approach used a finite number of 5 VNFs per SFC to analyze the complex problem efficiently. A VNF's processing resource (CPU core) initialization is based on a normal distribution with a mean of 3 and a standard deviation of 0.4, implying that each VNF wants 1 to 5 CPU cores.

Each run consists of 3000 episodes with 100 time-steps each, and as was indicated above, each time-step generates a unique VNF-FG with various resource

Table 8.2 DDQL model parameters

Parameters	
Hidden layers	6
Neurons per layer	300
Neural network	2
Target NN update (τ)	10,000
Activation function	ReLu
Optimizer	Adam
Learning rate	0.0001
Discounting factor	0.99
Batch size	64
Memory size	50,000
Epsilon decay	0.0001
Loss function	Mean square error (MSE)

Table 8.3 Topology-based nodal capacity

Topology	12-4 scenario	12-8 scenario
$\zeta_{Netrail}$	336 CPU cores	672 CPU cores
$\zeta_{BtEurope}$	1152 CPU cores	2304 CPU cores

requirements. The simulations are carried out on an Intel Core i7 processor and 64 GB of RAM, and the DDQL model was created in Python using the PyTorch module. The parameters used to create the DDQL model are listed in Table 8.2. Due to the growing complexity of the problem, we chose a DNN with six hidden layers for further analysis, as mentioned in Section 8.5.

To demonstrate the benefits of the DDQL paradigm for microservices embedding, we establish a bound known as the expected upper bound (EUB). An EUB specifies the maximum number of SFCs that a network topology can support based on the network architecture and nodal capacity scenario. The EUB is calculated as

$$EUB = \frac{\zeta_{topology}}{\vartheta \times \Upsilon} \tag{8.30}$$

where $\zeta_{topology}$ is the maximum nodal capacity of a topology, as described in Table 8.3. ϑ, Υ are the average CPU cores requested by a VNF and the number of VNF per SFC, respectively.

The outcomes are presented in two formats: (1) a moving average of normalized SAR over the last 50 episodes, and (2) a moving average of topology's nodal capacity over the last 50 episodes. In normalized SAR[†] graphs, the agent is more prone to investigating the environment at the beginning (episodes), and later, based on the collected data, it begins exploiting the environment. As a result, the acceptance rate for

[†]The SAR is normalized by the EUB values.

Table 8.4 Mean of normalized SAR

Topology	Scenarios (CPU cores)	DQL		Heuristic		DDQL	
		Mono	Decomp	Mono	Decomp	Mono	Decomp
Netrail	12-4	27.14	65.98	37.11	44.33	63.57	94.27
	12-8	27.51	60.47	33.06	39.68	67.27	95.40
BtEurope	12-4	8.98	17.16	19.93	25.46	34.99	74.83
	12-8	9.28	17.17	16.63	21.56	34.13	51.13

all cases has increased dramatically. Furthermore, when compared to other models, the nodal capacity graphs describe the proposed model's efficiency in terms of network resource usage. To analyze the constructed model's benefits, we compared the performance with the single estimator DQL model and a heuristic model. Table 8.4 highlights the models' performance in terms of service acceptance for all scenarios and topologies that have been investigated. The 'Mono' and 'Decomp' denote the attained mean of the normalized SAR for the monolithic and deconstructed models, respectively.

8.6.1 Heuristic model

We have proposed a heuristic model for substrate node selection, instead of random selection. In this model, the node selection is performed based on auxiliary variables, as described in Algorithm 3. On successful placement, lines 8 and 9 describe the update process of the auxiliary variables at every step. The auxiliary variables emphasize the significance of choosing a substrate node. The lower the auxiliary variable value, the more important it is to select that substrate node. The VLs/mVLs placement and decomposition are carried out in the same manner as in the suggested approach.

8.6.2 Time complexity

A number of factors determine the increase in time complexity in our suggested model. Begin with the number and quantity of episodes assumed for each run. Each episode, as mentioned in section IV, defines numerous arriving SFCs for the deployment; the complexity of these services is defined by the existence of the number of VNF and VL chained together. Furthermore, the introduction of decomposition (mVNFs/mVLs) based on the fine granularity factor increases the temporal complexity. We suggested a 'Granularity Criteria' and applied the 'Re-architecture' principle to govern this. Aside from the service type, the model's complexity is increased dependent on the density of the topology.

Our method is referred to as 'Dynamic Optimization', with the primary goal of dynamically incorporating the 'short-lived' services. Traditionally, supervised learning and other ML methods require pre-existing datasets for training, testing, and

Algorithm 3: VNF-FG placement based on heuristic

1 Initialize Latency
2 Initialize auxiliary variables
3 **foreach** *episode i = 1... epi* **do**
4   Reset the Environment
5   Initialize substrate node resource R_H, substrate link R_N
6   **foreach** *time-step t = 1... T* **do**
7   Observer the state s_t
8   Estimate the weights/auxiliary value for each substrate node
9   Arrange the Action space in ascending order
10   **while** *all VNFs are given to the agent* **do**
11   Using Weighted Action Selection Method a_t
12   a_t is the action with the lowest auxiliary value
13   **if** *Action a_t successfully embeds VNF v_t* **then**
14   Update the auxiliary variables
15   **else**
16   Check if *Decomposition* is possible
17   Initialize Decomposition Identifier
18   Initialize Granularity criteria
19   Observe decomposed v_t, choose actions **A**
20   Initialize Re-architecture model
21   end
22   end
23   Execute Algorithm 2
24 **end**
25 **end**

validation. However, anticipating the arrival service type is impossible in a realistic situation like this. This is one of the reasons for using the RL technique, which requires no pre-existing datasets but instead creates its own repository from all transitions and learns from it. The learning curve of the model changes depending on the relevant factors stated above. During the initial phase, the proposed model provides a sub-optimal solution to all the arriving services. The saved transitions train the model concurrently until it reaches a stable learning point. As soon as the stability phase has been acquired, the model begins to initiate the prediction phase by delivering the sub-optimal solution using the learned model (primary DNN). At the same time, the target DNN keeps track of new arrivals and updates the primary DNN on a regular basis, bringing stability to the learning graph. As a result, our algorithm is an online deployment process that learns and delivers the solution at the same time.

8.6.3 Netrail Topology

The normalized SAR moving average for the Netrail topology for the 12-4 and 12-8 nodal capacity scenarios, with the EUB as 22 and 44 SFCs, respectively, is shown in Figures 8.8 and 8.9. These normalized EUB values are represented in the figures as

Dynamic decomposition of service function chain 227

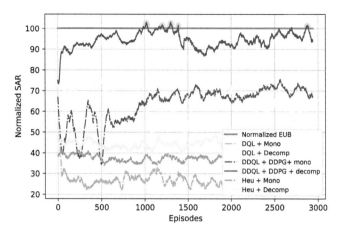

Figure 8.8 Netrail SAR: 12-4 scenario

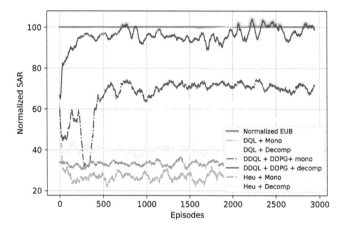

Figure 8.9 Netrail SAR: 12-8 scenario

'Normalized EUB'. Considering the DDQL model's performance for the aforementioned figures, the microservice (DDQL+Decomp) paradigm effectively maintained ~ 48% more SFCs than the monolithic (DDQL+Mono) paradigm for the 12-4 scenario. Similarly, the DDQL+Decomp model saved up to 42% more SFCs than the DDQL+Mono model at higher nodal capacity (12-8 CPU cores). This is so that the decomposed SFCs can still find a location even when the network availability is minimal. The proposed approach successfully decomposes resource-intensive VNFs into suitable mVNFs and efficiently deploys them, resulting in a lower rejection rate. In the case of monolithic VNFs, these resource-intensive VNFs failed to secure a substrate node for deployment during low network availability, resulting in a greater service rejection rate and poor performance. The results show that the proposed model incorporated much more SFCs than monoliths. These figures also show the benefit of using the re-architecture module, which efficiently managed to remove the redundant functionality and recovered enough link and nodal resources to deploy the SFCs more than anticipated. This is evident from the figures, for 12-4 scenarios, the normalized SAR was 2% higher than the normalized EUB and almost >2% higher for 12-8 cases (which has been highlighted in the figures). Therefore, the Netrail topology's resources have been efficiently used by the combination of microservices and re-architecture.

Figures 8.8 and 8.9 show performance comparisons of the DDQL, DQL, and heuristic. Although the heuristic decomposed (Heu+Decomp) model performs better than Heu+Mono and DQL+Mono; both DQL and heuristic models exhibit extremely unstable learning in general. Because the heuristic systems suffer from a greedy selection of highly favored nodes, leading to rapid depletion of their resources and a high proportion of underutilized nodes. The DQL model, on the other hand, has only one estimator that seeks to learn both the behavior and the target policies in order to find an optimal solution. This causes constant modifications to the weights, resulting in a considerable divergence in learning, as seen in all of the provided DQL findings. The learning in the DDQL model becomes more obvious with the double estimator (primary + target NNs), where the target network is changed at every τ iterations rather than every other step, enabling more stable learning significantly.

Figures 8.10 and 8.11 demonstrate the behavior of nodal capacity on the topology. The initialized nodal resource on topology is represented by the black trail, while the red trail shows the total processing resources (CPU cores) sought by the 100 SFCs across the episodes. For both the 12-4 and 12-8 scenarios, our proposed model (DDQL+Decomp) used available nodal resources more efficiently than the monolithic (DDQL+Mono) approach, resulting in greater in-service acceptance, as shown in Figures 8.8 and 8.9. As a result, this illustrates the efficacy of the recommended technique.

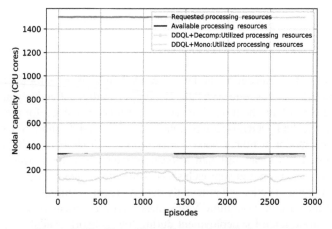

Figure 8.10 Netrail nodal capacity: 12-4 scenario

Dynamic decomposition of service function chain 229

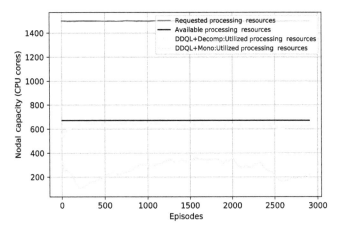

Figure 8.11 *Netrail nodal capacity: 12-8 scenario*

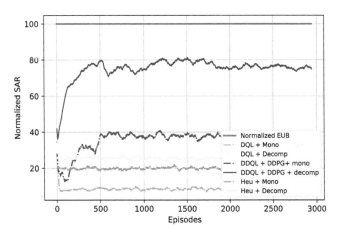

Figure 8.12 *BtEurope SAR: 12-4 scenario*

8.6.4 *BtEurope topology*

Figures 8.12 and 8.13 show the moving average of normalized SAR over the last 50 episodes for 12-4 and 12-8 nodal capacity scenarios in the BtEurope topology, with EUBs of 76 and 153 SFCs, respectively. In the 12-4 scenario with the denser topology, the proposed DDQL+Decomp successfully incorporated twice as many services as DDQL+Mono, as illustrated in Figure 8.12. Figure 8.13 shows that the suggested paradigm increases service acceptance by over 50% for bigger nodal capacity (12-8 CPU cores) than monolithic. Even with abundant nodal resources, the intended model could only achieve 50% of normalized SAR due to link resource depletion. Observing the DQL and heuristic models again, the performances remain unstable, with significant learning divergence. Even as topology complexity increased, the model learned to determine the decomposing VNFs and their granularity, as well as to restructure the SFCs, resulting in a more profitable normalized SAR.

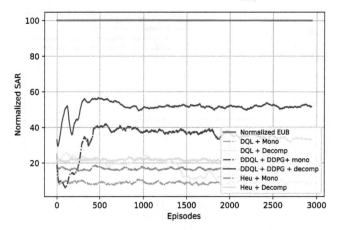

Figure 8.13   BtEurope SAR: 12-8 scenario

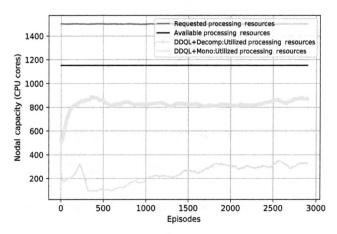

Figure 8.14   BtEurope nodal capacity: 12-4 scenario

8.6.5   Nodal capacity

Figures 8.10, 8.11, 8.14, and 8.15 describe the usage principle of how various models utilize nodal capacity based on topology. As demonstrated in Figures 8.10 and 8.11, the processing resources used by DDQL+Decomp approximately overlap with the total available processing resources in the topology, showing that the DDQL+Decomp model effectively used the available resources to deploy the arriving SFCs. However, in Figures 8.14 and 8.15, the suggested model DDQL+Decomp represents some residual resources in the topology, which might have enhanced the SARs with proper learning. This can be improved by employing denser NNs.

To summarize, we employed DDQL to address the microservices-based RA problem and compared its performance to DQL and a heuristic solution under various network density and nodal capacity scenarios. Microservices-based SFC indicated a considerable boost in service acceptability compared to monolithic SFC.

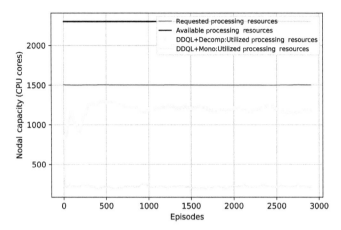

Figure 8.15   BtEurope nodal capacity: 12-8 scenario

The advantage of re-architecting is also perfectly evident, particularly in the Netrail topology.

8.7   Conclusions

In this study, a novel RA method based on microservices is examined for the effective deployment and decomposition of VNFs onto substrate networks. Our investigation revealed that decomposing each arriving VNF could be a resource- and time-intensive. The proposed algorithm strikes the proper balance between when and how much decomposition is advantageous. This study proposes a criterion for selecting prospective VNFs for decomposition and fine granularity technique. For deconstructed VNFs, the re-architecture concept is adopted.

We examined DDQL for the NFV-RA problem and compared its performance to DQL and a heuristic model. Under various network density and nodal capacity scenarios, the performance of DDQL for microservices and monoliths is comprehensively studied. By eliminating the over-estimation problem, the DDQL agent discovered the characteristics of the topologies under defining conditions and provided efficient solutions. The suggested methodology demonstrated a significant improvement in service acceptance for microservices-based SFCs over monolithic SFCs, with microservices recovering up to 50% more. The benefit of the re-architecture approach is noticeable, particularly in the Netrail topology, where the model recognizes the relevance of microservices and chooses to opt for them. For a smaller topology, the model exceeds the EUB; for a denser topology, it tries to approach the upper bound but is constrained by the exhaustion of the link resources. We recommend utilizing substantially denser NNs for better learning in denser topologies. As a result, our proposed model outperforms when resources are scarce.

The model requires a more extended learning period to thoroughly understand the environment as topology complexity increases, resulting in a longer calculation

Table 8.5 Notations

Notations	Descriptions
S	State-space
s_t	State at time-step t
A	Action-space
a_t	Action at time-step t
R_t	Discounted rewards at time-step t
π_t	Policy at time-step t
$Q^{\pi}_{s_t,a_t}$	Action value function
θ_t	Parameters for Q estimator
θ'_t	Parameters for Q' estimator
G	Directed graph for topology
N	Set of substrate nodes
L	Set of substrate links
n_y	Set of all nodal resource types in a substrate node y
l_u	Set of all link resource types in a substrate link u
$n_{y,w}$	Amount of resource type w in a substrate node y
$l_{u,w}$	Amount of resource type w in a substrate link u
$\Im_{node}$	Total nodal resource types
$\Im_{link}$	Total link resource types
G'	Directed graph for VNF-FG
V	Set of VNFs
J	Set of VLs
$\emptyset$	Potential/candidate VNF for decomposition
$G''_{\emptyset}$	Directed graph for mVNF-FG
$E_{\emptyset}$	Set of mVNFs
$F_{\emptyset}$	Set of mVLs
V_{ξ}	Set of VNFs in SFC ξ
E_{ξ}	Set of mVNFs in SFC ξ
R_{ξ}	Set of demanded resources by all VNFs in SFC ξ
H_{ξ}	Set of demanded resources by all mVNFs in SFC ξ
$r_{\xi,x}$	Set of requested resources by VNF x in SFC ξ
$e_{\xi,\emptyset}$	Set of finely decomposed mVNFs
$h_{\xi,\emptyset}$	Set of requested resources by candidate VNF $\emptyset$ in SFC ξ
$v_{\xi,x}$	VNF x in SFC ξ
$r_{\xi,x,w}$	Amount of requested resource type w in VNF x, SFC ξ
$h_{\xi,\emptyset,z,w}$	Amount of requested resource type w in mVNF z, SFC ξ
A^{vnf}_{ξ}	Set of action-space for all monolithic VNFs in SFC ξ
A^{mvnf}_{ξ}	Set of action-space for all decomposed VNFs in SFC ξ
$a^{vnf}_{\xi,x}$	Set of action-space for monolithic VNF x, SFC ξ
$a^{mvnf}_{\xi,x}$	Set of action-space for decomposed mVNF, SFC ξ
$R(\xi)$	Reward function for SFC ξ
$L_{reward}(\xi_{vnf/mvnf})$	Local reward for VNF or mVNF
$G_{reward}(\xi_{vnf/mvnf})$	Global reward for VNF-FG or mVNF-FG
$GR^{pt}_{G'}$	Global reward points for VNF-FG
$GR^{pt}_{G''}$	Global reward points for mVNF-FG

(continued)

Table 8.5 Continued

Notations	Descriptions
$R^{pt}_{vnf/mvnf}$	Reward points for VNFs or mVNFs
$P^{pt}_{vnf/mvnf}$	Penalty points for VNFs or mVNFs
$R^{pt}_{vl/mVl}$	Reward points for VLs or mVLs
$P^{pt}_{vl/mvl}$	Penalty points for VLs or mVLs
$\Xi(\cdot)$	Agent's function
$\mathcal{R}(\cdot)$	Currently available resource for selected action
K^{vnf}_{ξ}	Set of agent selected actions for all VNFs, SFC ξ
K^{mvnf}_{ξ}	Set of agent selected actions for all mVNFs, SFC ξ
$k^{vnf}_{\xi,x}$	Obtained action for VNF x, SFC ξ
$k^{mvnf}_{\xi,\emptyset}$	Decomposition status for VNF $\emptyset$, SFC ξ
$k^{mvnf}_{\xi,\emptyset,f}$	Obtained action for mVNF f, VNF $\emptyset$, SFC ξ
$\zeta_{topology}$	Maximum nodal capacity of a node in a topology
$\varphi_{topology,t}$	Currently available nodal resource in a topology at time-step t

time to solve the NFV-RA problem. To improvise the produced results, we will adopt the deep deterministic policy gradient (DDPG) approach, a model-free, actor-critic technique that deals with larger action spaces. Furthermore, applying the microservices model in real-time may generate several complicated concerns, such as security and increased operational complexity. In the future, we will focus on comprehending and solving such scenarios.

References

[1] Saad W, Bennis M, and Chen M. A vision of 6G wireless systems: applications, trends, technologies, and open research problems. *IEEE Networks*. 2019.
[2] Ahmadi H, Nag A, Khan Z, *et al*. Networked twins and twins of networks: an overview on the relationship between digital twins and 6G. *IEEE Commun Stan Mag*. 2021;5(4):154–160.
[3] Letaief KB, Chen W, Shi Y, *et al*. The roadmap to 6G: AI empowered wireless networks. *IEEE Commun Mag*. 2019;57(8):84–90.
[4] Network Functions Virtualisation – Introductory White Paper. Available from: https://portal.etsi.org/NFV/NFV_White_Paper.pdf.
[5] Rost M and Schmid S. On the hardness and inapproximability of virtual network embeddings. *IEEE/ACM Trans. Netw*. 2020;28(2):791–803.
[6] Fowler M. Microservices Guide. Available from: https://martinfowler.com/microservices/.

[7] Probst K and Becker J. Engineering Trade-Offs and the Netflix API Re-Architecture. Available from: https://netflixtechblog.com/engineering-trade-offs-and-the-netflix-api-re-architecture-64f122b277dd.
[8] Implementing Microservices on AWS.
[9] Nekovee M, Sharma S, Uniyal N, *et al.* Towards AI-enabled microservice architecture for network function virtualization. In: *IEEE ComNet*; 2020. p. 1–8.
[10] Rasheed A, San O, and Kvamsdal T. Digital twin: values, challenges and enablers from a modeling perspective. *IEEE Access*. 2020;8:21980–22012.
[11] Di Cicco N, Mercan EF, Karandin O, *et al.* On deep reinforcement learning for static routing and wavelength assignment. *IEEE J. Select. Top. Quantum Electron*. 2022; vol. 28, no. 4, pp. 1–12.
[12] Quang PTA, Hadjadj-Aoul Y, and Outtagarts A. A deep reinforcement learning approach for VNF forwarding graph embedding. *IEEE Trans. Netw. Serv. Manag*. 2019; 16(4):1318–1331.
[13] Dehury CK and Sahoo PK. DYVINE: fitness-based dynamic virtual network embedding in cloud computing. *IEEE J. Sel. Areas Commun*. 2019; 37(5):1029–1045.
[14] Agarwal S, Malandrino F, Chiasserini CF, *et al.* VNF placement and resource allocation for the support of vertical services in 5G networks. *IEEE/ACM Trans. Netw*. 2019;27(1):433–446.
[15] Nejad MAT, Parsaeefard S, Maddah-Ali MA, *et al.* vSPACE: VNF simultaneous placement, admission control and embedding. *IEEE Sel. Areas Commun*. 2018;36(3):542–557.
[16] Ruiz L, Barroso RJD, De Miguel I, *et al.* Genetic algorithm for holistic VNF-mapping and virtual topology design. *IEEE Access*. 2020;8:55893–55904.
[17] Habibi MA, Yousaf FZ, and Schotten HD. Mapping the VNFs and VLs of a RAN slice onto intelligent PoPs in beyond 5G mobile networks. *IEEE Open J. Commun. Soc*. 2022;3:670–704.
[18] Yuan Y, Tian Z, Wang C, *et al.* A Q-learning-based approach for virtual network embedding in data center. *Neural Comput. Appl*. 2020;32(7):1995–2004.
[19] Sciancalepore V, Yousaf FZ, and Costa-Perez X. z-TORCH: an automated NFV orchestration and monitoring solution. *IEEE Trans. Netw. Serv. Manag*. 2018;15(4):1292–1306.
[20] Sharma S, Uniyal N, Tola B, *et al.* On monolithic and microservice deployment of network functions. In:*IEEE NetSoft*; 2019. p. 387–395.
[21] Moro D, Verticale G, and Capone A. A framework for network function decomposition and deployment. In: *IEEE DRCN 2020*. IEEE; 2020. p. 1–6.
[22] Chowdhury SR, Anthony, Bian H, *et al.* A disaggregated packet processing architecture for network function virtualization. *IEEE J. Select. Areas Commun*. 2020;38(6):1075–1088.
[23] Fu X, Yu FR, Wang J, *et al.* Service function chain embedding for NFV-enabled IoT based on deep reinforcement learning. *IEEE Commun. Mag*. 2019;57(11):102–108.

[24] Luizelli MC, Bays LR, Buriol LS, *et al*. Piecing together the NFV provisioning puzzle: efficient placement and chaining of virtual network functions. In: *IFIP/IEEE International Symposium on Integrated Network Management*. IEEE; 2015. p. 98–106.

[25] Shihabur Rahman C, Reaz A, Nashid S *et al*. Revine: reallocation of virtual network embedding to eliminate substrate bottlenecks. In: *IFIP/IEEE International Symposium on Integrated Network Management*. IEEE; 2017. p. 116–124.

[26] Mijumbi R, Serrat J, Gorricho JL, *et al*. Design and evaluation of algorithms for mapping and scheduling of virtual network functions. In: *IEEE NetSoft*. IEEE; 2015. p. 1–9.

[27] Chetty SB, Ahmadi H, and Nag A. Virtual network function embedding under nodal outage using reinforcement learning. In: *IEEE International Conference on Advanced Networks and Telecommunications Systems*. IEEE; 2020.

[28] Chetty SB, Ahmadi H, Sharma S, *et al*. Virtual network function embedding under nodal outage using deep Q-learning. *Future Internet*. 2021;13(3):82.

[29] Sutton RS and Barto AG. *Reinforcement Learning: An Introduction*. Cambridge, MA: A Bradford Book; 2018.

[30] Van Hasselt H, Guez A, and Silver D. Deep reinforcement learning with double Q-learning. In: *Proceedings of the AAAI*. vol. 30; 2016.

[31] Gupta A, Habib MF, Mandal U, *et al*. On service-chaining strategies using virtual network functions in operator networks. *Comput. Netw.* 2018;133: 1–16.

[32] Chowdhury SR, Salahuddin MA, Limam N, *et al*. Re-architecting NFV ecosystem with microservices: state of the art and research challenges. *IEEE Netw.* 2019;33(3):168–176.

[33] Erdős P and Rényi A. On random graphs I. *Publ. Math. Debrecen*. 1959;6 (290–297):18.

Chapter 9

An Optimization-as-a-Service platform for 6G exploiting network digital twins

Oriol Sallent[1], José-Manuel Martínez-Caro[2],
Javier Baliosian[3], Luis Diez[4], Luis M. Contreras[5],
Jordi Pérez-Romero[1], Juan Luis Gorricho[1],
Matías Richart[3], Ramón Agüero[4], Joan Serrat[1],
Pablo Pavón-Mariño[2,6] and Irene Vilà[1]

The continuous evolution of the communications networks in the race to support 6G services will introduce important operational challenges for network operators, especially in terms of efficient usage of resources, near real-time performance, etc. Such advanced 6G services will necessarily leverage on existing brownfield networks complemented with novel technologies and supportive tools. In this context, the concept of *network* should be understood in a broad sense, including both networking and computing capabilities enabling the connectivity of the customers to service functions or applications in a very dynamic way.

Until recently, the main driver for network design and operation has been the guarantee of throughput since bandwidth-based services have been dominant so far. With the advent of 5G, delay-based services have started to become relevant as well, as a consequence of the offerings of low latency and interactive services, such as tactile Internet or augmented and virtual reality. Now, the advent of forthcoming beyond 5G (B5G) and 6G services is expected to introduce more stringent requirements motivating a new radical change in operations, adding new dimensions to service provision, such as delay-variation or precision-based services.

All this complexity triggers the need of evolving the mode of operation by introducing smarter, faster, and educated decisions on the operational processes, pursuing an overall optimization of the usage of both networking and compute resources.

[1]Signal Theory and Communications, Universitat Politècnica de Catalunya (UPC), Spain
[2]E-lighthouse Network Solutions, Cartagena, Spain
[3]Instituto de Computación, Facultad de Ingeniería, Universidad de la República, Uruguay
[4]Communications Engineering Department, Universidad de Cantabria, Spain
[5]Telefónica Innovación Digital, Madrid, Spain
[6]Department of Information and Communication Technologies, Technical University of Cartagena, Spain

In this context, the concept of network digital twins (NDTs) is gaining traction as a perfect complement to the foreseen complex operation. For instance, thanks to the paradigms of network softwarization and virtualization, the telco industry is now adopting the DevOps paradigm, supporting the continuous integration/continuous delivery (CI/CD) of services, introducing a dynamicity on the configuration actions realized over network and services never seen before. That high dynamicity of actions should be performed with extreme care for avoiding side effects and service affection on a critical infrastructure such as the telecom network. By means of the NDT, the expectation is to make available a replica of the network that can help to anticipate, validate, and predict actions in the real physical network but in a controlled environment. It should be noted that a key aspect for the operator is the efficient use of available network and compute assets and resources. It is here where the NDT can be exploited as an Optimization-as-a-Service (OaaS) platform for 6G.

The NDT is considered to be a replica of the physical network, existing side by side. In order to behave similarly, the NDT should ideally receive the same inputs as the physical counterpart, in terms of traffic profiles, configurations, and events, so that the observable outcomes could be comparable. This helps to understand what the expected behavior of the physical network can be by leveraging on the information obtained from the digital one. Those inputs can be provided in real-time, by appropriate monitoring and telemetry information, or off-line, as part of training processes based on datasets collected from real data. The quality of the used data (with respect to the real information) and the fidelity of the software models used in the NDT (as a representation of the network elements and the processing environments) will determine the accuracy of the information provided by the digital replica. With the aim of covering a variety of potential scenarios, the datasets used should be rich enough to ensure a proper training of the NDT, even though not all possible situations could be considered. In fact, the availability of a NDT will mostly help, precisely, to understand the behavior of the real system under, e.g., unexpected events or future traffic trends and patterns.

Apart from the data serving as input, one essential component of the NDT is related to the data analytics and processing, with the purpose of identifying the best operational decision at every moment. The adoption of artificial intelligence (AI) and machine-learning (ML) techniques, including their interaction across the different technological domains in the network (e.g., network-related, edge computing-related, and cloud computing-related), will permit to identify cognitive-based optimizations for later fulfillment on the real network. Thus, it allows us to predict and react to service degradations and failures in a prompt manner due to short- and mid-term future events, e.g., due to situations like rapid changes in the demand (flash crowd events), network congestion, etc.

In consequence, the integration of the NDT with the operational lifecycle management of the physical network can contribute as an efficient response system at run-time to spatially and temporally changing situations, such as traffic patterns, fluctuating network and computing conditions, and dynamic service contexts.

The use of digital twins (DTs) has recently received a lot of attention in the literature due to their applicability in different areas. The main challenges and enabling technologies related to DT are discussed in [1], which also includes some work in progress on three broad areas of DT applicability, namely manufacturing, healthcare, and smart cities. Similarly, a survey of the key technologies involved with DTs is presented in [2], particularly focusing on communications between components (physical and virtual), physical data processing, data twin modeling, and cloud and edge computing. Potential security threats in DTs were listed in [3] considering the different functionality layers where they take place and proposing some preliminary security recommendations and potential future approaches for general-purpose scenarios. The applicability of DT in the context of the Internet of Things (IoT) is surveyed in [4], discussing technical features, the main use case scenarios, and the architecture models for implementing DTs, while the authors of [5] discuss how to adopt DT as a key component of the IoT and analyze the possibility of integrating DT with technologies such as AI, 5G, or blockchain in the industry. In [6], they provide an overview of DT technology focusing on the importance of user-centered design and the need for effective data management. With respect to the NDTs, the development and uses of an AI-driven NDT referred to as B5GEMINI are presented in [7]. This is a 5G NDT oriented to optimize network performance and troubleshooting. Similarly, another NDT for 5G networks is presented in [8], putting the focus on the enabling technologies such as sensing technology, data analytics, and cloud computing and discussing potential applications for network optimization and troubleshooting. The paper also describes the main challenges in the application of NDT in 5G, including data quality, data security, and data integration. In turn, the authors of [9] describe the general architecture of an NDT focusing on the use of ML technologies for building some of its core components and presenting a case study that applies the NDT for routing optimization.

With all the above, this chapter introduces and describes an OaaS platform framed in the context of the OPTIMAIX project [10], which proposes an operational NDT integrated with ML and AI capabilities allowing the execution of advanced optimization algorithms in a flexible manner. Section 9.1 provides the functional architectural view of the platform and the main system APIs. Section 9.2 describes the data models considered for the platform design, permitting a smooth interaction in both the physical and the digital networks. Section 9.3 presents some examples of relevant use cases considered for assessing the validity of the platform.

9.1 OaaS platform: architectural overview

Existing telecommunication networks are complex, cost-intensive, and critical infrastructures. Optimizing their deployment is a paramount challenge, hindered by the number of network components, the utmost diversity of technological options, and the difficulties of facing a dynamic demand and traffic requisites, of very different nature.

In this context, an OaaS [11,12] platform provides a set of tools for assisting the automated planning and optimization of the network in a number of use cases, which can be classified as follows:

1. According to the application timeline: Offline or static use cases, typically conducted periodically (e.g., twice a year) by network carriers, like capacity planning and network dimensioning, demand for computationally intensive and potentially slow algorithms conceived for long running times. On the other hand, capacity-on-demand use cases like optimized path computations, require highly dynamic and agile algorithms able to adapt to the current needs of the networks. Both approaches require network validation functionalities that permit simulating the KPIs of computed optimized network configurations or deployments, and thus evaluate them from a performance point of view.
2. According to the network segment of the application: Network carriers typically organize their networks in different segments, like fixed-access, radio-access, metro-aggregation, and core. Different segments are commonly managed by different network planning personnel, which face different technical needs, and specialized in the segment equipment and services involved.
3. According to the service type: Network optimization is typically targeted to maximize telco operators profit in a number of so-called telco-level services, i.e., services representing aggregations of the traffic of multiple users, instead of user-level fine-grained services. Telco services have a prominent diversity, e.g., from point-to-point transport connections to full network slices.

The heterogeneity of use cases for an OaaS system, demands for:

1. A flexible and extensible architecture, which permits integrating multiple optimization algorithms, with arbitrary inputs and outputs, computational requisites, optimization targets, and application scopes.
2. A scalable scheme, applicable from small to carrier-size networks.
3. An open system, based on open application programming interfaces (APIs). This is an accepted community policy that permits the integration of contributions (e.g., optimization algorithms or computation resources), coming from third parties, which can retain the authorship and economic profit (if any) of their contributions to the whole system.

9.1.1 Functional architecture

Figure 9.1 illustrates an overview of the considered system. The left side of the figure represents a particular example of the target network of the OaaS platform. The network infrastructure consists of a radio access network (RAN) composed of a number of points of presences (PoPs) that include different resources (e.g., radio units, transport nodes, computational and storage resources, etc.) interconnected with transport links. Those PoPs handle many cells that provide radio coverage to the user equipment (UE) in specific geographical areas. The upper-left part represents different services to be allocated in the network infrastructure, consisting in this example of network slices, with their corresponding service level agreements (SLAs). Both the network infrastructure and the supported services are described according to a defined network model: a standards-based schema agreed upon the system actors, as described in Section 9.2.

OaaS platform for 6G exploiting NDTs 241

The right part of Figure 9.1 shows an overall perspective of the architecture of the proposed OaaS platform. It is conceived as a hybrid architecture where one logically centralized (centralized OaaS) and multiple federated (distributed OaaS) platforms coexist. These components are described in the subsections below. The logically centralized platform represents the entry point of the system, and a unified access point to the federated resources. The hybrid architecture, with a logically centralized plus federated platforms, has the advantage to cover two different profiles of third-party institutions/users potentially contributing to the OaaS system:

1. Small/scale contributors may be interested in integrating a small number of algorithms in a simple form to the centralized platform, making them available and runnable using its computing resources.
2. Special/large-scale contributors, willing to have full control not only on the algorithms/reports but also on the computing facilities where they will be running, and how the containers are stored and secured.

These functionalities are implemented as containers, as defined in Dockerfile, with one or more resources such as (i) optimization algorithms; (ii) NDT simulation software; or (iii) reporting schema running and waiting for requests. Each of these functionalities is implemented as a black box software that can be invoked by the network operator making use of the defined REST APIs. This strategy provides a useful level of isolation: each container can host algorithms implemented in different languages or libraries, which do not need to be known or have any impact on other algorithms. In addition, the API server is the binding block between the platform functionalities hosted in containers, their inputs from the hosted services, and the corresponding management actions on the network infrastructure.

9.1.1.1 Centralized OaaS platform

The system contains different interacting elements looking for an efficient use of the available infrastructure resources. The central unit, called the centralized OaaS platform, is the minimum required functionality and it is constituted by a set of generic and specific modules. Once the centralized platform is working, the system can be augmented according to the network operator needs.

The generic modules are constituted by Docker containers grouped into a computing cluster. Docker containers are instantiated as many times as necessary to materialize the required platform functionalities like placement algorithms, NDTs, reporting, or any other.

The computing cluster has to have the necessary computing power to run the Docker containers. This module can be hosted in local dependencies or in a cloud provider infrastructure (e.g., Amazon Web Services, Google Cloud Platform, Microsoft Azure, etc.).

On the other hand, the specific modules which are only present in the centralized OaaS platform are meant as a storage media to preserve and update network designs and the distributed resources metadata (URL, credentials, etc.), embracing:

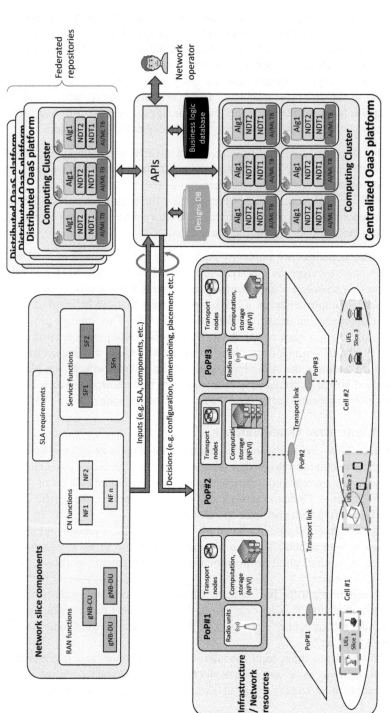

Figure 9.1 OaaS architecture

1. The Designs database contains the representation of the network infrastructure and the service requirements. The database is periodically updated with inputs from the infrastructure components and service requirements.
2. The business logic database stores metadata related to the different federated platforms, user permissions, containers allocation, etc.

9.1.1.2 Distributed OaaS platform

As part of the architecture, the system is targeted to permit the federation of an arbitrary number of distributed OaaS platforms. Each federated OaaS platform is accessible from the centralized unit and contains the generic modules available in the centralized OaaS platform such as a computing cluster, and one or more containers as a set of functionalities. The main advantages that bring a federation of distributed OaaS platforms are (i) simpler instances; (ii) independent platforms managed by users; (iii) enabling work concurrency between different platforms; and (iv) increasing the overall computational capacity of the system.

9.1.2 OaaS system APIs

Figure 9.2 illustrates the main APIs defined to govern the system. For all of them, the OpenAPI specifications [13] are adopted for the API description, due to its usefulness for the APIs development. The proposed APIs are implemented as REST APIs, with messages based on a human-readable format as the JavaScript Object Notation (JSON) format, pervasively used in these contexts. Lightweight implementations, robustness, fast processing, and easiness of documentation and testing process are the main advantages of the OaaS system APIs.

The proposed schema describes four key APIs for design handling, and others for the OaaS platforms, the repository, and the functionality management. These APIs may be extended for specific use cases or augmented requisites. The following subsections include the main details.

9.1.2.1 API1: design handling

Optimization algorithms, NDT simulations, and reporting algorithms require updated infrastructure information to get precise conclusions and decisions according to their status. Network and other infrastructure information is inserted into the Designs database and frequently updated using the data models presented in Section 9.2. This API, as depicted in Table 9.1 from top to bottom, allows the creation of new designs, the return of a list of all the available designs with their identifiers and descriptions, the details of a specific design, the update of the design parameters (topology, SLAs, demands, etc.), or the deletion of a specific design.

9.1.2.2 API2: OaaS platforms

As stated above, the system supports either a centralized or a federation of distributed OaaS platforms. As the number of federated platforms is dynamic, the system allows the enrolment of new platforms and the deletion of the existing ones according to the system, user, and infrastructure requirements. All platforms, either centralized or distributed, are made accessible through this API. The OaaS platform management

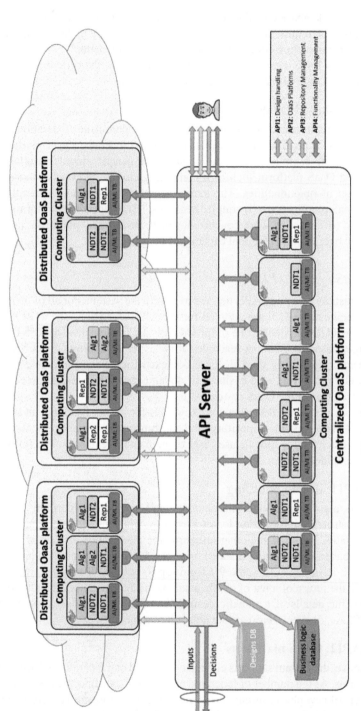

Figure 9.2 OaaS system APIs

Table 9.1 API1: design handling paths

Design handling paths	
POST/design Accepted: application/json Description: creates a new design	HTTPS/1.1
GET/designs Description: returns a list with the available designs (id and description)	HTTPS/1.1
GET/design/{design-id} Description: returns updated information of a design by id	HTTPS/1.1
PUT/design/{design-id} Accepted: application/json Description: updates the design and network information (topology information, SLAs, demands, etc.	HTTPS/1.1
DELETE/design/{design-id} Description: deletes a design by id	HTTPS/1.1

Table 9.2 API1: OaaS platform paths

OaaS platforms paths	
POST/oaas-platfom Accepted: application/json Description: registers a new OaaS platform	HTTPS/1.1
GET/oaas-platforms Description: returns a list with the federated OaaS platforms and (id description)	HTTPS/1.1
GET/oaas-platform/{*oaas-platform-id*} Description: returns the OaaS Platform details and status by id	HTTPS/1.1
PUT/oaas-platform/{*oaas-platform-id*} Accepted: application/json Description: updates a federated OaaS platform by id	HTTPS/1.1
DELETE/oaas-platform/{*oaas-platform-id*} Description: defederates an OaaS platform by id	HTTPS/1.1

adopts a set of available paths, as listed in Table 9.2, to carry on tasks such as registering a new OaaS platform, listing the identifiers and description of the existing federated OaaS platforms, returning detailed information of a particular OaaS platform, updating the specific parameters of an OaaS platform, or deleting an OaaS platform.

9.1.2.3 API3: repository management

Each OaaS platform (centralized or distributed) lays on a computing cluster, able to instantiate Docker containers with algorithms, NDTs, or reports (isolated or

Table 9.3 API3: repository management paths

Repository management paths	
POST/image Accepted: application/json Description: instantiates a new Docker image (centralized or distributed)	HTTPS/1.1
GET/containers Description: returns a list with all instantiated Docker containers (id, description, and OaaS platform)	HTTPS/1.1
GET/container/{*oaas-platform-id*} Description: returns a list with all instantiated Docker containers in a federated OaaS platform	HTTPS/1.1
GET/container/{*container-id*} Description: returns the Docker container details and status by id	HTTPS/1.1
DELETE/container/{*container-id*} Description: deletes an instantiated Docker container by id	HTTPS/1.1

combined). The API3, as shown in Table 9.3, performs basic operations for the successful operation of all the Docker containers. Sequentially, it deploys a new Docker image in an OaaS platform sending the URL and credentials (if required) from a public or private repository (e.g., DockerHub). Once deployed, the OaaS platform API returns the instantiated containers in the whole system or a specific OaaS platform (via *oaas-platform-id*). Making the queries with *container-id* in the path, the API returns the container details or deletes the instantiated container.

9.1.2.4 API4: functionality management

Enrolled Docker images implement a set of functionalities such as algorithms, NDTs, and reports defined in the Docker image. After the instantiation of the Docker container, the user may require the paths detailed in Table 9.4 for returning lists of available functionalities in all the architecture, in a given OaaS Platform or in a specific Docker container. In addition, other paths return all the details for a specific functionality and start/stop the execution of the running workflow.

9.1.3 A workflow example

In order to illustrate how the proposed system architecture evolves with the number of federated distributed OaaS platforms, Figure 9.3 depicts an initial OaaS system state where the centralized OaaS platform has two containers in its computing cluster. Then, the labels and arrows in the schema represent an example sequential system workflow detailed as follows:

1. The network operator inserts a new network design in the Designs database using the appropriate data model.

Table 9.4 API4: functionality management paths

Repository management paths	
GET/functionalities Description: returns a list with all functionalities (algorithms NDTs, and reports) available	HTTPS/1.1
GET/functionalities/{oaas-platform-id} Description: returns a list with all functionalities in a federated OaaS Platform	HTTPS/1.1
GET/functionalities/{container-id} Description: returns a list with all functionalities in a container	HTTPS/1.1
GET/functionalities/{functionality-id} Description: returns the functionality details and status by id	HTTPS/1.1
PUT/functionalities/{functionality-id}/start Accepted: application/json Description: runs a functionality (algorithm, NDTs, or report)	HTTPS/1.1
PUT/functionalities/{functionality-id}/stop Accepted: application/json Description: stops a functionality (algorithm, NDTs, or report)	HTTPS/1.1

2. Whenever the network infrastructure changes, notifications of these changes are sent to the API server via inputs updating the network design parameters (topology inventory, SLAs, etc.).
3. The network operator requests the federation of a new distributed OaaS platform in the system. The API server checks the availability of the platform and inserts a new entry in the Business logic database with the new OaaS platform parameters. If the federation is successful, the new platform and its computing cluster are ready to instantiate new repositories.
4. The previous step allows the deployment of containers on the new distributed OaaS platform. In the example, the network operator sends the instruction to implement a new container in the platform and updates the Business logic database with the changes.
5. Once the new container with its corresponding functionalities is implemented, it is ready to be launched via the API server. The API server detects the sender of the request, and it is informed about the functionality to be launched (e.g., a given algorithm) and the network design on which to apply the result.
6. Now, the API server retrieves the network parameters from the Designs database and the metadata of the OaaS platform and the requested container from the Business logic database and remits all the information to the container with the desired functionality.
7. The functionality executes the request(s) (e.g., an algorithm with a given purpose is executed). The execution time depends on the algorithmic complexity of the functionality and the dimension of the underlying network design.
8. When the execution is completed, the functionality returns a response to the API server, which will distribute it to the appropriate recipients in the OaaS system.

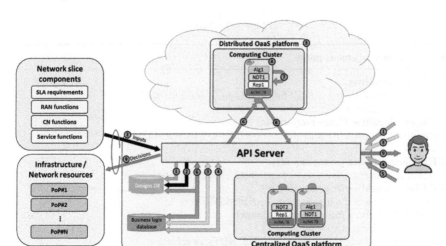

Figure 9.3 Simple OaaS system workflow example

9. For this particular example, the network operator receives a human-readable report with the conclusions of the execution and the infrastructure receives the appropriate actions derived from the execution to optimize the network operation.

9.2 OaaS platform: network model

Optimizing the deployment and operation of services raises the need for means to describe service requests, their architecture, as well as the physical resources where they will be allocated and executed. This inherent need is fulfilled through a network information model that provides resource descriptions at different levels of granularity, meeting the requirements of service specifications and infrastructure description. It is worth noting that the network information model needs to include not only connectivity functionality but also the computing functionality needed to host the service components. In the following subsections, we provide the grounding elements to build the network model of the OaaS platform.

9.2.1 Overview of standardized network models

3GPP, ETSI, and IETF are the SDOs that have promoted the majority of standards related to different types of networks and network resources, which are within the scope of the OaaS platform. The adoption of all these models in the project will be done up to the extent that is needed to support the target use cases. At the moment of writing this chapter, we summarize those that are looking more relevant.

9.2.1.1 Radio access network

A frame reference for specifying the model of the radio network is [14], which specifies the information model and solution set for the network resource model (NRM)

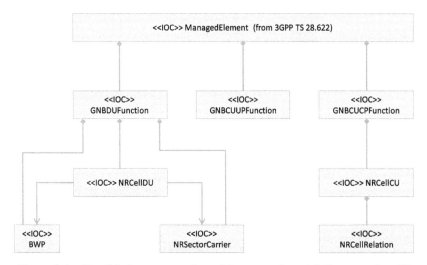

Figure 9.4   Simplified representation of the network model diagram of a gNB

definitions of NR, NG-RAN, 5G core (5GC) network, and network slice. The information model defines the semantics and behavior of the information object class attributes and relations that are visible on the management interfaces in a protocol and technology neutral way.

Focusing on the NG-RAN, Figure 9.4 depicts a simplified view with the set of classes (IOC: Information Object Class) that encapsulate the relevant information for a gNB according to the NRM of [14]. In 5G, a gNB is functionally split into different components, namely the gNB central unit (CU), which hosts the upper layers of the radio interface protocol stack, namely the service data adaptation layer (SDAP), radio resource control (RRC), and packet data convergence protocol (PDCP), and the gNB distributed unit (DU) that hosts the lower layers, namely radio link control (RLC), medium access control (MAC), and physical layer. Moreover, a gNB CU can be further split into a control plane function (gNB-CU-CP) that hosts the RRC and the control part of PDCP, and a user plane function (gNB-CU-UP) that hosts the SDAP and the user plane part of PDCP. The set of IOC classes of the gNB model reflects this functional split, so that a gNB is represented by a combination of a GNBCUCPFunction for modeling the gNB-CU-CP, one or more GNBCUUP-Functions for modeling the gNB-CU-UP and one or more GNBDUFunctions for modeling the gNB-DU.

One gNB handles one or more NR cells, where a cell provides radio coverage over a certain geographical area at a certain frequency carrier and is the radio network object that can be uniquely identified by a user equipment (UE) through a "cell Identifier" broadcast over a geographical zone. Correspondingly, the NRM of the gNB includes the NRCellDU class that represents the part of NR cell information that describes the specific resource instances, and the NRCellCU that represents the NR cell information responsible for the management of inter-cell mobility and neighbor relations. Each one of these classes has a specific list of attributes that can be found in [14]. A part from the NRCellDU, other IOCs that depend on the GNBDUFunction are the NRSectorCarrier, which represents the resources of each transmission point associated with corresponding cells, and the BWP, which represents a bandwidth part related to downlink, uplink, or supplementary uplink resource grids.

9.2.1.2 IP network domain

The IP SDN domain controller contains a set of network management datastores defined in Yang 1.1 [15] which will be accessible via a RESTCONF North Bound Interfaces (NBI) [16]. Based on these datastores, the Telecom Infra Project [17] proposes the mandatory technical requirements of the NBI to be exposed by IP domain controllers according to standards, or drafts, defined by the IETF. In particular, the document [18] defines abstract YANG data models for network and service topologies and inventories.

As depicted in Figure 9.5, the model in [18] starts with the class entitled ietf-network, which is a list of network nodes, supporting the concept of network hierarchy or network stack. In other words, the data model is a container with a list of networks, where each network is distinguished via the corresponding network identifier.

The above-mentioned ietf-network class is specialized by means of the ietf-network-topology class, which defines a generic topology model and their components like nodes, links, and termination points. Nodes and termination points are supported by underlying entities. Links, which are point-to-point and unidirectional, are distinguished by means of a link identifier within a given topology and establish a connection between a source and a destination node.

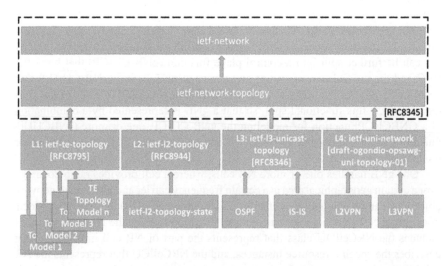

Figure 9.5 Augmented network model hierarchy in the IP network domain

The data model proposed in [18] allows a network to contain multiple topologies, for example, topologies at different layers and overlay topologies. The data model, therefore, allows relationships between topologies, as well as dependencies between nodes and termination points across topologies, to be captured.

Based on [17], the topological information has at least four topology levels, as depicted in Figure 9.5 and described hereafter.

1. Network #1 (L1) is the closest view of the physical representation of the network elements and applies to different L1 technologies with different TE topology models. The TE topology model is meant to be network technology agnostic and can be augmented using different technology-specific TE topology models (TE Topology Model n). The standard [19] supports L1 and the main parameters are network, network type, node, termination-point, inter-domain-plug-id, and links.
2. Network #2 (L2) represents the link-layer or Ethernet connections built with diverse L2 technologies. It is based on standard [20] and defines the YANG data model for physical and logical L2 network topologies by augmenting the abstract data models with L2-specific topology attributes built with different L2 technologies such as identifiers, identities (e.g., IEEE 802.1ah, IEEE 802.1ad, VXLAN), attributes and states of L2 networks, nodes, links, and termination-points. This L2 can be augmented with ietf-l2-topology-state that mirrors the ietf-l2-topology with read-only state data.
3. Network #3 (L3) illustrates the topology elements located at the network and IGP layer detailed in [21]. This data model introduces a holistic view of the L3 unicast network topology integrating different L3 unicast topology types under the specified ietf-l3-unicast-topology class that could be augmented to enhance the L3 unicast topology details (e.g., OSPF, IS-IS, etc.). This module augments the abstract data models with new L3 attributes in the tree at different leaves (network, network-type, node, termination-point, and link).
4. Network #4 (L4) is based on the latest version of the IETF draft-ogondio-opsawg-uni-topology [22] and augments the abstract data model by adding the concept of service attachment points. The service attachment points are an abstraction of the points where network services such as L3VPN or L2VPN can be attached allowing the information exchange between control elements and supporting VPN service provision and resource management. The service orchestration layer can determine which endpoint of interconnection to add to L2VPN or L3VPN service. With the help of other data models (e.g., L3SM and L3NM model) and mechanisms, hierarchical control elements could determine the feasibility of an end-to-end path and derive the sequence of domains and the points of interconnection to use.

9.2.1.3 Network slicing

Network slices are identified in 3GPP by means of the single network slice selection assistance information (S-NSSAI), which is defined in [23] and includes the slice/service type (SST) to reflect the network slice behavior in terms of features and services and the slice differentiator, used to differentiate amongst multiple network slices of the same SST. Up to date, 3GPP has defined five different values of SST, namely SST = 1 for handling enhanced Mobile BroadBand (eMBB) communication services, SST = 2 for Ultra Reliable and Low Latency Communication (URLLC) services, SST = 3 for Massive Internet of Things (MIoT), SST = 4 for Vehicle-to-Everything (V2X) communications services, and SST = 5 for supporting High Performance Machine Type Communications (HMTC).

In addition to the SST, a network slice should be characterized by a set of attributes that specify the SLA terms between the network slice provider (NSP) and the network slice consumer (NSC). In this respect, the GSMA generic slice template (GST) provides a standardized list of attributes (e.g., performance related, function related, etc.) that can be used to characterize different types of network slices [24]. GST is generic and is not tied to any type of network slice.

Using the different attributes of the GST, the GSMA NG.116 document also defines different cases of network slice type (NEST). A NEST is a GST filled with values or ranges of values. There are two kinds of NESTs: standardized NESTs (S-NEST), i.e., NESTs with values established by standards organizations, working groups, fora, etc. such as, e.g., 3GPP, GSMA, 5GAA, 5G-ACIA, etc.; and private NESTs (P-NEST), i.e., NESTs with values decided by the NSP. The NESTs defined in [24] are the NEST for eMBB with IP multimedia subsystem (IMS) support, the NEST for URLLC, the NEST for MIoT, the NEST for HMTC, and the NEST for public safety.

Another important aspect when modeling network slices is the way to specify the amount of resources allocated to each slice. In this respect, the 3GPP NRM of [14] has defined a set of classes and attributes that are relevant for the management of the resource allocation to network slices in the RAN through radio resource management (RRM) policies. A simplified view of the involved classes and attributes for characterizing RRM policies that allow configuring the way that resources are allocated to the slices is provided in Figure 9.6. Specifically, the RRMPolicyManagedEntity proxy class represents the different RAN-managed components (e.g., cell resources managed at DU, cell resources managed at CU functions, DU functions, etc.) that are subject to the RRM policies. These can be e.g., cell resources managed at DU, cell resources managed at CU functions, DU functions, etc. Correspondingly, the RRMPolicyManagedEntity can represent the different classes discussed in Section 9.2.1.1, such as an NRCellDU, an NRCellCU, and a GNBDUFunction. The RRMPolicy_IOC represents the properties of an abstract RRM policy that includes two attributes: the resourceType and the rRMPolicyMemberList.

On this basis, the RRMPolicyRatio IOC represents a particular realization of an RRM policy that establishes a resource model for resource distribution among slices based on three resource categories: shared resources among slices with no specific guarantees per slice, prioritized resources that are guaranteed for use by associated slices but still usable for other slices when free, and dedicated resources that are only used for the associated slices. Accordingly, the RRMPolicyRatio IOC includes three attributes for managing these categories, namely the rRMPolicyDedicatedRatio, the rRMPolicyMinRatio, and the rRMPolicyMaxRatio.

OaaS platform for 6G exploiting NDTs   253

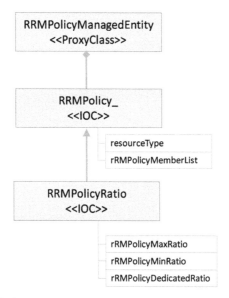

Figure 9.6 Simplified representation of the classes and attributes for configuration of RRM policies for slice management

9.2.2 A transport and computing infrastructure model

The physical infrastructure spans datacenters to provide computing resources and wide-area (or transport) networks that provide connectivity between the datacenters where resources will be allocated for the instantiation of slices.

Regarding datacenters, generally referred to as datacenter PoPs (DC-PoPs), "consist of a number of servers and networking elements for connecting all of them with the transport network. Depending on the number of servers in the data center, there will be distinct capabilities in terms of CPUs, memory, and storage available for deploying and running services" [25].

Figure 9.7 depicts a UML representation of the model proposed in [22]; it includes entities that abstract the datacenter elements needed for modeling the optimization problems addressed in Section 9.3. It is connected to the network topology model presented in Section 9.2.1.2 through one of the TE topology model classes that extend the L1:ietf-te-topology class of Figure 9.5. The dc object provides a general description of a datacenter, on top of which network slices will be instantiated. The relevant information for datacenter capabilities is described in terms of how much memory (the ram object), storage (the disk object), and computing power (the vcpu object) are available to the hypervisor in terms of total, used, and free capacity. The hypervisor is responsible for using those resources to instantiate particular images of virtualized computational entities (the image object). These computational entities are customized virtual machines or containers, or bundles usually referred to as flavors (the flavor object). As examples of those flavors, the authors of [25] reference bundles such as those proposed by the Common Network Function Virtualisation

254   Digital twins for 6G

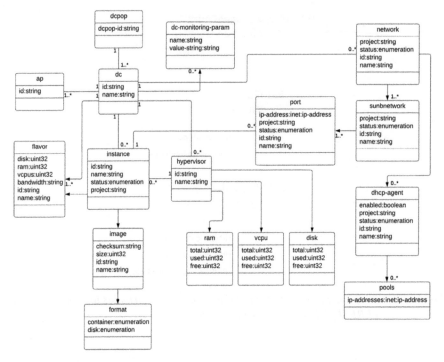

Figure 9.7   DC infrastructure model compiled on the basis of [22]

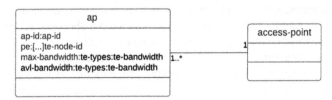

Figure 9.8   Access point model compiled on the basis of [26]

Infrastructure Telecom Taskforce [26]. This model can be extended with additional information referring to the management capabilities of the computing infrastructure, such as hypervisor details or virtualization technologies available. In addition, the network object adds information related to the internal networking details of the datacenter (IP addresses, Intradomain links bandwidth, etc.).

It is worth noting that the ap object binds this model with the transport IP network model mentioned before through the virtual network (VN) model presented in [27]. That VN model defines one or more access points (aPs) that describe the customers' end-point characteristics. Each AP models the termination point of a link between a particular DC and others outside the DC-POP, including its maximum and available bandwidth, as seen in Figure 9.8.

As an example of how this model works, let us suppose an optimization problem is meant to allocate service components or functions in a 6G network. In that context, the objective for a telco operator is to place those service components or functions in its different PoPs to distribute the workload by creating a topology that can satisfy service level objectives (SLOs), such as throughput, delay, and processing capacity. The computing PoPs are modeled here by means of the dcpop object in Figure 9.7. Thus, an optimization algorithm should decide to place each service component of a 6G service in a particular DC (modeled as a dc object). The algorithm must consider that the service components requirements should be satisfied by the available memory, CPU and storage of a DC; those availabilities are stored in the free attribute of the ram object, the vcpu object, and the disk object, respectively. Notice that those resources would be used through VMs or containers taken from the available flavors stored in the model of the DC that will host a particular service component. In order to consider constraints such as throughput and delay, the optimization algorithm should look at other parts of the model, such as the available bandwidth depicted in the avl-bandwidth attribute of the model in Figure 9.8, and at topology elements, such as tunnels modeled following the abstractions presented in [19].

9.3 OaaS platform: use cases

This chapter presents some relevant use cases that can be managed by exploiting the OaaS-NDT architecture. The use cases are presented as types of optimization problems involving different network segments and elements, according to the taxonomy described in Section 9.2. First, in Section 9.3.1, we will present static use cases mostly devoted to network dimensioning; later in Section 9.3.2 dynamic problems will be presented that aim to adapt the network operation on the fly.

It is worth noting that the OaaS-NDT is not tied to the problems presented below, but its architecture would permit addressing a larger catalog of problems and network configurations. In addition, the problems presented in this section could be merged together leading to richer management solutions.

9.3.1 *Dimensioning problems*

9.3.1.1 RAN placement and functional-split configuration

This type of problem deals with the configuration of the fronthaul network. In particular, over a substrate network, we aim to decide the best location of the elements that compose the base station, namely gNB-CU and gNB-DU. The former can be placed in any potential node of the network, provided that there are enough resources. On the other hand, the latter would be placed close to the RU. On top of it, we also decide on the functional split configuration, which in turn imposes bandwidth/delay requirements over the fronthaul network connecting the gNB-CU and gNB-DU. All in all, the problem presented below combines network embedding and resource allocation.

General problem statement

We consider a physical network made of devices and communication paths represented as a non-directed graph $G^p = (\mathcal{V}^p, \mathcal{E}^p)$, where the network nodes correspond to the graph vertices, $\mathcal{V}^p$, and the connections between the nodes correspond to the graph edges, $\mathcal{E}^p$. The set of physical nodes is defined as $\mathcal{V}^p = \mathcal{C}^p \cup \mathcal{D}^p$, where $\mathcal{C}^p$ corresponds to the set of possible gNB-CUs hosting nodes, $\mathcal{D}^p$ is the set of potential gNB-DUs hosting nodes (we assume that they have an association with RU).

A node i belonging to $\mathcal{C}^p$ has computational capacity ω_i^{pc}, or ω_i^{pd} when it belongs to $\mathcal{D}^p$. Furthermore, gNB-DUs are characterized by the location coordinates (x_i, y_i) of the RU, while we can add additional features η_i^{pd} to account for the number of antennas, physical resources, etc., depending on the use case.

Each link $l_{i,j}^p$ between two physical nodes $\{i, j\} \in \mathcal{V}^p$ is characterized by the transmission capacity $\beta_{i,j}^p$, delay $\delta_{i,j}^p$, and a generic cost of use $\gamma_{i,j}^p$. Similarly, a path between two nodes a and b, $P_{a,b}$ is represented by the transmission capacity of the bottleneck link $\beta_{a,b}^p$, aggregated delay $\delta_{a,b}^p$, and generic cost of use $\gamma_{a,b}^p$. We also define the set of all possible paths between nodes a and b, given by $\mathcal{P}_{a,b}^p$.

Over a common physical network, a set of virtual networks, $\mathcal{Q}$, are requested. We use the generic term virtual network to refer to either networks of VNOs or network slices. A request i is represented as a non-directed graph $G_i^r = \{\mathcal{V}_i^r, \mathcal{E}_i^r\} \in \mathcal{Q}$. Here, $\mathcal{V}_i^r = \mathcal{C}_i^r \cup \mathcal{D}_i^r$, where $\mathcal{C}_i^r$ and $\mathcal{D}_i^r$ denote the sets of requested CUs and DUs, of the ith request. Importantly, in each requested graph, the number of gNB-CUs, gNB-DUs, and links connecting them is the same and corresponds to the number of requested base-stations: $|\mathcal{C}_i^r| = |\mathcal{D}_i^r| = |\mathcal{E}_i^r|$. The jth base station of the ith request is defined by the desired location (x_i, y_i) and a location error $\rho_{i,j}^{rd}$, and generic fitting metric η_{ij}^{rd}.

It is worth noting that we do not impose requirements on gNB-CU nodes and links, but they will eventually appear, as a consequence of the split selection. Nevertheless, this model can be extended with constraints over the gNB-CUs, such as the co-location of gNB-CUs to foster coordinated multi-point (CoMP) techniques, imposing the same split level to certain base stations, etc.

For a given virtual request G_i^r, we define the set of candidate gNB-DU hosts that can support $d_{i,j}^r$ as $\Omega_j^i = \{d_k^p \in \mathcal{D}^p \mid \eta_{ij}^{rd} \leq \eta_k^{pd}, \text{dist}(d_j^r, d_k^p) \leq \rho_{ij}^r\}$. Note that all gNB-CU's hosting nodes are candidates to host the CUs. Regarding suitable assignments for a virtual link $l_{i,j}^r$, candidate paths comprise all possible routes that connect suitable physical nodes for $c_{i,j}^r$ and $d_{i,j}^r$, $\Lambda_j^i = \{R_{a,b} \in \mathcal{R}_{a,b}^p \mid a \in \Omega_j^i\}$.

Each requested base station will be configured with a given functional split from a set $\mathcal{S}$ of possible alternatives. The split selected for the jth base station of the ith request, $s_{i,j} \in \mathcal{S}$, is characterized by its transmission capacity β_{ij}^s, maximum admissible delay δ_{ij}^s, and gNB-CU and gNB-DU computational resources, ω_{ij}^{sc} and ω_{ij}^{sd}, respectively.

Altogether, for the ith request G_i^r, an assignment for the jth base station (made of c_{ij}^r, $d_{i,j}^r$ and $l_{i,j}^r$) is defined as a tuple $\alpha_{ij} = (c_{ij} \in \mathcal{C}^p, d_{ij} \in \Omega_{i,j}, l_{ij} \in \Lambda_j^i, s_{ij} \in \mathcal{S})$. It is worth noting that the goal is to find the assignments α_{ij} so that suitable physical nodes, physical paths, and functional split can be configured, according to the resource requirements.

A candidate solution, X, is given by an assignment for all requests in $\mathcal{Q}$, and we define as $\mathcal{F}$ the space of feasible solutions. Our goal is to find the solution x^* that optimizes the following problem:

$$x^* = \arg\max_{X \in \mathcal{F}} f(X) \tag{9.1}$$

where the objective function $f(X)$ can be defined according to the use case. For instance, we can consider the usage of links to foster exploiting some technologies, the degree of centralization of the RAN functions, particular features of each virtual network (or network slice), etc.

The abovementioned feasible solutions will be subject to a set of basic constraints, which could be extended according to the particular use case. We define the set of assignments for the ith request as $\mathcal{A}_i$. Similarly, $\mathcal{A}_{ab}^L$ holds for the set of assignments that use the physical link l_{ab}^p, and $\mathcal{A}_k^c$ and $\mathcal{A}_k^d$ are the sets of assignments that use a physical gNB-CU host c_k^p or gNB-DU host d_k^p, respectively. In any case, a feasible solution will be subject, at least, to the following constraints:

- The required capacity of the splits in an assigned physical link cannot exceed the physical link capacity: $\sum_{\alpha_{ij}} \beta_{ij}^s \leq \beta_{ab}^p \forall l_{ab}^p \in E^A$. Where E^A is the set physical links assigned.
- The delay of a selected path needs to satisfy the delay of the selected split: $\delta_{i,j}^P \leq \delta_{ij}^s \approx \forall \alpha_{ij} \in \mathcal{A}_i, 1 \leq i \leq |\mathcal{Q}|, 1 \leq j \leq |C_i^r|$.
- The aggregated capacity of assignments using a physical gNB-CU host cannot exceed the physical processing capacity: $\sum_{\alpha_{ij} \in \mathcal{A}_k^c} \omega_{ij}^{rc} \leq \omega_k^{pc} \forall 1 \leq k \leq |C^P|$. Similarly, the aggregated capacity of assignments using a gNB-DU host cannot exceed its physical capacity: $\sum_{\alpha_{ij} \in \mathcal{A}_k^d} \omega_{ij}^{sd} \leq \omega_k^{pd} \forall 1 \leq k \leq |\mathcal{D}^P|$.
- Constraint that involves a generic fitting feature in the gNB-DUs (e.g., number of antennas, number of PRBs, etc.): $\sum_{\alpha_{ij} \in \mathcal{A}_k^d} \eta_{ij}^{rd} \leq \eta_k^{pd} \forall 1 \leq k \leq |\mathcal{D}^P|$.

Finally, we need to ensure that all required base-stations are assigned: $|\mathcal{A}_i| = |C_i^r| \forall 1 \leq i \leq |Q|$.

Solving frameworks and approaches

In general, the resulting problems combine network embedding and route selection, which have been proved to be NP-hard. This type of problem has been traditionally addressed by employing branch-and-bound approaches, such as [29], which take advantage of the problem and network structure. In this type of solving approach, potential solutions are connected following a tree structure, where branches represent changes in the variables. Branch-and-bound algorithms travel the tree, discarding branches when the changes in such branches cannot improve the solution. For the described problem, branch-and-bound algorithms can benefit from the functional split structure, since increasing the centralization always imposes higher requirements. Also, the capillarity of the network can help to guide the search.

While branch-and-bound approaches are widely used, their performance would depend on the problem and network structure. More generic solvers can be leveraged using heuristics and meta-heuristics. For instance, in [30], the authors propose both a heuristic greedy algorithm and an evolutionary one to avoid the curse of dimensionality due to the large search space dimension. The results show that both approaches can provide results close to the optimum one, while the greedy approach allows affording the optimization of large networks. As an example, Figure 9.9 shows the average solution quality obtained by the proposed algorithms compared with the

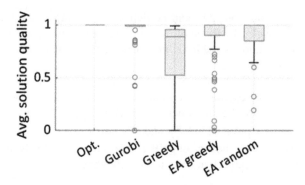

Figure 9.9 Average solution quality of algorithms in Figure 12(a) from [30]

optimum solution, where the quality of the solution is defined as the maximum centralization respecting the network constraints. As can be observed, the solutions based on evolutionary algorithms (EA) provide solutions close to the optimum.

Finally, it is worth noting the recent works that apply graph neural networks (GNNs) to virtual network embedding problems [31,32]. GNNs are a type of neural networks tailored to graph-structured data, which can learn the data features of the nodes and relationship with neighbors, and group together nodes with similar characteristics. Thus, by applying GNNs physical nodes can be clustered, thus reducing the dimensionality of the search space.

Within the considered OaaS architecture, the adopted solver would interact with the NDT that mimics the real network behavior of the RAN segment. Considering that we are dealing with a dimensioning problem, the evaluation of the solving technique over the NDT would follow a Monte Carlo approach. In this sense, the NDT would generate random network instances with certain characteristics according to the use case. For each instance, the network description would be passed to the solver that would return the embedding and functional split decisions. This way, the performance and execution time of solving techniques could be evaluated over the NDT before being applied to a real scenario.

9.3.1.2 Service function placement

The problem addressed in this section is concerned with end-user services, each constituted by a bundle of microservices to be deployed on top of cloud-network slices [29] and characterized by means of four technical dimensions: storage (a) required by each microservice; bandwidth (b) that captures the rate at which the microservice generates or consumes data; computing (c) capacity that is needed by the microservice to be executed, and the maximum delays allowed between pairs of microservices. As the microservices constituting an end-user service can have different requirements in terms of the above-mentioned four parameters, each may have to be deployed on top of different infrastructure elements, supporting a given cloud-network slice.

The modeling and solution approach is generic and not linked to the set of the four service characterization parameters above mentioned. The reason to select these ones is that they play a role in differentiating the main 5G application areas, namely Enhanced Mobile Broadband (eMBB), Ultra Reliable Low Latency Communications (URLLC), and Massive Machine Type Communications (mMTC). At the same time, these parameters can also be used to characterize the hosting infrastructure (both cloud and network infrastructures). In other contexts, this set could be augmented or shrunk as needed.

In that framework, the problem consists of finding the appropriate computational and networking resources in the substrate infrastructure to allocate a set of services fulfilling the requirements of their constituent microservices, and maximizing the profit derived from the allocation.

General problem statement

The problem can be modeled as an assignment problem, which can be stated in terms of a Mathematical Programming problem as follows.

There is a set of requested services $\mathcal{S}$, each of those services, $s_i : i \in 1,\ldots,n$, is decomposed into a set of microservices $\mathcal{F}_i = f_{ij} : j \in 1,\ldots,n_i$ each with requirements for storage s_{ij}, bandwidth b_{ij}, and computation c_{ij}. Each pair of linked microservices in a service, f_{ij} and f_{ik}, has a maximum communication delay requirement d_{ij-ik}. Those microservices are to be deployed in a set of infrastructure elements $\mathcal{R} = r_1,\ldots,r_m$. Each one of these elements has maximum capacities, σ_k, β_k, and k_k, corresponding to storage, bandwidth, and computing resources, with $k = 1,\ldots,m$.

Additionally, those elements are set on a network, and, for each pair of infrastructure elements r_o and r_p, there is a known average delay D_{op}. We call a subset $\widehat{\mathcal{F}} \subseteq \mathcal{F}$ a feasible solution, where $\mathcal{F}$ is the set of all possible deployments of the n services onto the infrastructure of m resources, if the items of $\widehat{\mathcal{F}}$ can be assigned to the infrastructure elements without exceeding their capacities, i.e., if $\widehat{\mathcal{F}}$ can be partitioned into m disjoint sets $\mathcal{F}_k$, such that $s(\mathcal{F}_k) \leq \sigma_k, b(\mathcal{F}_k) \leq \beta_k$, and $c(\mathcal{F}_k) \leq k_k$ with $k = 1, \ldots, m$. The objective is to select a feasible subset $\widehat{\mathcal{F}}$, so that the profit of deploying the n services is maximized. The problem is formulated as follows:

$$\max \sum_{k=1}^{m} \sum_{i=1}^{n} \sum_{j=1}^{n_i} p_{ij} x_{ijk} \qquad (9.2)$$

$$\text{s.t.} \quad \sum_{i=1}^{n} \sum_{j=1}^{n_i} s_j x_{ijk} \leq \sigma_k, \ k = 1,\ldots,m \qquad (9.3)$$

$$\sum_{i=1}^{n} \sum_{j=1}^{n_i} b_j x_{ijk} \leq \beta_k, \ k = 1,\ldots,m \qquad (9.4)$$

$$\sum_{i=1}^{n} \sum_{j=1}^{n_i} c_j x_{ijk} \leq \kappa_k, \ k = 1,\ldots,m \qquad (9.5)$$

$$\sum_{k=1}^{m} x_{ijk} \leq 1, \ i = 1,\ldots n; j = 1,\ldots n_i \qquad (9.6)$$

$$x_{iro}x_{itp}D_{op} \leq d_{ir_it}, i = 1\ldots,n; r,t = 1,\ldots,n_i; o,p = 1,\ldots,m \qquad (9.7)$$

$$x_{ijk} = \begin{cases} 1 & \text{if } f_{ij} \text{ in element } r_k \\ 0 & \text{otherwise} \end{cases} \qquad (9.8)$$

Solving frameworks and approaches

The above problem is NP-hard. Therefore, it is not possible to find a solution making use of conventional solvers, unless the size of the problem is relatively small. More efficient approaches have been envisaged, and one of them is based on flow networks [34].

A flow network [35] is a directed graph $G = (V, E)$ with a source vertex $s \in V$ and a sink vertex $t \in V$, where each edge $(u, v) \in E$ has capacity $c(u, v) > 0$ and cost $a(u, v)$. The cost of sending a flow $f(u, v)$ along an edge (u, v) is: $f(u, v) \cdot a(u, v)$. In the above flow network, the objective is to minimize the total cost through all edges $\sum_{(u,v) \in E} a(u, v) \cdot f(u, v)$, maximizing the ingress flow to the network: $\sum_{w \in V} f(s, w)$.

The rationale behind the use of a flow network to solve an allocation problem is quite straightforward. To simplify the description process, assume there are N monolithic service types, each characterized by a unique parameter requirement, i.e., the required computation, to be mapped to a set of M servers, each characterized by its maximum available computation capacity. The flow network would consist of four steps. The first one containing the source s, the second containing the N service types, the third containing the M servers, and the forth the target t. In addition, assume that we know the number of services of each type to be allocated. The modeling of link capacity would be as follows. The capacity of each link (s, u) would be the total number of services characterized by the node u and the capacity of the link (v, t) would be the available capacity of the server characterized by node v. In addition, the capacity of the links (u, v) would be infinite. Likewise, the modeling of link costs would be as follows. The cost of links (s, u) and (v, t) would be zero. In addition, the cost of links (u, v) would be determined by the specific cost of allocating a service represented by node u on the server represented by node v. The algorithm that computes the minimum cost maximum flow ends up with a solution consisting of integer flows between pairs of nodes u and v, representing how many services of each type are allocated on each server.

Within the considered OaaS architecture, an algorithm that deals with the solution to this problem consists of a modified version of the Successive Shortest Path algorithm (SSP) [27], which is commonly used to compute the minimum cost (maximum profit) flow in a flow network. This algorithm would be properly fed with the network state provided by an online monitoring system and the output would drive the microservices allocation onto the substrate network. The processing may be for an entire batch of service requests or incrementally based on a continuous stream of service requests. In the latter case, the solver should cope with microservice migrations and sub-optimal solutions. Therefore, those solutions should be validated on the NDT before they are effectively applied to the infrastructure.

9.3.2 Operational problems
9.3.2.1 Capacity sharing for RAN slicing

Network slicing is a key feature of the 5G system enabling the creation of multiple end-to-end logical networks, referred to as network slices, on top of the same physical infrastructure, each one optimized to the requirements of specific services and application domains. A network slice includes a 5G core subnet instance and a RAN subnet instance, denoted as RAN slice. The deployment of RAN slices has to deal with the management of the radio resources available in the existing cells in accordance with the requirements of each slice. For this purpose, capacity-sharing mechanisms are needed to dynamically modify the amount of resources allocated to each RAN slice in each cell, ensuring an efficient use of the available radio resources and, at the same time, the fulfillment of the RAN slice requirements. This sub-section addresses the optimization of this capacity-sharing problem.

General problem statement

This problem assumes a RAN infrastructure composed of N cells, each one associated to a radio unit, gNB-DU and gNB-CU, and having different characteristics, e.g., cell radius, transmission power, frequency, etc. The cell n has a total of W_n Physical Resource Blocks (PRBs) with PRB bandwidth B_n depending on the used numerology (i.e., subcarrier spacing). The RAN infrastructure is shared among K tenants by providing each one with a RAN slice instance. Assuming communication in the downlink direction, the SLA of the kth RAN slice is specified through a set of requirements on specific attributes denoted as $SLA_k = SLA_{k,1},\ldots,SLA_{k,M}$. In the most general case, the fulfillment of the mth SLA attribute for tenant k requires that the value of this attribute is within a specific range, generally denoted as $[SLA_{k,m,min}, SLA_{k,m,max}]$.

Let us define the capacity share or resource quota of tenant k in time t as $\sigma_k(t) = [\sigma_{k,1}(t),\ldots,\sigma_{k,n}(t),\ldots,\sigma_{k,N}(t)]$, where each component $\sigma_{k,n}(t)$ corresponds to the proportion of the total PRBs W_n in cell n allocated to the slice of the tenant k in time t. The resource quota $\sigma_{k,n}(t)$ of the kth tenant in the nth cell can be configured through the *rRMPolicyDedicatedRatio* attribute defined in the network resource model of 3GPP TS 28.541, as described previously in Section 9.2.1.3.

Under the above considerations and assuming that the target of the optimization is to allocate the minimum amount of resources to the different slices that allow satisfying the SLA requirements, the optimization problem can be formally defined as:

$$\arg\min_{\sigma_{k,n}(t)} \sum_{n=1}^{N} \sum_{k=1}^{K} \sigma_{k,n}(t) W_n \qquad (9.9)$$

$$\text{s.t.} \quad 0 \leq \sigma_{k,n}(t) \leq 1 \quad \forall n \in \{1,\ldots,N\}, \forall k \in \{1,\ldots,K\} \qquad (9.10)$$

$$\sum_{k=1}^{K} \sigma_{k,n}(t) \leq 1 \quad \forall n \in \{1,\ldots,N\} \qquad (9.11)$$

$$SLA_{k,m,min} \leq SLA_{k,m} \leq SLA_{k,m,max} \quad \forall k \in \{1,\ldots,K\}, \forall m \in \{1,\ldots,M\} \qquad (9.12)$$

Constraint (9.10) ensures that the allocated resource quota values $\sigma_{k,n}(t)$ for any slice and cell are between 0 and 1. In turn, Constraint (9.11) ensures that the aggregate of resource quotas allocated to all the slices of any cell n does not exceed 1, because otherwise the total allocation would be larger than the maximum amount of resources in the cell. Finally, Constraint (9.12) intends to ensure that all the SLA attributes for all the tenants are within their specified bounds.

Solving framework and approaches

The problem of capacity sharing in RAN slicing has been addressed in the literature using different techniques and under different assumptions. For example, different works have considered the problem in single-cell scenarios applying different approaches, such as the exponential smoothing model in [38], the Karush Kuhn Tucker conditions in [39], or market-oriented models in [40,41]. Other works have considered multi-cell scenarios through different heuristic algorithms, such as dynamic programming in [42], integer mathematical programming in [43], or a fisher market game in [44].

Overall, given the complexity and the inherent dynamic uncertainty of the wireless environment, machine learning (ML) tools and, in particular, reinforcement learning (RL) methods are relevant candidates to deal with the capacity-sharing problem, as they allow optimizing complex dynamic decision-making problems depending on a multiplicity of inputs with large dimension. An example of the application of RL for capacity sharing is presented in [45] through a multi-agent reinforcement learning (MARL) algorithmic solution based on deep Q-network (DQN).

The DQN-MARL solution includes one DQN agent per tenant in charge of determining the resource quota values $\sigma_{k,n}(t)$ to be applied for each cell in the RAN slice of the tenant. For this purpose, the DQN agent of a tenant includes a deep neural network that takes as an input a vector with different components of the state (including the fraction of resources of the tenant for each cell, the available resources not used by any tenant in each cell, and the SLA metric) and provides as an output a selected action that determines, for each cell, whether the resource quota $\sigma_{k,n}(t)$ has to be increased by a factor Δ, has to be decreased by a factor Δ, or has to remain unaltered.

The policy for obtaining the actions based on the input state is determined by the weights of the deep neural network and is learned as a result of a training process. In the context of the OaaS architecture, the training process is conducted based on interactions between a DQN agent and the NDT that mimics the behavior of the real network. In each interaction, the DQN agent generates an action that leads to a configuration of the resource quota values in the different cells of the NDT. As a result of this action, the NDT generates a reward that measures how good or bad the action has been in accordance with the desired performance. This reward is used to update the weights of the deep neural network following the procedure of [46]. The interactions between the DQN agent and the NDT and the subsequent policy updates are repeated until reaching a termination condition. Once the training process is completed, the learnt policy can be stored as an algorithm on the OaaS platform. Then, during the ML inference stage, the capacity-sharing function at the management system will

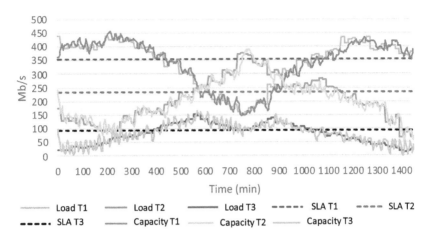

Figure 9.10 Illustration of the operation of the DQN-MARL capacity-sharing solution

apply the learnt policy to dynamically adjust the resource quota values in the cells of the real network.

The reader is referred to [45] for all the specific details of the DQN-MARL capacity-sharing solution. Then, just to illustrate its operation, Figure 9.10 plots an example of the evolution of the offered or demanded load per tenant in one cell and the assigned capacity (i.e., the assigned resource quota $\sigma_{k,n}(t)$ multiplied by the cell capacity) for a scenario with three tenants T1, T2, and T3 during one day. The SLA of each tenant is defined in terms of a certain aggregate data rate to be provided to the tenant as long as it has sufficient demand. The considered cell has a total of $W_n = 73$ PRBs of 360 kHz bandwidth and the cell capacity is approximately 700 Mb/s. The figure shows that since most of the time, there is enough capacity in the cell to fulfill the offered load of the three tenants, the assigned capacity matches the offered load nearly all the day, even when the offered load of a tenant exceeds its SLA (e.g., for tenant T1 between time 0 and 400 min). In turn, during the time when the aggregate offered load of all tenants exceeds the cell capacity, the algorithm limits the capacity of those tenants that demand more than their SLA. For example, this occurs with tenant T2 approximately between times 1000 min and 1100 min, when the offered load of this tenant exceeds the SLA and the algorithm assigns a capacity close to the SLA.

9.3.2.2 Dynamic functional split selection

In this case, we consider a scenario where service verticals are deployed and connectivity is provided by a number of physical base stations by means of network slices. For that, the physical resources of the base stations are distributed between slices. As described in [48], the different centralization options enable different coordination techniques which improve the SINR received by end users.

Having that in mind, we aim to adapt to changes in the services' traffic and wireless conditions by adapting both the centralization level along with the distribution of physical resources among slices. Although the underlying problem formulation presented below is generic, the decision time scale is considered large enough so that the time elapsed for the network reconfiguration is negligible.

General problem statement

We consider a set of base stations $\mathcal{B}$ and service verticals or slices $\mathcal{S}$. Each base station $b \in \mathcal{B}$ has a number of physical wireless resources η_b, and each slice $s \in \mathcal{S}$ has a capacity demand $d_s(t)$, where t is a particular time instant, measured in traffic rate units (typically Mb/s).

For each service vertical $s \in \mathcal{S}$, we can use the Shannon's formula to estimate the experienced average rate in a time instant t, $\rho_s(t)$, as follows:

$$\rho_s(t) = B_s(t) \cdot \log_2\left(1 + \frac{S_s(t)}{\sigma^2 + I_s(t)}\right) \tag{9.13}$$

where $B_s(t)$ is the number of physical resources granted to the slice, while $S_s(t)$ and $I_s(t)$ hold for the average signal and interference experienced by the devices of the service vertical.

Let $\mathcal{F}$ be the set of possible functional splits configurable in the base stations, and $f_b(t)$ the functional split selected for the base station $b \in \mathcal{B}$ at time t. In addition, $\alpha_s^b(t)$ hold for the number of physical resources granted by base station $b \in \mathcal{B}$ to slice $s \in \mathcal{S}$.

It follows that the variables in (9.13) may depend on both random and decision variables. If we consider a generic random variable $\omega(t)$, we can express the variables in that equation as functions as follows: $B_s(t) = \widehat{B}_s(\alpha(t))$; $I_s(t) = \widehat{I}_s(\alpha(t), \omega(t))$; and $S_s(t) = \widehat{S}_s(\omega(t))$.

In the general case, an optimization problem would aim to maximize a generic function $F(\cdot)$ of the rates experienced by the services over time as follows:

$$\max_{f(t),\alpha(t)} F\left(\sum_t \sum_{s \in \mathcal{S}} \rho_s(t)\right) \tag{9.14}$$

s.t. $\quad G(\rho_s(t)) \geq d_s \quad \forall s \in \mathcal{S} \tag{9.15}$

$$\sum_s \alpha_s^b(t) \leq \eta_b \quad \forall b \in \mathcal{B}, \forall t \tag{9.16}$$

$$f_s(t) \in \mathcal{F} \quad \forall s \in \mathcal{S}, \forall t \tag{9.17}$$

where (9.15) ensures that a generic function $G(\cdot)$ of the services' rate satisfies the demand. This generic formulation can be particularized to different use cases. In the case of a strict demand constraint, $G(\cdot)$ becomes the identity function applied to every time instant t. Other use cases can apply averages and expectations of the rate (i.e., time average). Besides, (9.16) and (9.17) ensure that assigned resources do not exceed the amount of physical ones and that the selected split belongs to the set of possible functional splits, respectively.

Solving frameworks and approaches

Although the particular type of problem would depend on the definition of the functions $F(\cdot)$ and $G(\cdot)$, we envisage two main groups of problems. In the first group, the mentioned functions would apply to every time instant independently. In such cases, independent problem instances would be posed every time instant, lying in the realm of integer programming. According to the particular definitions of $G(\cdot)$ and $F(\cdot)$, the corresponding sequence of problems might belong to different types, such as linear, convex or generic ones, and so different techniques should be used to solve them.

In a second group of problems, $G(\cdot)$ and $F(\cdot)$ would consider the performance over time, for instance, time averages. In this case, the problem would correspond to a stochastic program, which can be addressed by control techniques, such as those rooted in Lyapunov theory. In this case, distributed implementations could help to reduce the burden to the solver.

In either case, within the OaaS framework, the solver would interact with the NDT in a time-based fashion. In this sense, the NDT would periodically monitor the network and service state and send such scenario status to the solver. Then, the solver's response would be used to update the network configuration.

Acknowledgments

This work is funded by the Spanish Ministry of Economic Affairs and Digital Transformation and the European Union – NextGenerationEU under projects OPTIMAIX_OaaS (Ref. TSI-063000-2021-34) and OPTIMAIX_NDT (Ref. TSI-063000-2021-35).

References

[1] A. Fuller, Z. Fan, C. Day, and C. Barlow, "Digital twin: enabling technologies, challenges and open research," *IEEE Access*, vol. 8, pp. 108952–108971, 2020, doi: 10.1109/ACCESS.2020.2998358.

[2] Y. Wu, K. Zhang, and Y. Zhang, "Digital twin networks: a survey," *IEEE Internet of Things Journal*, vol. 8, no. 18, pp. 13789–13804, 2021, doi: 10.1109/JIOT.2021.3079510.

[3] C. Alcaraz and J. Lopez, "Digital twin: a comprehensive survey of security threats," *IEEE Communications Surveys & Tutorials*, vol. 24, no. 3, pp. 1475–1503, 2022, doi: 10.1109/COMST.2022.3171465.

[4] R. Minerva, G. M. Lee, and N. Crespi, "Digital twin in the IoT context: a survey on technical features, scenarios, and architectural models," *Proceedings of the IEEE*, vol. 108, no. 10, pp. 1785–1824, 2020, doi: 10.1109/JPROC.2020.2998530.

[5] S. Mihai, M. Yaqoob, D. V. Huang, *et al.*, "Digital twins: a survey on enabling technologies, challenges, trends and future prospects," *IEEE Communications Surveys & Tutorials*, vol. 24, no. 4, pp. 2255–2291, 2022, doi: 10.1109/COMST.2022.3208773.

[6] B. R. Barricelli, E. Casiraghi, and D. Fogli, "A survey on digital twin: definitions, characteristics, applications, and design implications," *IEEE Access*, vol. 7, pp. 167653–167671, 2019, doi: 10.1109/ACCESS.2019.2953499.

[7] A. Mozo, A. Karamchandani, S. Gómez-Canaval, M. Sanz, J. J. Moreno, and A. Pastor, "B5GEMINI: AI-driven network digital twin," *Sensors*, vol. 22, pp. 4106, 2022. https://doi.org/10.3390/s22114106

[8] R. Ramirez, C.-Y. Huang, and S.-H. Liang, "5G digital twin: a study of enabling technologies," *Appllied Science*, vol. 12, pp. 7794, 2022. https://doi.org/10.3390/app12157794

[9] P. Almasan, M. Ferriol-Galmés, J. Paillisse, *et al.*, "Network digital twin: context, enabling technologies, and opportunities," *IEEE Communications Magazine*, vol. 60, no. 11, pp. 22–27, 2022, doi: 10.1109/MCOM.001.2200012.

[10] OPTIMAIX project, https://optimaix.upc.edu/.

[11] M. Garrich, C. San Nicolas Martinez, F. Javier Moreno-Muro, M. V. Bueno Delgado, and P. Pavon Marino, "Network optimization as a service with Net2Plan," European Conference on Networks and Communications 2019 (EuCNC'19).

[12] M. Garrich, J. L. Romero-Gázquez, F. J. Moreno-Muro, M. Hernandez-Bastida, and P. Pavon Marino, "Joint optimization of IT, IP and WDM layers: from theory to practice," in *International Conference on Optical Network Design and Modeling 2020 (ONDM'20)*.

[13] OpenAPI Initiative (Online). https://www.openapis.org.

[14] 3GPP TS 28.541 v18.1.2, "Management and orchestration; 5G Network Resource Model (NRM); Stage 2 and stage 3 (Release 18)," September, 2022.

[15] M. Bjorklund (Ed.), "The YANG 1.1 Data Modeling Language," IETF RFC 7950, August 2016.

[16] A. Bierman, M. Bjorklund, and K. Watsen, "RESTCONF Protocol," IETF RFC 8040, January 2017.

[17] Telecom Infra Project, "Open Transport SDN Architecture Whitepaper," 2021.

[18] A. Clemm, J. Medved, R. Varga, R. Bahadur, N. Ananthakrishnan, and X. Liu, "A YANG data model for network topologies," *IETF RFC*, vol. 8345, 2018.

[19] X. Liu, I. Bryskin, V. Beeram, T. Saad, H. Shah, and O. Gonzalez de Dios, "YANG data model for Traffic Engineering (TE) Topologies," *IETF RFC*, 87952020.

[20] J. Dong, X. Wei, Q. Wu, M. Boucadair, and A. Liu, "A YANG data model for layer 2 network topologies," *IETF RFC*, vol. 8944, 2020.

[21] A. Clemm, J. Medved, R. Varga, X. Liu, H. Ananthakrishnan, and N. Bahadur, "A YANG data model for layer 3 topologies," *IETF RFC*, vol. 8346, 2018.

[22] O. Gonzalez de Dios, S. Barguil, Q. Wu, and M. Boucadair, "A YANG model for User-Network Interface (UNI) Topologies," 2020. https://datatracker.ietf.org/doc/draft-ogondio-opsawg-uni-topology/01/.

[23] 3GPP TS 23 501 v17.6.0. "System architecture for the 5G System (5GS); Stage 2 (Release 17)," September, 2022.

[24] GSMA "Generic Network Slice Template Version 7.0," Official Document NG.116, June, 2022.
[25] Y. Lee, X. Liu, L. and M. Contreras, "DC aware TE topology model," 2020. https://datatracker.ietf.org/doc/draft-llc-teas-dc-aware-topo-model/02/.
[26] "Cloud iNfrastructure Telco Taskforce Reference Model, Reference Architectures," https://cntt.readthedocs.io/en/stable-elbrus/ref_arch/README.html.
[27] Y. Lee, D. Dhody, D. Ceccarelli, I. Bryskin, and B. Y. Yoon, "A YANG data model for virtual network (VN) operations," 2022 (Issue draft-ietf-teas-actn-vn-yang-16). https://datatracker.ietf.org/doc/draft-ietf-teas-actn-vn-yang/16/.
[28] F. J. Moreno-Muro, M. Garrich, I. Iglesias-Castreno, S. Zahir, and P. Pavon Marino, "Emulating software-defined disaggregated optical networks in a containerized framework," *Applied Sciences*, vol. 11, no. 5, p. 2081, 2021, doi: 10.3390/app11052081.
[29] A. Garcia-Saavedra, J. X. Salvat, X. Li, and X. Costa-Perez, "WizHaul: on the centralization degree of cloud RAN next generation fronthaul," *IEEE Transactions on Mobile Computing*, vol. 17, no. 10, pp. 2452–2466, 2018, doi: 10.1109/TMC.2018.2793859.
[30] C. C. Erazo-Agredo, M. Garza-Fabre, R. A. Calvo, L. Diez, J. Serrat, and J. Rubio-Loyola, "Joint route selection and split level management for 5G C-RAN," *IEEE Transactions on Network and Service Management*, vol. 18, no. 4, pp. 4616–4638, 2021, doi: 10.1109/TNSM.2021.3091543.
[31] A. Rkhami, T. A. Quang Pham, Y. Hadjadj-Aoul, A. Outtagarts, and G. Rubino, "On the use of graph neural networks for virtual network embedding," in *2020 International Symposium on Networks, Computers and Communications (ISNCC)*, 2020, pp. 1–6, doi: 10.1109/ISNCC49221.2020.9297270.
[32] F. Habibi, M. Dolati, A. Khonsari, and M. Ghaderi, "Accelerating virtual network embedding with graph neural networks," in *2020 16th International Conference on Network and Service Management (CNSM)*, 2020, pp. 1–9, doi: 10.23919/CNSM50824.2020.9269128.
[33] "Novel Enablers for Cloud Slicing – NECOS Project." http://www.h2020-necos.eu.
[34] M. C. Luizelli, D. Raz, and Y. Sa'ar, "Optimizing NFV chain deployment through minimizing the cost of virtual switching," in *IEEE INFOCOM 2018 – IEEE Conference on Computer Communications*, 2018, pp. 2150–2158.
[35] D. R. Ford and D. R. Fulkerson, *Flows in Networks*. Princeton, NJ: Princeton University Press, 2010.
[36] R. Ahuja, T. Magnanti, and J. Orlin, *Network Flows: Theory, Algorithms and Applications*. Englewood Cliffs, NJ: Prentice-Hall, 1993.
[37] J. Baliosian, L. M. Contreras, P. Martinez-Julia, and J. Serrat. "An efficient algorithm for fast service edge selection in cloud-based Telco networks," *IEEE Communications Magazine*, vol. 59, no. 10, pp. 34–40, 2021. doi: 10.1109/ MCOM.111.2001195.
[38] A. S. D. Alfoudi, S. H. S. Newaz, A. Otebolaku, G. M. Lee, and R. Pereira, "An efficient resource management mechanism for network slicing in a LTE network," *IEEE Access*, vol. 7, pp. 89441–89457, 2019.

[39] D. Marabissi and R. Fantacci, "Highly flexible RAN slicing approach to manage isolation, priority, efficiency," *IEEE Access*, vol. 7, pp. 97130–97142, 2019.

[40] J. Pérez-Romero, O. Sallent, R. Ferrús, and R. Agustí, "Profit-based radio access network slicing for multi-tenant 5G networks," in *2019 European Conference on Networks and Communications (EuCNC)*, Valencia, Spain, 2019, pp. 603–608.

[41] Ö. U. Akgül, I. Malanchini, and A. Capone, "Dynamic resource trading in sliced mobile networks," *IEEE Transactions on Network and Service Management*, vol. 16, no. 1, pp. 220–233, 2019.

[42] J. Shi, H. Tian, S. Fan, P. Zhao, and K. Zhao, "Hierarchical auction and dynamic programming based resource allocation (HA&DP-RA) algorithm for 5G RAN slicing," in *2018 24th Asia-Pacific Conference on Communications (APCC)*, Ningbo, China, 2018, pp. 207–212.

[43] J. Gang and V. Friderikos, "Optimal resource sharing in multi-tenant 5G networks," in *2018 IEEE Wireless Communications and Networking Conference (WCNC)*, Barcelona, 2018, pp. 1–6.

[44] P. Caballero, A. Banchs, G. De Veciana, and X. Costa-Pérez, "Network slicing games: enabling customization in multi-tenant mobile networks," *IEEE/ACM Transactions on Networking*, vol. 27, no. 2, pp. 662–675, 2019.

[45] I. Vilà, J. Pérez-Romero, O. Sallent, and A. Umbert, "A multi-agent reinforcement learning approach for capacity sharing in multi-tenant scenarios," *IEEE Transactions on Vehicular Technology*, vol. 70, no. 9, pp. 9450–9465, 2021.

[46] V. Mnih, K. Kavukcuoglu, D. Silver, *et al.*, "Human-level control through deep reinforcement learning," *Nature*, vol. 518, no. 7540, pp. 529–533, 2015.

[47] S. J. Pan and Q. Yang, "A survey on transfer learning," *IEEE Transactions on Knowledge and Data Engineering*, vol. 22, no. 10, pp. 1345–1359, 2010, doi: 10.1109/TKDE.2009.191.

[48] A. M. Alba, S. Janardhanan, and W. Kellerer, "Enabling dynamically centralized RAN architectures in 5G and beyond," *IEEE Transactions on Network and Service Management*, vol. 18, no. 3, pp. 3509–3526, 2021, doi: 10.1109/TNSM.2021.3071975.

Chapter 10
Robotics digital twin for 6G

Milan Groshev[1], Carlos Guimarães[2] and
Antonio de la Oliva[1]

10.1 Introduction

Over the last two centuries, several industrial revolutions have been witnessed. The first industrial revolution introduced the use of steam power and the mechanization of production, marking the transition from handmade production methods to machines, machines started to play a paramount role in many industrial environments. The first industrial revolution was followed by the second and third revolutions which, respectively, introduced the use of electricity and the mass production of goods, and incorporated the use of computers and automation in industrial processes. The fourth industrial revolution, commonly referred to as Industry 4.0, is currently underway and puts together Artificial Intelligence (AI), robotics, and the Internet of Things (IoT). Notwithstanding, the next revolution has already begun, focusing on how to bring back the human presence to achieve human–robot interaction, collaboration, and long-term sustainability within industrial environments.

10.2 The Shift from Industry 4.0 to Industry 5.0

Industry 4.0 offers a new way of understanding and organizing distinct industrial processes, pushed by the increasing interest to accomplish a digital transformation of many existing industries. It focuses on smart interconnections between humans, machinery, and products to build an efficient, agile, and more autonomous industrial environment. To achieve this vision, flexible factories, intelligent production lines, and automation of processes appeared as key features to fulfill the increased demands from customers [2,3]. Although Industry 4.0 is still underway and has not yet been implemented in a significant percentage of industrial setups, the promise of complete process automation and cost savings has clearly captured the industry's attention. Also contributing to this was the fact that the information and communication technologies (ICTs), such as cloud computing, Internet of Things (IoT), artificial

[1]Telematics Department, Universidad Carlos III de Madrid, Spain
[2]ZettaScale Technology, Saint-Aubin, France

intelligence and machine learning (AI/ML), and Big data, that form the backbone of Industry 4.0, are becoming technologically mature and seeing widespread and growing deployments.

Even so, the research community is already looking for the next industrial revolution. Industry 5.0 is currently envisioned to bring back the human presence on the factory floor [4,5], where the unique creativity of human experts will be used in collaboration with powerful, smart, and extremely accurate robots. Such collaboration will enable a new conceptual shift in the industrial processes—from mass customization to mass personalization—allowing highly personalized and unique products to be easily produced in mass. Humans are given tasks that require critical and creative thinking for the personalization aspects, while robots are given the monotonous and repetitive tasks required for mass production. In other words, humans will increase the creativity, diversity, and personalization of production, while robots ensure the high quality and precision of manufactured products. Industry 5.0 will continue to significantly increase industrial efficiency and sustainability, and create a symbiosis between humans and machines, enabling responsibility for interaction and constant monitoring activities.

Compared to Industry 4.0, Industry 5.0 will shift from mass customization toward mass personalization, and humans will be involved in more skilled and creative jobs that require tight collaboration and cooperation with robots. While Industry 4.0 focuses on the creation of Cyber-Physical Systems (CPSs) and bringing together the IT and OT worlds, Industry 5.0 will enhance industrial processes and applications with real-time, accurate, and predictive operations. To do so, Industry 5.0 will leverage on predictive analytics and operational intelligence to create autonomous robotic systems that can make extremely accurate decisions in real-time, a paramount aspect to ensure the safety of the humans on the factory floor at all times [4]. In addition, Industry 5.0 aims at adopting greener and more sustainable solutions within the industrial environment, going beyond the limited approaches of Industry 4.0.

10.3 Digital Twin as the Pillar of Industry 5.0

Industry 5.0 has the potential to provide revolutionary solutions for the factories of the future not only by exploiting the best of all the previous revolutions but also by effectively supporting emerging mission-critical and real-time robotic systems. Autonomous robotic systems are serving the manufacturing industry for more than 40 years focusing on applications like spot and arc welding, painting, material handling, and pick and place of objects. These industrial systems have always been controlled and/or monitored by specialized human workers where, due to strict industrial safety policies, the robot and human workspaces were mostly separated.

However, to effectively enable an efficient and optimized collaboration between human workers and robots, these physical or virtual barriers must be eliminated [6]. Having this in mind, the research community is centering efforts on proposing and providing robotic services to support industrial activities through human–robot interactions [6]. Their goal is to develop mission-critical and real-time robotic systems that can behave in natural, predictable, and trustworthy ways. Only by doing so, human-workers can intuitively and safely work with robots in close proximity. Popular applications of collaborative robots (also known as *cobots*) include robotic arms that support human workers in handling heavy parts when loading machines or picking components out of bins, or cognitive robots for knowledge-intensive industrial work [7]. Although the applicability of cobots on the industrial floor remains limited today, their ability to perform daily industrial tasks is rapidly improving and opening new pathways for the future.

Digital twin concepts regain importance as the key pillar to support these novel and collaborative applications. Digital twins are nowadays enabling a real-time, remote, and coordinated control and monitoring of all the industrial processes, including robots, as a way to optimize the entire production chain [8]. In its essence, digital twin aims at bridging both physical and virtual worlds through the integration of physical processes, computing, and communications technologies [9]. It defines a *"a virtual representation of a physical asset enabled through data and simulators for real-time prediction, optimization, monitoring, controlling, and improved decision-making"* [10]. This concept truly embodies the integration of CPS within Industry 5.0, combining any industrial process achieved through closed-loop feedback mechanisms, thus allowing virtual objects in a virtual world to replicate the behavior of the physical devices in the physical world (or vice versa) through connected data [11].

Nevertheless, the adaptation of digital twin in Industry 5.0, in what can be referred to as a full-fledged end-to-end (E2E) digital twin system, is inseparable from recent advances in ICT, such as next-generation networks, computational technologies, and cloud-Native approaches, and CPSs and networked robots.

10.4 ICT technologies and adaptation for Industry 5.0

Industry 5.0 will heavily rely on the integration of advanced information and communications technology (ICT) technologies in order to enable its vision and to optimize its efficiency and performance. The adoption and implementation of these technologies may require significant adjustments and adaptations in order to fully integrate them into the digital twin processes. The following is a list of key enabling technologies that will support Industry 5.0 and help in achieving an E2E digital twin system:

NextG wireless networks (NGWNs): Wireless technologies provide a number of benefits for industrial environments, such as greater flexibility for connecting physical devices, reductions of installation and maintenance costs, support for mobility, and improved safety for the employees [12]. With the deployment of the fifth generation of mobile networks (5G), Industry 4.0 recognized the opportunity for a unified communication interface that can support the stringent requirements of real-time industrial applications. In parallel, a broad spectrum of recent research focuses on defining the next-generation mobile networks or 6G systems that envision close integration of artificial intelligence into every aspect of networking systems to support hyper-flexible intelligent E2E network architecture [13–15]. Moreover, the 6G systems will feature the integration of sensing and communication functions that will provide high-resolution network sensing capabilities, high-accuracy localization and tracking, imaging, and environment reconstruction capabilities to improve communication performance and, at the same time, set up the data foundation for building an intelligent digital world [16]. Primary features of the 6G networks include (i) AI-native service-based architecture (SBA) to provide hybrid 6G services; (ii) integrated sensing and communication functions at different levels; (iii) sub-ms data-transportation and computation networks; and (iv) support for reliable collaborative human-centric systems. In this regard, computational technologies (CT) integration will play a big part in realizing the 6G vision.

Computational technologies (CT): One of the main technologies behind Industry 4.0 is cloud computing [17]. It offers cloud computing resources to expand the industrial systems and to increase their accessibility, reconfigurability as well as to increase the visibility of the data that is generated on the factory floor (the so-called *data lakes*). Similar to cloud computing, edge computing [18] and fog computing [19] are envisioned as the main computational technologies behind Industry 5.0. These technologies will extend the computing resources of industrial systems by placing them near, or even beyond, the edges of the network. While the edge computing paradigm covers the far edge or edge-cloud concept where the static computation resources (e.g., servers) with pre-defined locations (e.g., edge data centers) are pushed deeper in the network infrastructure, the fog computing paradigm covers the near edge concept where volatile, mobile and constrained devices (e.g., access points, robots) that are in the factory floor are integrated into the enhanced computing infrastructure. Thus, the edge computing and fog computing paradigms extend cloud computing to form the so-called cloud-to-robot continuum (hereinafter referred to only as *continuum*) making available a truly distributed, decentralized, and holistic computational infrastructure. Different industrial services, such as anomaly detection, fault prediction, remote control, or robot functionalities, can be distributed anywhere in the continuum. Moreover, real-time context information, like connectivity metrics, is expected to be available at the edge of the network, enabling a dynamic adaptation of the applications' logic to the actual conditions of the communication (e.g., radio channel) [20]. The robots interact with the network and computing infrastructure to realize an integrated E2E system that provides seamless support in daily industrial activities using the available resources on demand.

cloud-Native NextG wireless networks: NGWNs are adopting cloud-native guidelines for their deployment, management, and control processes, thus becoming inseparable from software-defined networking (SDN) and network function virtualization (NFV) enabled communication networks and implemented as a decentralized and distributed microservice architecture. By doing so, NGWNs are empowered with new capabilities that can sustain the needs of Industry 5.0 vision. In particular, this integration will enable automated and intelligent network resource management and operations to deploy dynamic network functions and industrial services, which provides a high degree of flexibility in numerous demanding industrial use cases. The concept of cloud-native NGWNs will include the convergence of (i) AI-native service-based architecture; (ii) network and industrial applications virtualization; (iii) federation and orchestration of services; (iv) E2E digital twin system implementation for human-robot interactions; and (v) networked robotics.

CPS and networked robots: The rapid advancements in ICTs are transforming the industrial sector toward a full digitalization and integration concept. This industrial transformation enhances industrial systems with the ability to make decentralized and autonomous decisions through the use of CPSs. Consequently, the industrial world can improve productivity, logistics, and lower production costs [21]. CPSs are the main linchpin for the industrial world to move toward a fully automated industrial infrastructure that relies on real-time capabilities, distributed control systems, virtualization, service orientation, and modularity [22]. The concept of a networked robot truly embodies the cyber-physical integration where a networked robot is defined as a *robotic device connected to a wireless or wired communications network (e.g., Internet or LAN) and based on the available protocols such as TCP, UDP, or IEEE 802.11. There are two subclasses in networked robots: (i) Tele-operated, where the human operator sends commands and receives feedback via the network; and (ii) autonomous, where robots and sensors exchange data via the network* [23]. Digital twins concepts are a true embodiment of the CPS and networked robots by combining the operation and monitoring of any industrial process or robotic system through closed-loop feedback mechanisms, thus paving the way for more advanced and collaborative applications or the integration of AI/ML mechanisms within such processes or systems [1].

10.5 Unified role of Industrial E2E digital twin systems in Industry 5.0

An E2E digital twin system for robotic systems is an emerging concept that integrates all of the aforementioned information and communication technologies to create a highly consistent, accurate, and synchronized virtual representation of its physical counterpart, anywhere, anytime, and in any conditions. However, an E2E digital twin system is commonly mistaken for a digital twin application. The latter, which is already tackled by Industry 4.0, only refers to the virtual representation of a physical robot that is used for real-time prediction, optimization, monitoring, controlling, and improved decision-making. In turn, the E2E digital twin system, which is being pushed by Industry 5.0, continuously orchestrates, manages, and controls the complete system that includes the physical robot, its virtual replica but also the available computation and communication infrastructures throughout its entire lifecycle. Moreover, since Industry 5.0 brings several opportunities to collaborative robots, Industrial E2E digital twin systems will not comprise a single robotic twin but an entire industrial environment twin, unifying its view on robots, human-personal, and all the existing processes.

Essentially, the network infrastructure and the digital twin application are split into different virtual functions and they can be distributed anywhere in the continuum (i.e., fog, edge, or cloud). Industrial E2E digital twin systems can then be implemented at various levels of the layered communication pyramid, which resources or services will be orchestrated and federated according to their computational (i.e., computing, storage, and communication) requirements and available resources. Moreover, the cyber-space mirrored through the Industrial E2E digital twin system arises as the perfect playground for the development of AI agents [24] and new sensing capabilities. Digital twin systems already provide the tools for transferring the domain expertise of specialized personnel in raw data in the cyber-space, which can be later used to train and cross-validate different ML algorithms used in AI agents, as an alternative to heuristic or decision-tree-based solutions. These AI agents will not only develop expertise in specific tasks but also extend and optimize it beyond human capability due to the volume of data they can handle to make decisions. Ultimately, safer, smarter, and extremely accurate digital twins can be devised where autonomy is achieved through AI-controlled processes that operate in all types of environments and conditions, including in the presence of humans nearby the physical robots.

The unified view of the Industrial E2E digital twin system together with the opportunity for smarter and autonomous robots and new sensing capabilities are paving the way for human–robot symbiosis envisioned by Industry 5.0. The implementation of these systems involves the use of various critical enablers, such as NGWNs, computational technologies, AI/ML, and CPS and Networked Robots, to ensure that industrial large amounts of data are collected, processed, and analyzed in real-time. This allows for greater degrees of visibility and control over industrial processes and systems, leading to improved efficiency, productivity, and overall performance, while ensuring a symbiosis between human workers and running robots due to the increased safety for both to share the same coworking space.

10.6 Fundamentals and challenges of digital twins for robotic systems

In recent years, the use of digital twins has gained significant attention in various industrial sectors, to the point of being placed as a key component in manufacturing processes and robotic applications. In what concerns its use in the context of CPS, digital twins have a broad range of applications, from design, simulation, and testing to performance optimization, monitoring, and predictive maintenance. However, the implementation of digital twin technology in the industrial setting is not achieved without its challenges. This section will provide an in-depth analysis of the fundamentals and challenges of digital twins for robotics.

10.7 Digital twins in real industrial environments

The world of manufacturing is always looking for a way to improve time to market and save money. It is also an area that has many sub-fields, from large factories to small machines and sensors. For this reason, the majority of existing research studies revolve around the main applications for digital twins in different manufacturing sectors, arguing on the importance of emerging technologies for the development of smart manufacturing [11,25–27]. Digital twins are applied in the automotive industry to monitor the production of automobiles, making punctual modifications, and optimize processes [28]. Moreover, they facilitate the adjustment of the industrial production operations through autonomous factory planning [29,30] as well as the lifecycle management of the production phase in small- and medium-sized enterprises (SME) [31–33]. Lastly, digital twins also serve as a simulation-based environment that helps realize complex control algorithms for real-time production control [34].

Industrial cloud [35] and edge computing [36] are known concepts in the manufacturing world. They enable computation and control to be offloaded to a computing infrastructure, making digital twins more scalable, real-time and ensuring accessibility to physical devices anytime and from anywhere. Currently, cloud-based digital twin frameworks for industrial robotics [37], manufacturing execution systems [38,39], and smart product-service systems [40] have been developed to provide an insight into smart manufacturing. edge-based digital twins have also been studied in the literature with their benefits for manufacturing such as anomaly detection for automation systems [41] and improving production quality, efficiency, and costs for metal additive manufacturing system [42].

Another key driving feature of digital twins in the automation and manufacturing industries is industrial wireless networks. The potential advantages of wireless technologies are significant and they have been studied through papers that elaborate on the importance of the industrial wireless [12] and 5G environment [43–45] for smart manufacturing. Besides conceptual works, some recent study investigates Digital Twin solutions over IoT communication technologies (LTE-M, LoraWan, and Sigfox) for smart manufacturing assembly systems [46]. A digital twin prototype for mission-critical application over 4G was developed to identify the main obstacles and cyber-security issues in the realization of an Industry 4.0 vision [47].

10.8 From digital twins in Industry 4.0 to Industrial E2E digital twin systems in Industry 5.0

Industry 4.0 is increasingly adopting digital twin concepts as a way to integrate any industrial process achieved through closed-loop feedback mechanisms. To do so, virtual factories are conceived which include geometrical and virtual models of tools, machines, operatives, finished products, etc., as well as behaviors, rules, physics, and analytic models. As such, based on their functionality, the digital twin can be classified into [48]: (i) monitoring, (ii) simulation, and (iii) operational digital twins. While the monitoring digital twin mostly gathers data about the operational states or behaviors of the physical device (e.g., production line dashboard), the simulation digital twin contains 3D models, simulation tools, and machine-learning models to describe, understand, and predict the future behavior of the physical device. Finally, the operational digital twin enables the human worker to interact with the physical device and execute different recommended actions (e.g., remote control) to enhance the industrial process.

Several standardization bodies and alliances, such as ETSI, 3GPP, 5G Alliance for Connected Industries and Automation (5G ACIA), and Next-Generation Mobile

Table 10.1 Connectivity requirements of digital twin

Digital twin	Latency	Data rate	Reliability	Scalability
Monitoring	50–100 ms	0.1–0.5 Mbps	99.9%	100–1000 nodes
Simulation	20–50 ms	1–1000 Mbps	99.99%	1–100 nodes
I4.0 operation	0.5–20 ms	1–100 Mbps	99.9999%	1–50 nodes
I5.0 operation	0.1–0.5 ms	1–10 Gbps	99.999999%	1–500 nodes

Networks Alliance (NGMN), defined new use cases for such vision in Industry 4.0 [49–52]. For example, use cases where every industrial process or sensor is continuously monitoring and their information pushed to the cloud (i.e., monitoring digital twin); use cases focusing on factory remote maintenance (i.e., simulation digital twins); or use cases that include remote operation, motion control, safe control, and closed-loop control (i.e., operational digital twins). Each one of them has distinct connectivity requirements that mainly depend on how objects, processes, systems, and humans interact with the digital twin system [48].

Although the Industrial E2E digital twin systems will comprise any of the three types of digital twins, Industry 5.0 is expected to take operational digital twins to even greater demands and stringent requirements. When robots are operating together or in the proximity of human workers, the safety of the latter is of vital importance. Thus, E2E digital twin systems in industrial environments must operate over the same assumptions as mission-critical systems where the failure or misoperation of any of the E2E components can result in significant harm to people or even the environment. Because of the high level of importance and potential consequences associated with a particular situation or task, robots must meet strict real-time and availability requirements to ensure that they can operate safely and reliably by means of their digital twin counterpart, either by a human worker or an AI agent. Moreover, they might require to perform time-critical tasks, on which controlling commands must be completed within a specific time frame in order to ensure its safety and success. Other aspects cannot be overlooked, like stringent reliability and scalability requirements, that acquire additional importance when put into the scope of an E2E digital twin system. Finally, the Industrial E2E digital twin systems must operate continuously and without fail for long periods of time, while being able to quickly recover from any failures.

Table 10.1 summarizes the most critical connectivity requirements for Industrial E2E digital twin systems in comparison to those from Industry 4.0.

10.9 The infrastructure behind industrial digital twins

The infrastructure to support industrial digital twins systems is responsible for extending the capabilities of the robotic systems by providing access to *computing*, *storage*, and *connectivity*. In doing so, industrial digital twins can be expanded with "unlimited" resources that are not available locally in the devices themselves, while making them "virtually" available in any geographic location. If the former is relevant for e.g. high performance and scalability reasons, the latter might prove to be beneficial in what concerns data protection and privacy.

10.9.1 Computing and storage

Currently, cloud-based digital twin frameworks for industrial robotics [37], manufacturing execution systems [38,39], and smart product-service systems [40] have been developed to provide an insight into smart manufacturing. edge-based digital twins have also been studied in the literature with their benefits for manufacturing such as anomaly detection for automation systems [41] and improving production quality, efficiency, and costs for metal additive manufacturing system [42]. The choice between using the cloud or the edge to support the digital twins is mostly due to the connectivity requirements highlighted in the previous sections, or due to the amount of data and its privacy.

For example, edge-based digital twins might prove to be more suitable to guarantee the lower latency of the *Operation Digital Twins*, while cloud-based digital twins might prove to cope better with monitoring digital twins and the subsequent application of AI/ML mechanisms for, e.g. predictive maintenance decisions. Nonetheless, each digital twin deployment must be assessed on a case-by-case since the available infrastructure and computing resources are highly dependent on the geographic location.

10.9.2 Connectivity

Industrial wireless networks are another key driving feature of industrial digital twins systems in terms of automation and flexibility of the manufacturing industries. The potential advantages of wireless technologies are significant and they have been studied through papers that elaborate on the importance of industrial wireless [12], 5G environment [43–45], WiFi 6E [53], and low power wide area (LPWA) networks for smart manufacturing.

5G brings powerful mobile connections on several frequency bands to industrial digital twins, providing benefits not only in terms of achievable throughput, low latency, device scalability, reliability, and availability but also in terms of an interference-free operation due to the licensed spectrum. In addition, 5G enables the deployment of industrial 5G network private, while still interconnected to the public 5G network [54]. In some specific cases, 4G technologies might also be adopted as a complemented technology. For example, a digital twin prototype for mission-critical application over 4G was developed to identify the main obstacles and cyber-security issues in the realization of an Industry 4.0 vision [47].

WiFi 6E besides the two commonly used 2.4 GHz and 5 GHz bands also introduces up to 1.2 GHz of Wi-Fi spectrum available in the 6 GHz band. With WiFi 6E, devices can achieve higher throughput, lower latency, and higher density of connected nodes, while at the same time creating a powerful combination with 5G systems [55]. However, industrial digital twins research did not yet cover such a combination and its clear benefits for their performance and efficiency.

Besides conceptual works, smart manufacturing assembly systems have been adopting LPWA connectivity on digital twin solutions [46]. Their application is mostly limited to Internet of Things (IoT) applications and monitoring-related tasks in industrial digital twins where node scalability and low-power consumption are the main requirements.

10.10 Enablers for industrial digital twins

Industrial digital twin solutions in manufacturing need to integrate network virtualization, computing, and wireless technologies so that the most stringent requirements of real-time applications are met. Moreover, the adoption of such technologies is changing the way solution providers build digital twin systems and the way users interact with them, by following cloud-native [56,57] and edge-native [58] oriented approaches.

10.10.1 Cloud-to-robot continuum for digital twins

The underlying network infrastructure of the digital twin comprises different dynamic and heterogeneous typologies. It can be divided into three segments, as shown in Figure 10.1: (i) Aggregation Ring; (ii) Access Ring; and (iii) (Radio) Access Network ((R)AN). The Aggregation Ring resides far from the physical objects, relying on wired connectivity to connect cloud-based digital twins that are suitable for human-scale responsive services and delay-tolerant tasks (e.g., monitoring). The Access Rings go closer to the physical objects, interconnecting multiple (R)ANs. The Access Rings are locally present and expose radio network information (e.g., radio channel) to edge-based Digital Twins, namely for time-sensitive tasks (e.g., remote manipulation). Finally, the (R)AN is in the vicinity of the factory floor, providing connection to the physical objects using both wired and wireless connectivity. Different Radio Access Technologies (RATs) are available (e.g., WiFi, LTE, and 5G), differing in their capabilities with respect to latency, range, data rate, power profile, and scalability.

Wired technologies are most suitable for fulfilling the communication requirements of digital twins. Due to their limitations in terms of flexibility, mobility, and

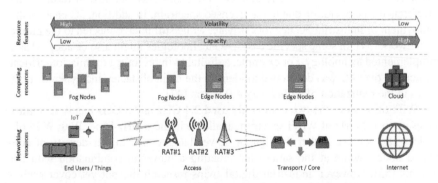

Figure 10.1 Cloud-to-Things continuum characterization

high-density connections, wireless technologies are becoming more appealing in the (R)AN. However, the critical processes within industrial environments are sensitive to radio-frequency interference, requiring RATs to be interference-free, to work on licensed bands, and to provide an extremely controlled environment. Industry 4.0 claims 5G as a key enabler to fulfill the communication requirements set by digital twin [45], not only through radio enhancements but also by employing network slicing and virtualization as core features. At the same time, WiFi 6E appears as another candidate for Industry 4.0, with trials already showcasing its capability to sustain the presence of interference and noise and to meet the stringent requirements of most use cases.

10.10.2 Computation offloading for digital twin

In Industry 4.0, physical objects are composed of either low-performance and constrained hardware or hardware tailored to a specific task. Owing to the development of virtualization, software components of the physical object are represented as modular virtualized functions, which execution is outsourced into more powerful computing resources.

Cloud-based solutions have been initially exploited for implementing such concepts [59], by providing elastic and powerful computing capabilities required to support the digital twin. However, cloud providers cannot ensure the performance of the network between the physical object and its digital replica, worsening with their network distance and the number of providers in between. As a result, cloud-based digital twins suffer from time-varying network delays, unpredictable jitter, limited bandwidth, or data loss. These drawbacks prevent time-sensitive tasks, including real-time remote control, to be fully supported by the cloud computing substrate.

To overcome the shortcomings of cloud computing, edge and fog emerged as a natural extension. While edge computing provides computing capabilities near the physical objects via static substrates, fog computing also integrates volatile, constrained, or mobile resources (including the physical objects). By exploiting edge and fog computing, the digital twin can offload time-sensitive processing from the physical object, which in turn contributes toward further optimizations of the hardware costs. Additionally, new algorithms for efficient data filtering, envisioning privacy and security improvements [60], can be applied and the data can be restricted within a trusted private infrastructure. Finally, due to the close proximity, edge-based digital twins can use the available radio network information to adapt the physical objects' operations or to optimize resource allocation in order to improve the Quality of Experience (QoE).

10.10.3 Digital twin as a service

Virtualized network functions (VNFs), also referred to as service functions, are widely embedded in network operators' deployments. It allows them to migrate network functionalities from costly vendor hardware to general-purpose resources. Network services are usually formed as a composition of VNFs, each providing a specific functionality of the whole service. In addition, functional network modules that compose the network service are split, centralizing a subset of functions into a data center. This functional split was proposed as a feature for the next generation of radio access networks (RANs), where the processing of all the base stations has been centralized into a data center. Pooling computational resources reduces the cost of 5G deployments and centralization enables easier coordination between next-generation base stations [61].

The adaptation of industrial processes toward a concept based on digital twin is facing a major challenge with respect to the unification and exchange of information by different applications [62]. To make use of the digital twin, applications must have access to information from different systems (as shown in Figure 10.2), making abstractions a very important topic to handle the underlying heterogeneity. Moreover, manufacturing service encapsulation, composition, and publication are identified as key technologies that need to be studied [63].

Applying the network service and functional split concepts to digital twin gives the factory operator a unified and modular view of the system and extends the management capabilities beyond the networking aspects to implement a digital twin as a service. These concepts enhance the flexibility of the infrastructure as well as monitoring and management operations, contributing to fulfilling the scalability and availability requirements.

10.10.4 Robot operating system framework

Robot operating system (ROS) is an open-source software development kit for robotics applications that have been gaining popularity and adoption, being used for a variety of robotic applications and supported by many companies in the robotics field. It provides a set of tools, libraries, and protocols to aid in the design of robust and complex robot behavior across a range of robotic platforms. Thus, ROS is mostly used to create robot software, which is one of the foundations for Industry 4.0 and 5.0.

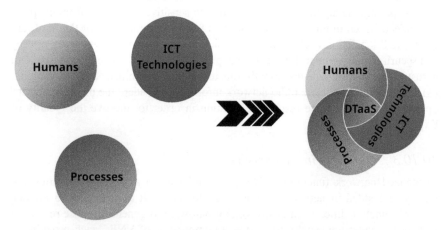

Figure 10.2 *Digital twin as a service concept*

Today, ROS makes available ROS1, at the end of its lifetime but still widely used, and its successor ROS2 aims to provide several improvements and innovations. Despite the benefits introduced by its modular design, ROS1 has several limitations, including meeting real-time latency and throughput constraints in non-ideal networking conditions, and on scalability in multi-robot deployments. ROS2 aims to tackle some of these issues by introducing data distribution service (DDS) as the communication layer and enhancing real-time communication features. Despite the performance and scalability benefits in most of today's realistic robotic deployment, a constrained battery capacity still represent a major drawback in robotic applications, in particular those characterized by heavy computations.

To address some of these limitations, the research community started to integrate ROS-based applications with both the cloud and edge, as a way to offload part of the computational load out of the robotic devices. Still, both ROS1 and ROS2 follow different approaches for their communication capabilities. On the one hand, ROS1 follows a centralization of the entire system information on a master node and then relies on peer-to-peer connections over TCPROS/UDPROS for component communication. On the other hand, ROS2 removes the need for a master node and incorporates a DDS abstraction layer to achieve higher levels of decentralization but restricts its usage within the multicast domain. Thus, their integration with the cloud and the edge cannot be tackled in a unified way, since each imposes a different set of requirements as well as intrinsic advantages and drawbacks.

10.10.5 Resource and service federation

In order to satisfy both vertical customers and end-users, digital twin service providers dynamically dimension and manage their infrastructure to fit the expected demand. In the case of an infrastructure failure (e.g., burned data center) or a sudden increase in the number of users for a given service (e.g., number of robots), digital twin providers are often not able to react on time to keep a service operational while guaranteeing all the requirements defined in the service-level agreement (SLA). In these situations, virtualization is less costly and more time-efficient to enable a provider to use services and/or resources owned by other operators/providers, thus expanding its infrastructure. This process, defined as a federation, enables digital twin service providers to orchestrate services and resources toward external administrative domains.

In an NFV environment, the federation mechanism is used by administrative domains (ADs) to deploy network services or allocate resources over the infrastructure of an external domain. The federation procedure is triggered by a consumer domain as a need of extending a specific service (e.g., at a specific geo-location) or the lack of constituent resources. The federation requires all collaborative administrative domains to have mutual cooperation agreements. Typically, these agreements are signed by business executives. Upon agreement, the agreed ADs trust each other and establish peer-to-peer connectivity among themselves or define a trusted centralized entity to manage their interactions. Therefore, federation interactions can be executed in a centralized, decentralized, or distributed manner. In a centralized solution, all involved ADs have to sign mutual agreements and trust a centralized entity located in a neutral location. The centralized entity is managed by the joint community of involved ADs and oversees the federation interaction, acting as a neutral "middle-man." The positive side is that it is a highly trusty and scalable solution; however, the disadvantage is the set-up time, joint effort, continuous maintenance, and that it is a single point-of-failure system. A decentralized solution is the simplest to employ, at the cost of the lowest scalability. Each AD establishes peer-to-peer connectivity with each external AD. This implies that each newly created (federation-ready) connection between ADs is followed by a business meeting and inter-domain connectivity setup (e.g., 50 connections between different ADs would take at least 50 or more days). A distributed solution is a hybrid approach, more similar to the centralized solution, where the central entity is distributed in each AD.

10.11 Open challenges to achieve Industrial E2E digital twin systems

Despite the recent advances for Industry 4.0, several challenges still remain to be addressed, especially if Industry 5.0 also comes into play. Below are identified five open challenges that require further research for their practical application in Industry 5.0.

Challenge C1: How to use the real-time application and network contextual information to optimize the Industrial E2E digital twins?

Context-aware mechanisms arise as a unique opportunity for digital twin services not only to optimize the E2E performance of provided service but also to enable their mutability in case of a changing environment. Solutions that exploit contextual information from the network and radio links are very well defined in the existing literature [64]. Nevertheless, the challenges still remain on how Industrial E2E digital twin systems in Industry 5.0 can exploit application-context information not to optimize their overall connectivity or E2E performance but instead to adapt their operations when detecting changing environments. For example, when a human worker is detected approaching a robot on a manufacturing floor, how can the Industrial E2E digital twin system be capable of changing the operation of the latter with safer controls and movements?

Challenge C2: How can Industrial E2E digital twins systems predict and optimize the entire physical systems in order to make manufacturing industries more sustainable and green?

The manufacturing sector, as the largest world energy consumer (55% of the world's energy [65]), is in a constant search for energy-efficient optimization strategies that will reduce energy consumption. Cloud robotics started to contribute toward energy optimization in the manufacturing sector by offloading computation from the factory floor to the cloud, even though more energy consumption has been shifted to communication. Still, the computation offloading was only suitable for delay-tolerant, long-term, and high-computation robotic applications, which started to be addressed by offloading to the edge. The challenge still remains on how to achieve optimal and intelligent orchestration of the Industrial E2E digital twin system that not only fulfills the strict safety, latency, and reliability requirements but also contributes toward global energy optimization. For example, how components can be deployed in such a way that there is a reduction of energy consumption and an increase in the usage of green energy throughout the time, without impacting the E2E performance?

Challenge C3: How can an Industrial E2E digital twin systems extend its service footprint in a fast and efficient way over infrastructures that are owned by different stakeholders?

By placing computing, networking, and storage resources near the edge, Industrial E2E digital twin systems can execute their application components closer to the physical system resulting in more predictable communication and overall better system performance. Often enough, an edge digital twin service requires fast and short-lasting expansion of the service footprint over infrastructures that are owned by different stakeholders. Resource and service federation have been already studied over the years, but the challenge still remains on how to perform the federation of services in a secure and trusted way while not interrupting the operation of the digital twin service. Examples of open challenges in a multi-domain federation [66,67] in dynamic environments include (i) the trade-off between the administrative domains openness in admission control and preserving the privacy, security, and trust; (ii) monitoring the availability of the participating administrative domains over time with a resilient solution; (iii) dynamic pricing and billing for federation services with secure and trustful agreements in the form of dynamic SLAs; (iv) guaranteeing the requirements across federating domains in dynamic environments; and (v) achieving high security and privacy interconnection schema between administrative domains.

Challenge C4: How can AI/ML assist the Industrial E2E digital twin systems in its integration with the network and computing infrastructures to satisfy the expected real-time performance?

The cyber-physical integration in industrial environments mirrored through the networked robots arises a perfect opportunity for the development of AI agents to devise smarter and more accurate robots. ML is a strong candidate to implement such AI agents, as an alternative to heuristic or decision-tree-based solutions, among others. Digital twin systems acquire a huge amount of raw information about their physical counterpart and its surrounding, which can be later used to train and cross-validate different ML algorithms and AI agents. These agents not only learn about the details of the specific industrial tasks but also extend and optimize them beyond human capability due to the volume of data they can handle to make decisions. However, the challenge still relies on how to build AI agents that benefit from ML algorithms to exploit the exciting data sources with context information at both the application and infrastructure levels.

Challenge C5: How to achieve accurate and cost-effective localization in industrial digital twin systems?

With the advances of industrial digital twins, localization and positioning features are demanded from wireless technologies to ensure that machinery, robots, and equipment are operating correctly, hence, preventing accidents and reducing downtime. State-of-the-art localization solutions use optical sensors, such as laser range scanners (LIDAR), RGB cameras, and stereo cameras to produce accurate indoor maps. However, not only are optical sensors impaired by the presence of dust and smoke, but their use is also drastically limited by poor illumination (e.g., dimness or darkness). Acoustic sensor-based localization, such as ultrasonic and microphones, are resistant to lighting dynamics, but they suffer from short sensing range or become ineffective in noisy environments such as the factory floor. The challenge of having accurate and cost-effective positioning solution for industrial digital twin systems still exist that will be able to consider also the above-mentioned challenging situations.

10.12 6G enablers and their applicability to E2E digital twins systems

With the current development of 6G, new key enabling technologies are arising that have the potential to address the above-mentioned challenges in current Industry 4.0, and upcoming Industry 5.0, E2E digital twins systems. This section will provide an in-depth analysis of key enabling technologies for E2E digital twins systems.

10.13 Context awareness

The adaptation of E2E digital twins systems in Industry 5.0 is inseparable from the implementation of 5G technologies, such as 5G connectivity and edge computing, and upcoming 6G technologies, such as extreme connectivity and network as a sensor. These networks are architected to simultaneously support different types of real-time network service profiles in the shared infrastructure through the ultra-reliable low latency communication (URLLC). Together with edge computing, they provide a context-aware communication link with low end-to-end latency, low jitter, and localization awareness. Context awareness refers to the capability of a digital twin system to link changes in the network or in the industrial environment with the operation of the digital twin system. Context information related to digital twin systems is structured into two main categories: network information (surrounding computation resources and communication) and application information (location, sensors, cameras, and user state). This potential of E2E digital twins systems to support context-aware services for managing automation or computing best real-time decisions is still to be exploited.

The network contextual information provides information regarding the user connectivity (e.g., radio channel status) or available computing resources that can be used for dynamic adaptation of the digital twin application's logic to the actual status of the network (e.g., radio channel) [20]. On the other side, the application contextual information provides information regarding the surrounding industrial environment of the physical object via different sources of information such as IoT sensors, cameras, and noise sensors. In this view, information from different applications on the factory floor can be used as context for the digital twin system as long as there is a set of relationships that links the digital twin application with its environment. ETSI MEC defines the concept of services that are envisioned to expose contextual information. Radio network information service (RNIS) [68] and WLAN network information service (WNIS) [69] are such examples that provide radio network-related information, such as up-to-date radio network conditions, measurement and statistics information related to the user plane, and information related to users served by the radio nodes. Another example is given by the location service [70] which provides location-related information about the users (e.g., all of them or a subset) currently served by the radio nodes. The location information can be geolocation, Cell ID, etc. A mobile robot digital twin application can use the network contextual information about the wireless connectivity (e.g., signal level, number of retransmissions, and packet losses at the data link level) to adapt the speed of the mobile robot. This information is first consumed by an adaptive control algorithm that is tested on the virtual model of the mobile robot to ensure smooth operation. Once the adaptive control algorithm is tuned correctly, it can be used to perform adaptive remote control of the physical robot.

The natural evolution of computing infrastructures from the cloud toward the near and extreme edge of the network enables Digital Twin systems to be locally present and have access to the locally available network context information from the radio access network. The wide acceptance of SBA in 5G eases the implementation of context-aware E2E Digital Twins by placing heavy emphasis on services as primary architecture components when building Digital Twin applications. Such Digital Twin applications can consume services that provide the real-time status of the radio access network and adapt their operations accordingly.

10.14 Joint communication and sensing

The enhancement of communications networks with sensing capabilities is a very promising area that presents many opportunities and challenges. The main benefit of the communication networks in the context of sensing is that most of the infrastructure is already in place, providing full coverage of the factory floor which makes the sensing capabilities to be provided for free. The latest trends in wireless communication systems have leaned toward providing more bandwidth, and, with this, raised the need for higher carrier frequencies. This was noticeable in 5G and Wi-Fi, where mmWave bands around 24–28 GHz and 60 GHz, respectively, were considered to provide high bandwidths. The directional mmWave transmission and high temporal resolution from the multi-GHz bandwidth provide accurate localization (positioning). This localization potential has to stimulate the standardization bodies to propose standards improvements for the next-generation of positioning systems. Due to these performance boosts, many novel use cases and application improvements have emerged including the E2E digital twins systems.

The virtual representation of the physical factory floor requires accurate localization and mapping of any moving or stationary object in the physical manufacturing environment. These localization capabilities can be used to improve the performance of the factory floor network by providing optimizations input for network maneuvers. For example, sensing would be able to detect objects that will block the direct communication between a physical object and a point of access. This sensing information can be useful for beam steering or switching to a different point of access for communication with this physical object. Another example of sensing applicability in the digital twin system would be the precise position estimation around the manufacturing robot that can help to determine if a robot is located in near proximity of any interfering physical object such as another robot or a human. The accurate estimation of an object that a robot should pick is also a useful feature.

The communication networks are very well positioned for providing sensing capabilities in the manufacturing environment. They are composed of network nodes that provide full coverage of the factory floor. In this context, the infrastructure for sensing is basically already in place. The physical objects are already equipped with communication hardware (e.g., antennas) that has the possibility to use the reflections in different directions from one or more transmitting network nodes. Moreover, the network nodes that are present on the factory floor are already connected via the backhaul and core network, so the sensing information and possibly local pre-processing are already available.

10.15 Semantic orchestration

There is a rising interest in the networking community for providing support to digital twin robotic use cases (e.g., security/surveillance, cleaning, delivery of goods, and collecting products) in industrial deployments. The goal is always to meet the key performance indicators (KPIs) that mission-critical digital twin applications require as it is in the case of remotely controlled mobile robots where the goal is to provide uninterrupted connectivity over the radio access network. Table 10.1 in the introduction summarizes the key connectivity KPIs of different industrial use cases mapped into the digital twin categorization. For example, monitoring digital twins can be applied to use cases such as industrial condition monitoring and process automation based on sensors. Simulation digital twins can be applied to factory remote maintenance, whereas the operational digital twins include remote operation, motion control, safety control, and closed-loop control use cases. However, ensuring these KPIs over wireless networks and heterogeneous cloud-to-robot infrastructure is very challenging and many times it requires the correct selection of the radio points of attachment, the proper configuration parameters, and the use of resources that may span from the cloud to the edge of the network infrastructure (including MEC), to the fog (e.g., connected robots).

To fulfill these KPIs and provide the needed QoS, the digital twin system needs to provide the selection of the radio points of attachment and the use of resources that may span from the cloud to the edge of the network infrastructure (including MEC), to the fog (e.g., connected robots). Also, it is critical that robot VNFs are placed and connected, so as to (i) meet the target KPIs values, (ii) make efficient use of the different available resources, thus avoiding resource shortage, and (iii) minimize the service deployment cost, one of the main concerns for both mobile operators and vertical industries [71–74]. Traditionally, this problem is solved by optimization, heuristic, and game theoretic-based approaches. However, the networks evolving toward 6G become increasingly complex, and it is very challenging to formulate an accurate mathematics model and solve the large-scale problem to get global optimal results in a short time. Moreover, existing state-of-the-art solutions pre-define the number and type of resources needed to perform a given digital twin task or define them in a monolithic fashion, which leads to sub-optimal performance. Therefore, semantic (task-oriented) orchestration is explored as an alternative solution, where the semantics of the different digital twin tasks are considered to reduce the network overhead and allocate resources in a flexible way. Flexibility allows for consideration of multiple cloud-to-thing allocations and RAN slices to the same task-related performance, ultimately improving the end-to-end system performance.

The 6G data-driven architecture will be designed to support different purposes such as exposure of the data-driven architecture to verticals (e.g., AI as a service (AIaaS), sensing as a service (SaaS), and compute as a service (CaaS)). This data presents an opportunity for industrial digital twin systems to develop and integrate a set of novel mechanisms extensively exploiting the semantics from the digital twin applications for internal resource management and orchestration in order to optimize performance, reduce energy consumption, and ensure the strict KPI/QoS of the end-to-end digital twin system.

10.16 Distributed ledger technology federation

In recent years, distributed ledger technology (DLT) (also known as blockchain technology) gained significant applicability in the networking community to address different challenges in network slicing [75], wireless networking technologies [76], and network distribution [77]. In the context of the federation, DLT has the potential to be the solution for multi-domain federation in a dynamic environment.

Every administrative domain can deploy a blockchain node connected to an orchestrator. Just by deploying a blockchain network, most of the challenges enumerated above can be resolved. In such deployment, the admission control will be dependent on the blockchain governance policy. For example, in a permissioned blockchain, a common approach is to accept members via voting. Although domains may act maliciously and reject the entry of new members, domains typically have the incentive to increase participants. The availability will be guaranteed by the incentive of each domain to maintain an active blockchain node. Therefore, this improves the blockchain network security (avoiding 51 percent attacks) and increases the domain's usage budget (e.g., gas in Ethereum [78]). Security and privacy will be established by limiting the usage budget and the use of cryptography. Newly joined domains will have a lower limited usage budget or a limited number of federation announcements and thus are unable to spoof or spam the participating domains. Communications between domains will be recorded and validated as immutable transactions on the ledger where cryptography will be used to preserve the privacy of the data in the transactions exchanged. Finally, dynamic pricing and billing, and multi-domain QoS will be achieved by the implementation of dynamic SLAs and QoS monitoring.

The use of smart contracts is a promising solution for the integration of both dynamic SLAs and QoS monitoring. Smart contracts are deterministic and independent applications that reside on the blockchain ledger. The ETSI PDL specification [79] provides a hint of how to employ QoS through an example scenario of using smart contracts. It envisions a marketplace of SLAs where each smart contract represents a specific service offered with QoS metrics. Customers ready to deploy a service from the marketplace need to send a payment blockchain transaction to the specific smart contract. A third-party entity is used (as an oracle) to monitor the QoS metrics and record the SLA fulfillment directly in the smart contract. If QoS is not satisfied, the smart contract automatically sends back a blockchain payment transaction to the customer blockchain address with the penalty amount. Additionally, service providers as smart contract owners can dynamically change the prices in every smart contract, of course, prior to customers making a deposit.

The combination of context information and location coordinates allows the mobile robot to move within the boundaries of a single administrative domain (AD). Through the application of service federation, a mobile robotics service would not be limited (to a single AD) and it would be able to extend the desired service footprint at anytime, anywhere.

10.17 Artificial intelligence

The adaptation of digital twin in Industry 5.0 is inseparable from recent advances in ICT, such as wireless connectivity and supporting technologies. 6G networks are going to be designed to simultaneously support different types of digital twin services in the shared resilient, energy-efficient, and reliable infrastructure. Together with the edge [80] and fog [81] computing, they provide a communication link with low E2E latency, low jitter, and localization awareness to industrial Digital Twin services. Still, by themselves, these technologies cannot efficiently manage automation or compute the best decisions to address the existing industrial environmental problems, such as carbon emission and nuclear pollution. This existing environmental problem forced industries to shift from extensive economic growth to sustainable development. The key to realizing comprehensive sustainability is in balancing the financial, environmental, social, and governance dimensions. This process increases manufacturing costs and introduces several challenges for the industrial verticals.

In this sense, the cyber space mirrored through E2E digital twins systems arises as the perfect playground for the development of AI-based agents [24]. Moreover, ML is a strong candidate to implement such agents, as an alternative to heuristic or decision-tree-based solutions, among others. E2E digital twins systems provide the tools for transferring the domain expertise of specialized personnel into raw data in the cyber space, which can be later used to train and cross-validate different ML algorithms used in AI agents. These agents not only develop expertise in specific tasks but also extend and optimize them beyond human capability due to the volume of data they can handle to make decisions. Ultimately, smarter, more accurate, and cost-efficient E2E digital twins systems can be devised where autonomy is achieved through AI-controlled processes that operate in all types of environments and conditions.

Table 10.2 presents exemplary AI agents for E2E digital twins systems that can be executed in the infrastructure or application domain and have the potential to impact and optimize the performance factory floor. The *In-Network* or *On-Device*

Table 10.2 Example of AI agents for industrial E2E digital twin systems (adapted from [1])

	AI agent	Input data	Outcomes
Application	Movement Prediction	Historic of commands, real-time commands	Predictions on the N next commands
	Task learning	Demonstrations of the task from different knowledge domains (e.g., physical object states)	Generalized task policy
	Risk reduction	Sensor data, video streams, localization data, and machinery states	Identification and forecasting unsafe situations
	Predictive Maintenance	Machinery and environmental sensor data (e.g., motors status, vibration, and temperature)	Failure predictions
Infrastructure (Computing and networking)	Dynamic Scaling	Resource usage, date and time, task, number of instances, application KPIs and SLAs	Scale in/out or up/down suggestions
	Privacy, security, and intrusion detection	Infrastructure and network context information, traffic flows patterns, service and infrastructure KPIs	Security breaches and suspicious flows
	Heterogeneous RAT selection	Radio network information, available resources, mobility patterns, application KPIs, and SLAs	RAT and handover candidate selection

deployment strategies are envisioned while leveraging on the pervasiveness of the complete cloud-to-things continuum (i.e., fog, edge, and cloud). Moreover, AI agents can be trained in the cloud (cloud learning) for computation-intensive training, or, in the edge (edge learning), for local training considering enormous real-time and private data generated by industrial processes. As an example of an application-related AI agent, the predictive maintenance AI agent is a suitable candidate to decrease unplanned downtime and emergency maintenance by preemptively detecting failures or repair needs. It checks if the available sensor data might lead to failure situations and, if so, it schedules the maintenance of the physical object. At the infrastructure level, the dynamic scaling AI-agent uses data such as resource consumption, date and time, task, number of instances, and sessions and can then compute scaling decisions in order to fulfill KPIs, SLAs and optimize the resource usage.

For resilient, sustainable, and reliable development an end-to-end digital twin system integration, which has been increasingly considered and recommended worldwide, AI is a powerful tool. Through consistently linking, processing, and analyzing all available data across the entire digital twin system, a comprehensive evaluation of sustainability, resilience, and reliability can be achieved so that industrial verticals can develop their systems in that direction. For Industry 5.0, which is facing new challenges from mass product customization to mass product personalization, AI-driven digital twin systems are expected to provide the additional manufacturing understanding that enables a demand-oriented and real-time system, considering the resilience, sustainability, and reliability factors in all the different layers of the digital twin system.

10.18 Industrial E2E digital twin systems in collaborative robotic applications

Industry 5.0 is paving the way for a new set of robotics applications that aim at creating new opportunities in terms of human–robot cooperation [82]. The integration of recent ICT technologies will expand the level of automation in areas that have been difficult to automate via traditional processes. Moreover, the significant advances in Industrial E2E digital twin systems are key to unlocking the full potential of human–robot collaboration by contributing to higher degrees of adaptability and coping with the dynamics of human presence.

10.19 Manufacturing: localization and material inspection

The use of higher frequency bands from mmWave, wider bandwidth, and denser distribution of massive antenna arrays in future 6G systems will enable the integration of wireless signal communication and sensing in a single system to mutually enhance each other. Such capabilities are opening new opportunities regarding collaboration between humans and cobots, providing a broad range of new services to the human worker, like detecting human worker presence and tracking his location, recognizing human worker activities and movements, and even inspecting the environment and objects.

The communications system as a whole can serve as a sensor by exploring radio wave transmission, reflection, and scattering to sense and better understand the physical world. In other words, the same wireless communication system is simultaneously used for communication and sensing purposes. These two-fold capabilities are important in scenarios where cost is a limiting constraint, or in extremely harsh conditions that make the use of other types of sensors unfeasible.

On the one hand, accurate human detection, and its subsequent tracking, allows the cobot to estimate the position of the human worker in an indoor environment while tracking tries to determine the trajectory of movements by monitoring its change of position over time. Moreover, the cobot must identify human behaviors and intentions in order to accurately predict his movements. By combining all this information, the cobot has a more accurate view of its surroundings, which gains greater importance in realizing autonomous and safe collaborative tasks. On the other hand, the capabilities of high-accuracy localization, imaging, and environment reconstruction obtained from sensing could help improve communication performance such as more accurate beamforming, faster beam failure recovery, and less overhead to track the channel state information.

E2E industrial digital twin systems will become a paramount tool in manufacturing by providing a holistic view of the entire factory plant. They will expose new services, like indoor navigation, localization, and tracing solutions, thus better adapting to dynamic changes in the environment and manufacturing setup. Rather than relying completely on manually discovering the indoor environment with their mobile or sensors, E2E industrial digital twin systems will make use of any available wireless communication system to continuously create indoor maps in a rapid manner and keep human workers localized and traced accurately within the map.

Overall, the integration of cobots in manufacturing will enable energy savings and increased efficiency and safety in the workplace. Human workers and cobots will be able to locate and sense one another more accurately, while also providing the former added value in terms of novel services for material inspection.

10.20 Warehouse: material handling and logistics

Pick and place, assembly, packaging, and palletizing are some of the existing tasks in manufacturing industries that require material handling [83]. They are exhausting and dangerous tasks since workers might handle heavy or hazardous materials, thus likely to induce injuries and loss of items if mistakes are made.

Although large manufacturing plants have the capacity to automate all the aforementioned tasks with a traditional industrial robot, small- and medium-sized are expected to leverage cobots to partially automate their plants regarding any of these heavy and repetitive tasks. The cobot would work alongside human workers, either by doing complementary activities or by actually interacting in the same activity. For example, human workers can be responsible for tasks such as loading and unloading items while the cobot would execute precise and repetitive tasks such as assembling products, welding, and packaging the final product. Still, cobots could also coordinate with human workers and support them in loading and unloading heavy or dangerous items.

In such an environment, cobots must be designed to be safe to work alongside human workers. Not only do they need to interact with one another via a control plane or voice commands but also cobots must support adjustable capabilities (such as speed and force) and sensors (such as force sensors and distance limits) that can detect and adapt to human presence. The cobot would be programmed to slow down if it detects a human worker nearby to ensure a safe and effective collaboration as well as to entirely stop in case of a critical condition [84].

E2E industrial digital twin systems have a two-fold objective in material handling and logistics use cases. On the one hand, they will optimize the movement and storage of materials by monitoring and predicting the flow of materials across the entire manufacturing chain. Such capability will allow not only to identify of bottlenecks but also to balance production demands and green-energy and sustainable opportunities. On the other hand, they will be used to simulate cobots' movements, and to optimize their performance in a particular task or environment, ensuring that the cobot is able to work safely and efficiently alongside human workers.

Overall, the integration of cobots in material handling and logistics can greatly enhance the efficiency and safety of the workplace, especially for smaller businesses. Cobots will work alongside human workers in a collaborative manner, improving the flexibility and adaptability of the production processes. They will also enable human workers to focus on more complex and value-added activities. Nevertheless, safety protocols must be in place at all times to ensure people's physical well-being, safety, and health [85].

10.21 Construction: safety takeover

The collaboration between humans and cobots can also embody a more passive interaction, especially for tasks that require human intervention or are hazardous or physically demanding for humans. The human worker is responsible for executing the task, while the cobot is maintained on stand-by and only takes control in the case of incapacity of the former. Such cooperation is essential to ensure that the human worker's health and safety are not compromised during the task.

The human worker performing the task is continuously monitored in terms of his physiological signals (such as heart rate, respiration rate, arterial pressure, and body temperature), which are analyzed in real time to measure his well-being and awareness. If any of the physiological signals indicate fatigue or become critical, the robot takes over to finish the task in order to prevent any harm to the operator. To do so, the cobot has access to the same controlling mechanism as the human worker and, when activated, takes priority over the commands executed by the human operator.

In this scenario, E2E industrial digital twin systems must have two virtual representations: one of the human workers, representing all his physiological signals, and another of the entire environment where the task is being executed, including all the controlling entities. While the former will allow the cobot to make an assessment of the human worker's condition at every point in time, the latter will allow the cobot to make an assessment of the task progress and to continuously compute preventive measures to be executed only if needed.

Moreover, the human worker can perform tasks in a virtual environment, where he can test and experiment with different scenarios before executing them in the physical world, while the cobot can provide feedback on better-suited actions. This reduces the risk of errors and accidents and can help to optimize task performance.

Overall, human–robot collaboration leverages the strengths of both humans and robots to improve the efficiency and safety of their operations. They can take not only a more active collaboration but also a passive collaboration triggered only in case of need.

10.22 Healthcare: patient rehabilitation

Industry 5.0 is having a ripple effect in various industries, including healthcare. Healthcare 5.0, the Industry 5.0 revolution applied in healthcare, is introducing new developments for patient rehabilitation [86] where cobots are used to assist with exercises and therapy. While healthcare workers are going to assess the extent of the injury and determine the best treatment, cobots will be used to assist patients with their rehabilitation exercises and to provide guidance and support.

Physical therapy and rehabilitation are usually required during patients' recovery from injuries, surgeries, or any other conditions that affect mobility. It requires the correct and consistent execution of range-of-motion exercises, which in case of limited mobility or pain can be difficult to perform. A cobot can then be programmed to help and guide patients performing such exercises while monitoring and assessing the patient's progress. For example, the cobot can provide resistance based on the patient's strength, grip, and ability, or to guide and teach the movement as needed to perform the exercises correctly and achieve the best results. Additionally, the cobot can adapt resistance or support parameters according to the patient's recovery progress.

Moreover, cobots can provide immediate feedback on the patient's performance to the healthcare worker, ensuring the best treatment is provided at all times. The combination with augmented and virtual reality (AR/VR) technology will create immersive rehabilitation environments [87], expanding the interaction capabilities between the healthcare worker and the patient. The former will be able to remotely take control of the cobot and perform a similar in-person evaluation of the patient's condition while also instructing new movements to be supported by the cobot for the next recovery stage.

E2E industrial digital twin systems will be used to provide virtual representations of both the patient [88], generated from multimodal patient data and real-time updates on the patient and the cobot. These two-fold twins will allow the healthcare workers to have a complete and accurate view of the patient's health and his progress on the treatment as well as to have an entire view of the cobot while controlling it from a remote location.

Overall, the use of cobots in patient rehabilitation will improve the effectiveness and efficiency of rehabilitation therapy, while providing healthcare workers and patients with a more positive and empowering experience during their recovery. However, further research is needed to fully understand the potential of cobots in inpatient rehabilitation and to identify the most effective, secure, and safe ways to integrate them into clinical practice.

10.23 Conclusions

The current trends in manufacturing systems are the result of a long evolution of ICT. The current and future industrial revolutions focus on digitalization and the creation of cyber-physical systems (CPSs) that aim to bring together the IT and OT worlds. Digital twin is a concept that truly embodies cyber-physical integration, combining any industrial process achieved through closed-loop feedback mechanisms. In robotics terms, an E2E digital twin system is an emerging concept that integrates information and communication technologies to create a highly consistent, accurate, and synchronized virtual representation of its physical counterpart, anywhere, anytime, and in any conditions.

This chapter starts with an in-depth analysis of the fundamental 5G-related technologies behind the digital twins for robots today, followed by a clear image of the open challenges that remain to be addressed in order for the robotic digital twins to find practical applicability on the factory floor. Namely, the following five challenges have been identified:

- How E2E digital twin systems can exploit application-context information not to optimize their overall connectivity or E2E performance but instead to adapt their operations when detecting changing environments?
- How to achieve optimal and intelligent orchestration of the E2E digital twin system that not only fulfills the strict safety, latency, and reliability requirements but also contributes toward global energy optimization?
- How can E2E digital twin systems extend its service footprint in a fast and efficient way over infrastructures that are owned by different stakeholders?
- How can AI/ML assist the E2E digital twin systems in its integration with the network and computing infrastructures to satisfy the expected real-time performance?
- How to achieve accurate and cost-effective localization in E2E digital twin systems?

With the current development of 6G, new key enabling technologies are arising that have the potential to address the above-mentioned challenges, and be the driving factor for the upcoming Industry 5.0, E2E digital twins systems. This chapter also provides insight into the 5 key enabling technologies for E2E digital twins systems:

- **Context awareness**: The natural evolution of the computing infrastructure from the cloud toward the near and extreme edge of the network enables E2E digital twin systems to have access to the locally available network and application context information. Such digital twin applications can adapt their operations to the real-time status of the network and the application.
- **Joint communication and sensing**: The communication networks provide full coverage of the factory floor and are very well positioned for providing sensing capabilities in the manufacturing environment. The robots are already equipped with communication hardware (e.g., antennas) that has the possibility to use the reflections in different directions from one or more transmitting network nodes for sensing the environment.
- **Semantic orchestration**: The 6G data-driven architecture presents an opportunity for E2E digital twin systems to develop and integrate a set of novel mechanisms extensively exploiting the semantics from the digital twin applications for network and computation resource management and orchestration in order to optimize performance, reduce energy consumption, and ensure the strict KPI/QoS of the E2E digital twin system.
- **Blockchain technology**: Blockchain technology brings a unique combination of characteristics, namely secure data sharing, data logging, and new incentive mechanisms, which makes it an ideal candidate to provide a multi-domain federation for digital twin systems.

- **Artificial Intelligence**: E2E digital twins systems provide the tools for transferring the domain expertise of specialized personnel into raw data in cyber space, which can be later used to train and cross-validate different ML algorithms used in AI agents in the application or infrastructure domain.

The advent of the robotic digital twin era is yet to come and a set of powerful ICTs have the potential to lead the new generation of robotics services. This chapter contributes toward the realization of such a digital twin service by addressing some of the challenges faced by current digital twin solutions. This new set of innovative 6G concepts such as semantic orchestration, DLT federation, predictive control, or joint communication and sensing in E2E digital twins has the potential to lead toward future efficient, collaborative, and automated industries.

Acknowledgments

This work has been partially founded by Horizon 2020, project DAEMON (Grant No. 101017109), by the Horizon Europe research and innovation program, project ICOS (Grant No. 101070177), by the European Commission Horizon Europe SNS JU PREDICT-6G (GA 101095890) and the Spanish Ministry of Economic Affairs and Digital Transformation and the European Union-NextGenerationEU through the UNICO 5G I+D 6G-EDGEDT.

References

[1] M. Hermann, T. Pentek, and B. Otto, "Design principles for industrie 4.0 scenarios," in *2016 49th Hawaii International Conference on System Sciences (HICSS)*, 2016, pp. 3928–3937. doi: https://doi.org/10.1109/HICSS.2016.488.

[2] A. A. F. Saldivar, Y. Li, W. Chen, Z. Zhan, J. Zhang, and L. Y. Chen, "Industry 4.0 with cyber-physical integration: a design and manufacture perspective," in *2015 21st International Conference on Automation and Computing (ICAC)*, 2015, pp. 1–6. doi: https://doi.org/10.1109/ IConAC.2015.7313954.

[3] S. Nahavandi, "Industry 5.0—a human-centric solution," *Sustainability*, vol. 11, no. 16, 2019. doi: https://doi.org/10.3390/su11164371.

[4] M. Golovianko, V. Terziyan, V. Branytskyi, and D. Malyk, "Industry 4.0 vs. Industry 5.0: co-existence, transition, or a hybrid," *Procedia Computer Science*, vol. 217, pp. 102–113, 2023, *4th International Conference on Industry 4.0 and Smart Manufacturing*.

[5] S. Robla-Gómez, V. M. Becerra, J. R. Llata, E. González-Sarabia, C. Torre-Ferrero, and J. Pérez-Oria, "Working together: a review on safe human–robot collaboration in industrial environments," *IEEE Access*, vol. 5, pp. 26754–26773, 2017. doi: https://doi.org/10.1109/ ACCESS.2017.2773127.

[6] S. El Zaatari, M. Marei, W. Li, and Z. Usman, "Cobot programming for collaborative industrial tasks: an overview," *Robotics and Autonomous Systems*, vol. 116, pp. 162–180, 2019.

[7] F. Tao, J. Cheng, Q. Qi, M. Zhang, H. Zhang, and F. Sui, "Digital twin-driven product design, manufacturing and service with big data," *The International Journal of Advanced Manufacturing Technology*, vol. 94, Feb. 2018. doi: https://doi.org/10.1007/s00170-017-0233-1.

[8] R. Rajkumar, I. Lee, L. Sha, and J. Stankovic, "Cyber-physical systems: the next computing revolution," in *Design Automation Conference*, 2010, pp. 731–736. doi: https://doi.org/10.1145/1837274.1837461.

[9] A. Rasheed, O. San, and T. Kvamsdal, "Digital twin: values, challenges and enablers from a modeling perspective," *IEEE Access*, vol. 8, pp. 21980–22012, 2020. doi: https://doi.org/10.1109/ACCESS.2020.2970143.

[10] B. R. Barricelli, E. Casiraghi, and D. Fogli, "A survey on digital twin: definitions, characteristics, applications, and design implications," *IEEE Access*, vol. 7, pp. 167653–167671, 2019. doi: https://doi.org/10.1109/ACCESS.2019.2953499.

[11] A. Willig, K. Matheus, and A. Wolisz, "Wireless technology in industrial networks," *Proceedings of the IEEE*, vol. 93, no. 6, pp. 1130–1151, 2005. doi: https://doi.org/10.1109/JPROC.2005.849717.

[12] W. Jiang, B. Han, M. A. Habibi, and H. D. Schotten, "The road towards 6G: a comprehensive survey," *IEEE Open Journal of the Communications Society*, vol. 2, pp. 334–366, 2021. doi: https://doi.org/10.1109/OJCOMS.2021.3057679.

[13] K. B. Letaief, Y. Shi, J. Lu, and J. Lu, "Edge artificial intelligence for 6G: vision, enabling technologies, and applications," *IEEE Journal on Selected Areas in Communications*, vol. 40, no. 1, pp. 5–36, 2022. doi: https://doi.org/10.1109/JSAC.2021.3126076.

[14] C. D. Alwis, A. Kalla, Q.V. Pham, *et al.*, "Survey on 6G frontiers: trends, applications, requirements, technologies and future research," *IEEE Open Journal of the Communications Society*, vol. 2, pp. 836–886, 2021. doi: https://doi.org/10.1109/OJCOMS.2021.3071496.

[15] D. K. Pin Tan, J. He, Y. Li, *et al.*, "Integrated sensing and communication in 6G: motivations, use cases, requirements, challenges and future directions," in *2021 1st IEEE International Online Symposium on Joint Communications & Sensing (JC&S)*, 2021, pp. 1–6. doi: https://doi.org/10.1109/JCS52304.2021.9376324.

[16] G. Aceto, V. Persico, and A. Pescapé, "Industry 4.0 and health: Internet of Things, big data, and cloud computing for Healthcare 4.0," *Journal of Industrial Information Integration*, vol. 18, p. 100–129, 2020. doi: https:/doi.org/10.1016/j.jii.2020.100129. [Online]. Available: https://www.sciencedirect.com/science/article/pii/S2452414X19300135.

[17] I. Sittón-Candanedo, R. S. Alonso, S. Rodrí-guez-González, J. A. Garcí-a Coria, and F. De La Prieta, "Edge computing architectures in Industry 4.0: a general survey and comparison," in F. Martíadnez Álvarez, A. Troncoso Lora, J. A. Sáez Muñoz, H. Quintián, and E. Corchado, Eds., *14th International Conference on Soft Computing Models in Industrial and Environmental*

[18] S. Rani, A. Kataria, and M. Chauhan, "Fog computing in Industry 4.0: applications and challenges—a research roadmap," in R. Tiwari, M. Mittal, and L. M. Goyal, Eds., *Energy Conservation Solutions for Fog-Edge Computing Paradigms*, Singapore: Springer Singapore, 2022, pp. 173–190. doi: https://doi.org/10.1007/978-981-16-3448-2_9. [Online]. Available: https://doi.org/10.1007/978-981-16-3448-2_9.

[19] M. Peng, S. Yan, K. Zhang, and C. Wang, "Fog-computing-based radio access networks: issues and challenges," *IEEE Network*, vol. 30, no. 4, pp. 46–53, 2016.

[20] K. Henning, W. Wolfgang, and H. Johannes, *Securing the future of German manufacturing industry: Recommendations for implementing the strategic initiative INDUSTRIE 4.0 (Final report of the Industrie 4.0)*, Apr. 2013.

[21] P. Leitão, A. Colombo, and S. Karnouskos, "Industrial automation based on cyber-physical systems technologies: prototype implementations and challenges," *Computers in Industry*, vol. 81, 2015. doi: https://doi.org/10.1016/j.compind.2015.08.004.

[22] V. Kumar, D. Rus, and G. S. Sukhatme, "Networked robots," in B. Siciliano and O. Khatib, Eds., *Springer Handbook of Robotics*, Berlin, Heidelberg: Springer Berlin Heidelberg, 2008, pp. 943–958. doi: https://doi.org/10.1007/978-3-540-30301-5_42.

[23] M. Groshev, C. Guimarães, J. Martíadn-Pérez, and A. de la Oliva, "Toward intelligent cyber-physical systems: digital twin meets artificial intelligence," *IEEE Communications Magazine*, vol. 59, no. 8, pp. 14–20, 2021. doi: https://doi.org/10.1109/MCOM.001.2001237.

[24] K. Xia, C. Sacco, M. Kirkpatrick, *et al.*, "A digital twin to train deep reinforcement learning agent for smart manufacturing plants: environment, interfaces and intelligence," *Journal of Manufacturing Systems*, vol. 58, pp. 210–230, 2021. doi: https:/doi.org/10.1016/j.jmsy.2020.06.012.

[25] F. Tao, H. Zhang, A. Liu, and A. Y. C. Nee, "Digital twin in industry: state-of-the-art," *IEEE Transactions on Industrial Informatics*, vol. 15, no. 4, pp. 2405–2415, 2019. doi: https://doi.org/10.1109/TII.2018.2873186.

[26] A. Fuller, Z. Fan, C. Day, and C. Barlow, "Digital twin: enabling technologies, challenges and open research," *IEEE Access*, vol. 8, pp. 108952–108971, 2020. doi: https://doi.org/10.1109/ACCESS.2020.2998358.

[27] F. Pires, A. Cachada, J. Barbosa, A. P. Moreira, and P. Leitão, "Digital twin in industry 4.0: technologies, applications and challenges," in *2019 IEEE 17th International Conference on Industrial Informatics (INDIN)*, vol. 1, 2019, pp. 721–726. doi: https://doi.org/10.1109/INDIN41052.2019.8972134.

[28] S. Weyer, T. Meyer, M. Ohmer, D. Gorecky, and D. Zühlke, "Future modeling and simulation of CPS-based factories: an example from the automotive industry," *IFAC-PapersOnLine*, vol. 49, no. 31, pp. 97–102, 2016, 12th IFAC Workshop on Intelligent Manufacturing Systems IMS

2016. doi: https:/doi.org/10.1016/j.ifacol.2016.12.168. [Online]. Available: https://www.sciencedirect.com/science/article/pii/S2405896316328397.

[29] R. Rosen, G. von Wichert, G. Lo, and K. D. Bettenhausen, "About the importance of autonomy and digital twins for the future of manufacturing," *IFAC-PapersOnLine*, vol. 48, no. 3, pp. 567–572, 2015, 15th IFAC Symposium on Information Control Problems inManufacturing. doi: https:/doi.org/10.1016/j.ifacol.2015.06.141. [Online]. Available: https://www.sciencedirect.com/science/article/pii/S2405896315003808.

[30] J. Vachálek, L. Bartalsky, O. Rovný, D. Sismisova, M. Morhac, and M. Loksik, "The digital twin of an industrial production line within the industry 4.0 concept," in *2017 21st International Conference on Process Control (PC)*, Jun. 2017, pp. 258–262. doi: https://doi.org/10.1109/PC.2017.7976223.

[31] S. Konstantinov, M. Ahmad, K. Ananthanarayan, and R. Harrison, "The cyber-physical e-machine manufacturing system: virtual engineering for complete lifecycle support," *Procedia CIRP*, vol. 63, pp. 119–124, 2017, Manufacturing Systems 4.0 – Proceedings of the 50th CIRP Conference on Manufacturing Systems. doi: https:/doi.org/10.1016/j.procir.2017.02.035. [Online]. Available: https://www.sciencedirect.com/science/article/pii/S2212827117301324.

[32] T. H.-J. Uhlemann, C. Lehmann, and R. Steinhilper, "The digital twin: realizing the cyber-physical production system for industry 4.0," *Procedia CIRP*, vol. 61, pp. 335–340, 2017, *The 24th CIRP Conference on Life Cycle Engineering*. doi: https:/doi.org/10.1016/j.procir.2016.11.152. [Online]. Available: https://www.sciencedirect.com/science/article/pii/S2212827116313129.

[33] R. Söderberg, K. Wärmefjord, J. S. Carlson, and L. Lindkvist, "Toward a digital twin for real-time geometry assurance in individualized production," *CIRP Annals*, vol. 66, no. 1, pp. 137–140, 2017. doi: https:/doi.org/10.1016/j.cirp.2017.04.038. [Online]. Available: https://www.sciencedirect.com/science/article/pii/S0007850617300380.

[34] M. Schluse, M. Priggemeyer, L. Atorf, and J. Rossmann, "Experimentable digital twins—streamlining simulation-based systems engineering for industry 4.0," *IEEE Transactions on Industrial Informatics*, vol. 14, no. 4, pp. 1722–1731, 2018. doi: https://doi.org/10.1109/TII.2018.2804917.

[35] O. Givehchi, H. Trsek, and J. Jasperneite, "Cloud computing for industrial automation systems—a comprehensive overview," in *2013 IEEE 18th Conference on Emerging Technologies Factory Automation (ETFA)*, 2013, pp. 1–4. doi: https://doi.org/10.1109/ETFA.2013.6648080.

[36] W. Dai, H. Nishi, V. Vyatkin, V. Huang, Y. Shi, and X. Guan, "Industrial edge computing: enabling embedded intelligence," *IEEE Industrial Electronics Magazine*, vol. 13, no. 4, pp. 48–56, 2019. doi: https://doi.org/10.1109/MIE.2019.2943283.

[37] W. Xu, J. Cui, L. Li, B. Yao, S. Tian, and Z. Zhou, "Digital twin-based industrial cloud robotics: framework, control approach and

implementation," *Journal of Manufacturing Systems*, vol. 58, pp. 196–209, 2021, Digital Twin towards Smart Manufacturing and Industry 4.0. doi: https:/doi.org/10.1016/j.jmsy.2020.07.013. [Online]. Available: https://www.sciencedirect.com/science/article/pii/S0278612520301230.

[38] P. D. Urbina Coronado, R. Lynn, W. Louhichi, M. Parto, E. Wescoat, and T. Kurfess, "Part data integration in the shop floor digital twin: mobile and cloud technologies to enable a manufacturing execution system," *Journal of Manufacturing Systems*, vol. 48, pp. 25–33, 2018, Special Issue on Smart Manufacturing. doi: https:/doi.org/10.1016/j.jmsy.2018.02.002. [Online]. Available: https://www.sciencedirect.com/science/article/pii/S027861251830013X.

[39] L. Hu, N.-T. Nguyen, W. Tao, *et al.*, "Modeling of cloud-based digital twins for smart manufacturing with MT connect," *Procedia Manufacturing*, vol. 26, pp. 1193–1203, 2018, 46th SME North American Manufacturing Research Conference, NAMRC 46, Texas, USA. doi: https:/doi.org/10.1016/j.promfg.2018.07.155. [Online]. Available: https://www.sciencedirect.com/science/article/pii/S235197891830831X.

[40] C. Zhang, G. Zhou, J. He, Z. Li, and W. Cheng, "A data- and knowledge-driven framework for digital twin manufacturing cell," *Procedia CIRP*, vol. 83, pp. 345–350, 2019, 11th CIRP Conference on Industrial Product-Service Systems. doi: https:/doi.org/10.1016/j.procir.2019.04.084. [Online]. Available: https://www.sciencedirect.com/science/article/pii/S2212827119306985.

[41] H. Huang, L. Yang, Y. Wang, X. Xu, and Y. Lu, "Digital twin-driven online anomaly detection for an automation system based on edge intelligence," *Journal of Manufacturing Systems*, vol. 59, pp. 138–150, 2021. doi: https:/doi.org/10.1016/j.jmsy.2021.02.010. [Online]. Available: https://www.sciencedirect.com/science/article/pii/S0278612521000467.

[42] C. Liu, L. Le Roux, C. Körner, O. Tabaste, F. Lacan, and S. Bigot, "Digital twin-enabled collaborative data management for metal additive manufacturing systems," *Journal of Manufacturing Systems*, 2020. doi: https:/doi.org/10.1016/j.jmsy.2020.05.010. [Online]. Available: https://www.sciencedirect.com/science/article/pii/S0278612520300741.

[43] H. X. Nguyen, R. Trestian, D. To, and M. Tatipamula, "Digital twin for 5G and beyond," *IEEE Communications Magazine*, vol. 59, no. 2, pp. 10–15, 2021. doi: https://doi.org/10.1109/MCOM.001.2000343.

[44] J. Cheng, W. Chen, F. Tao, and C.-L. Lin, "Industrial IoT in 5G environment towards smart manufacturing," *Journal of Industrial Information Integration*, vol. 10, pp. 10–19, 2018. doi: https:/doi.org/10.1016/j.jii.2018.04.001. [Online]. Available: https://www.sciencedirect.com/science/article/pii/S2452414X18300049.

[45] L. Girletti, M. Groshev, C. Guimarães, C. J. Bernardos, and A. de la Oliva, "An intelligent edge-based digital twin for robotics," in *2020 IEEE Globecom Workshops (GC Wkshps)*, 2020, pp. 1–6. doi: https:/doi.org/10.1109/GCWkshps50303.2020.9367549.

[46] K. Židek, J. Pitel', M. Adámek, P. Lazorík, and A. Hošovský, "Digital twin of experimental smart manufacturing assembly system for industry 4.0 concept," *Sustainability*, vol. 12, no. 9, 2020. doi: https://doi.org/10.3390/su12093658. [Online]. Available: https://www.mdpi.com/2071-1050/12/9/3658.

[47] H. Laaki, Y. Miche, and K. Tammi, "Prototyping a digital twin for real time remote control over mobile networks: application of remote surgery," *IEEE Access*, vol. 7, pp. 20325–20336, 2019. doi: https://doi.org/10.1109/ACCESS.2019.2897018.

[48] M. Groshev, C. Guimarães, A. De La Oliva, and R. Gazda, "Dissecting the impact of information and communication technologies on digital twins as a service," *IEEE Access*, vol. 9, pp. 102862–102876, 2021. doi: https://doi.org/10.1109/ACCESS.2021.3098109.

[49] ETSI, "Digital Enhanced Cordless Telecommunications (DECT); study on URLLC use cases of vertical industries for DECT evolution and DECT-2020," European Telecommunications Standards Institute (ETSI), Technical Report (TR) 103 515 v1.1.1, Mar. 2018.

[50] 3GPP, "Service requirements for cyber-physical control applications in vertical domains," 3rd Generation Partnership Project (3GPP), Technical Specification (TS) 22.104 v17.4.0, Sep. 2020.

[51] 5G Alliance for Connected Industries and Automation (5G ACIA), *5G for Connected Industries and Automation, Second Edition*, Feb. 2019.

[52] NGMN, "5G E2E technology to support verticals URLLC requirements," Next Generation Mobile Networks (NGMN), Final Deliverable v2.5.4, Feb. 2020.

[53] P. Drahoš, E. Kučera, O. Haffner, R. Pribš, and L. Beňo, "Trends in industrial networks including APL, TSN, WiFi-6E and 5G technologies," in *2022 Cybernetics & Informatics (K&I)*, 2022, pp. 1–7. doi: https://doi.org/10.1109/KI55792.2022.9925965.

[54] C. Guimarães, X. Li, C. Papagianni, *et al.*, "Public and non-public network integration for 5Growth industry 4.0 use cases," *IEEE Communications Magazine*, vol. 59, no. 7, pp. 108–114, 2021. doi: https://doi.org/10.1109/MCOM.001.2000853.

[55] G. Naik and J.-M. J. Park, "Coexistence of Wi-Fi 6E and 5G NR-U: can we do better in the 6 GHz bands?" In *IEEE INFOCOM 2021 – IEEE Conference on Computer Communications*, 2021, pp. 1–10. doi: https://doi.org/10.1109/INFOCOM42981.2021.9488780.

[56] A. Balalaie, A. Heydarnoori, and P. Jamshidi, "Migrating to cloud-native architectures using microservices: an experience report," in A. Celesti and P. Leitner, Eds., *Advances in Service-Oriented and Cloud Computing*, Springer International Publishing, 2016, pp. 201–215.

[57] D. Gannon, R. Barga, and N. Sundaresan, "Cloud-native applications," *IEEE Cloud Computing*, vol. 4, no. 5, pp. 16–21, 2017.

[58] Eclipse Foundation Edge Native Working Group, "From DevOps to EdgeOps: A Vision for Edge Computing," Eclipse Foundation, Tech. Rep., Apr. 2021.

[59] K. Borodulin, G. Radchenko, A. Shestakov, L. Sokolinsky, A. Tchernykh, and R. Prodan, "Towards digital twins cloud platform: microservices and computational workflows to rule a smart factory," in *Proceedings of The 10th International Conference on Utility and Cloud Computing*, Austin, TX, USA, 2017. doi: https://doi.org/10.1145/3147213.3149234.

[60] M. De Donno, A. Giaretta, N. Dragoni, A. Bucchiarone, and M. Mazzara, "Cyber-storms come from clouds: security of cloud computing in the IoT era," *Future Internet*, vol. 11, no. 6, 2019.

[61] A. M. Alba, J. H. G. Velásquez, and W. Kellerer, "An adaptive functional split in 5g networks," in *IEEE INFOCOM 2019 - IEEE Conference on Computer Communications Workshops (INFOCOM WKSHPS)*, 2019, pp. 410–416. doi: https://doi.org/10.1109/INFCOMW.2019.8845147.

[62] M. Grieves and J. Vickers, "Digital twin: mitigating unpredictable, undesirable emergent behavior in complex systems," in Springer Nature, Aug. 2017, pp. 85–113. doi: https://doi.org/10.1007/978-3-319-38756-7_4.

[63] F. Tao and M. Zhang, "Digital twin shop-floor: a new shop-floor paradigm towards smart manufacturing," *IEEE Access*, vol. 5, pp. 20418–20427, 2017. doi: https://doi.org/10.1109/ACCESS.2017.2756069.

[64] M. N. Tehrani, M. Uysal, and H. Yanikomeroglu, "Device-to-device communication in 5g cellular networks: challenges, solutions, and future directions," *IEEE Communications Magazine*, vol. 52, no. 5, pp. 86–92, 2014.

[65] EIA, "International Energy Outlook 2017," U.S. Energy Information Administration's (EIA), Report (R) IEO2017, Sep. 2017.

[66] R. B. Uriarte and R. DeNicola, "Blockchain-based decentralized cloud/fog solutions: challenges, opportunities, and standards," *IEEE Communications Standards Magazine*, vol. 2, no. 3, pp. 22–28, 2018. doi:https://doi.org/10.1109/MCOMSTD.2018.1800020.

[67] K. Katsalis, N. Nikaein, and A. Edmonds, "Multi-domain orchestration for NFV: challenges and research directions," in *2016 15th International Conference on Ubiquitous Computing and Communications and 2016 International Symposium on Cyberspace and Security (IUCC-CSS)*, 2016, pp. 189–195. doi: https://doi.org/10.1109/IUCC-CSS.2016.034.

[68] ETSI, "Mobile Edge Computing (MEC); Radio Network Information API," European Telecommunications Standards Institute (ETSI), Group Specification (GS) 012 v1.1.1, Jul. 2017.

[69] ETSI, "Multi-access Edge Computing (MEC); WLAN Information API," European Telecommunications Standards Institute (ETSI), Group Specification (GS) 028 v2.0.1, Jul. 2018.

[70] ETSI, "Mobile Edge Computing (MEC); Location API," European Telecommunications Standards Institute (ETSI), Group Specification (GS) 013 v1.1.1, Jul. 2017.

[71] ETSI, "MEC in 5G Networks," European Telecommunications Standards Institute (ETSI), White Paper 28, Jun. 2018.

[72] P. Zhao and G. Dán, "A Benders decomposition approach for resilient placement of virtual process control functions in mobile edge clouds," *IEEE Transactions on Network and Service Management*, 2018.

[73] F. Malandrino and C.-F. Chiasserini, "Getting the most out of your VNFs: flexible assignment of service priorities in 5G," in *IEEE WoWMoM*, 2019.
[74] Y. Sang, B. Ji, G. R. Gupta, X. Du, and L. Ye, "Provably efficient algorithms for joint placement and allocation of virtual network functions," in *IEEE INFOCOM*, 2017.
[75] F. Javed, K. Antevski, J. Mangues-Bafalluy, L. Giupponi, and C. J. Bernardos, "Distributed ledger technologies for network slicing: a survey," *IEEE Access*, vol. 10, pp. 19412–19442, 2022. doi: https://doi.org/10.1109/ACCESS.2022.3151150.
[76] P.-H. Kuo, A. Mourad, and J. Ahn, "Potential applicability of distributed ledger to wireless networking technologies," *IEEE Wireless Communications*, vol. 25, no. 4, pp. 4–6, 2018. doi: https://doi.org/10.1109/MWC.2018.8454517.
[77] Y. Zhou, A. N. Manea, W. Hua, *et al.*, "Application of distributed ledger technology in distribution networks," *Proceedings of the IEEE*, vol. 110, no. 12, pp. 1963–1975, 2022. doi: https://doi.org/10.1109/JPROC.2022.3181528.
[78] G. Wood, "Ethereum: a secure decentralised generalised transaction ledger," *Ethereum Project Yellow Paper*, vol. 151, no. 2014, pp. 1–32, 2014.
[79] ETSI, "Permissioned Distributed Ledgers (PDL) Smart Contracts System Architecture and Functional Specification," European Telecommunications Standards Institute (ETSI), TSI ISG PDL 004 V1.1.1, Feb. 2021.
[80] W. Shi, J. Cao, Q. Zhang, Y. Li, and L. Xu, "Edge Computing: Vision and Challenges," *IEEE Internet of Things Journal*, vol. 3, no. 5, pp. 637–646, 2016.
[81] C. Mouradian, D. Naboulsi, S. Yangui, R. H. Glitho, M. J. Morrow, and P. A. Polakos, "A comprehensive survey on fog computing: state-of-the-art and research challenges," *IEEE Communications Surveys Tutorials*, vol. 20, no. 1, pp. 416–464, 2018.
[82] A. A. Malik and A. Brem, "Digital twins for collaborative robots: a case study in human-robot interaction," *Robotics and Computer-Integrated Manufacturing*, vol. 68, p. 102 092, 2021. doi: https:/doi.org/10.1016/j.rcim.2020.102092.
[83] S. Comari, R. Di Leva, M. Carricato, *et al.*, "Mobile cobots for autonomous raw-material feeding of automatic packaging machines," *Journal of Manufacturing Systems*, vol. 64, pp. 211–224, 2022. doi: https:/doi.org/10.1016/j.jmsy.2022.06.007.
[84] C.-C. Chan and C.-C. Tsai, "Collision-free speed alteration strategy for human safety in human–robot coexistence environments," *IEEE Access*, vol. 8, pp. 80120–80133, 2020. doi: https://doi.org/10.1109/ACCESS.2020.2988654.
[85] M. Valori, A. Scibilia, I. Fassi, *et al.*, "Validating safety in human–robot collaboration: standards and new perspectives," *Robotics*, vol. 10, no. 2, 2021. doi: https://doi.org/ 10.3390/robotics10020065.
[86] D. Townsend and A. MajidiRad, "Trust in human–robot interaction within healthcare services: a review study," in *International Design Engineering*

Technical Conferences and Computers and Information in Engineering Conference, American Society of Mechanical Engineers, vol. 86281, 2022, V007T07A030.

[87] S. Ali, Abdullah, T. P. T. Armand, *et al.*, "Metaverse in healthcare integrated with explainable AI and blockchain: enabling immersiveness, ensuring trust, and providing patient data security," *Sensors*, vol. 23, no. 2, p. 565, 2023.

[88] R. Sahal, S. H. Alsamhi, and K. N. Brown, "Personal digital twin: a close look into the present and a step towards the future of personalised healthcare industry," *Sensors*, vol. 22, no. 15, 2022. doi: https://doi.org/10.3390/s22155918.

Index

accessing network data 196
acoustic sensor-based localization 284
action space 215
actor-critic (A3C) learning 92, 110
Adafruit ESP32 60
additive white Gaussian noise (AWGN) 32
administrative domains (ADs) 281, 288
AI as a service (AIaaS) 287
air-to-ground (ATG) channels 30
amazon web services (AWS) 72, 193
application layer 187
   device life-cycle monitoring and maintenance 188
   hardware configuration 188–9
   intent-based service provisioning 187
   network attack failure detection and mitigation 187
application programming interfaces (APIs) 177, 240
artificial intelligence (AI) 4, 7–8, 134, 155, 196, 238, 269, 288–90, 295
   based communication enhancement 98–100
   based fault and anomaly detection 102
   based network management 94
     planning and monitoring 94–5
     resource allocation 96–8
   based network performance enhancement 91
     energy efficiency 93
     latency optimization 91–3
     task offloading and content delivery optimization 93–4
   based prediction analysis 100–1
   based security and privacy preservation 102–5
augmented reality (AR) 8, 133, 156, 204
authentication 75
auto encoders 162
automatic gain control (AGC) 183
automatic repeat request (ARQ) mechanisms 6
autonomous systems 204
autonomous transportation 8
autonomous vehicle (AV) 105
Average Wasserstein GAN with Gradient Penalty (AWGAN-GP) 102, 109–10

base band processing unit (BBU) 136
Bayesian convolutional neural network (BCNN) 95
BeagleBone Board 60
beyond 5G (B5G) 159, 237
Bi-directional GAN (BiGANs) 109
Bidirectional Gated Recurrent Unit (BiGRU) 104
Bidirectional Generative Adversarial Networks (BiGAN) 102
Bin-Packing problem 207
bit error rate (BER) 179
blockchain technology 75, 287, 294
branch-and-bound algorithms 257
BtEurope topology 229–30

cell densification 177
cellular networks 87
centralized OaaS platform 241–3
centralized RAN (C-RAN) 134, 137–8
central units (CU) 138
centre of mass (CM) model 181

channel ageing 15
channel model 30–2
channel parameter estimation 147
Channel State Information (CSI) 12
classic Shannon's equation 32
cloud-based storage 72
cloud computing 269
cloud-Native NextG wireless networks 272–3
cloud radio access networks (C-RAN) 16
cloud-to-robot continuum 272
cloud twin (CT) 14
clustering techniques 147
cobots 271, 292
commercial off-the-shelf (COTS) 138
complementary metal-oxide semiconductor (CMOS) 5
component-level models 182
   complex physical phenomena 182
   heterogeneity 182–3
   limited data availability 183
   scalability 184
computational technologies (CT) 272
computation-oriented communications (COC) 204
compute as a service (CaaS) 287
conditional GAN (C-GAN) 101
conditional generative adversarial network (C-GAN) 14
connected robotics 204
constraints 209
   latency 211
   seamless path 211
   VL mapping 210
   VN mapping 210
context awareness 284–5, 294
contextually agile eMBB communications (CAeC) 204
continuous integration and continuous deployment (CI/CD) methodology 159, 238

convolutional LSTM (C-LSTM) 109
convolutional neural networks (CNNs) 99, 108, 162
coordinated multi-point (CoMP) techniques 256
COSMOS 193
cross phase modulation (XPM) 181
cyber-physical systems (CPSs) 72, 270, 273, 293

data acquisition 55
   contextual relations 58–9
   physical features 56
   telemetry data 56–7
data annotation 151
data anonymization 76
data availability 113
data cleaning 65–6
data collection layer 178–9
data collection mechanisms 196
data distribution service (DDS) 281
data fusion layer 179
data integration 66
data lakes 272
data minimization and masking 76
data mining 52
data modeling languages 68–70
data ownership 19
data privacy 65
data quality 113
data reduction 66
data transformation 66
data uncertainty 52–3
data variety 52
decision-tree-based solutions 274
decomposition identifier 217–18
deep deterministic policy gradient (DDPG) 111
Deep Double Q-Network (DDQN) 95
deep learning (DL) 83, 147
deep neural networks (DNNs) 182
   architecture 222–3
deep Q learning (DQL) 101, 111, 209, 262

deep reinforcement learning (DRL) 110, 188, 207
deep transfer learning 65
denial of service (DoS) attacks 187
digital data 54
digital twin-definition language (DTDL) 68
digital twin entity manager (DTEM) 189
digital twin networks (DTNs) 9, 11, 49, 83, 156, 161
  AI-enabled data analysis and interpretation 72–4
  challenges in data management 52–3
  development efforts 86
    cellular networks 87
    industrial IoT networks 89
    networks in general 86–7
    optical networks 88
    satellite and aeronautic networks 88–9
    vehicular networks 89
    wireless networks 87
  ecosystem 83–6
  ethical considerations of using AI in data management 74–5
  explanation of AI's role in data management for 51
    data acquisition, preprocessing, modeling, and data storage 51
    data analysis and predictive modeling 51–2
    ethical considerations and security aspects 52
  importance of data management in 49–51
  key tasks in 89
    AI-based communication enhancement 98–100
    AI-based fault and anomaly detection 102
    AI-based network management 94–8
    AI-based network performance enhancement 91–4
    AI-based prediction analysis 100–1
    AI-based security and privacy preservation 102–5
  main AI models and tools harnessed by 105
    DL models and techniques 108–10
    FL and collaborative learning 111–12
    graph and network analysis techniques 112–13
    ML tools and models 106–8
    RL and optimization techniques 110–11
  main challenges in AI-based 113
    key challenges 113–14
    responsible AI considerations 114–15
  privacy-preserving techniques for protecting sensitive data in 75–6
  remaining challenges and open research questions in AI-enabled data management for 76–7
  security aspects, including data privacy and protection 75
  three states of 53–5
  twinning process 55
    data acquisition 55–9
    data preprocessing methods for cleaning, filtering, and transforming raw data 65–6
    digital twin ontology and data modeling 67–70
    edge computing and distributed data collection 63–5
    network topology-based data collection 61–3
    sensor-based data collection 59–61

storage architectures for
    managing large-scale data in
    70–2
digital twin optical network (DTON)
    189
digital twins (DTs) 27, 135, 144,
    155, 171, 204, 206–7, 238
  6G applications 8
    autonomous transportation 8
    DT 9
    evaluate 6G KPIs and
      applications 9–10
    extended reality 8–9
    rescue communications 9
    tele-medicine and remote
      robotic surgery 8
  6G KPIs 3–4
  6G technologies 4
    artificial intelligence and
      machine learning 7–8
    design of fresh waveform,
      modulation, coding, and
      information theory methods
      5–6
    integrated sensing and
      communication 7
    intelligent reflective surfaces 7
    non-orthogonal multiple
      access 6
    optical wireless
      communication 5
    semantic communication 7
    terrestrial and non-terrestrial
      networks collaboration 7
    THz communication 4–5
    ultra-massive MIMO 6
  as accelerator toward digitalization
    145–6
  application of 171
  cloud-to-robot continuum for
    278–9
  computation offloading for 279
  definition of 144
  deployment challenges 19–20
  general architecture of 144–5

and higher frequency technologies
    11
    THz 11–12
    visible light communications
      12
and intelligent networks 17
and Internet of Things (IoT)
    17–18
low latency DT 18–19
and new network technologies 15
    network slicing 16–17
    new radio access network
      (RAN) technologies 16
    new physical layer and
      multi-antenna techniques 14
    massive MIMO and cell-free
      massive MIMO 14–15
    RISs 15
of non-terrestrial networks 12
    satellite communications 13
    UAV communications and
      networks 13–14
on O-RAN architecture 146
    channel modeling for RAN
      optimization 147
    network fault detection 149
    network traffic forecasting and
      mobility management 147–8
    security and threat detection
      148–9
dimensioning problems 255
  RAN placement and
    functional-split configuration
    255–8
  service function placement
    258–60
discretization 66
distributed data collection 63–5
distributed ledger technology (DLT)
    287
  federation 287–8
distributed OaaS platform 243
distributed RAN (D-RAN) 137
distributed units (DUs) 15, 138, 163
document database 72

domain knowledge expertise 53
double-deep Q learning (DDQL) model 188, 208–9
Double deep Q-network 111
double deep Q-network (DDQN) algorithm 92, 96, 111
DT empowered task offloading model 32
  edge processing 33
  local processing 33
DT-empowering URLLC-based edge networks model 30
DT virtualization layer 145

edge AI models 74
edge computing 63–5, 75
edge processing 33
electromagnetic information theory (EIT) 4–5
elliptic curve cryptography (ECC) 148
end-to-end (E2E) network 159
energy and power consumption model 34
energy efficiency 113
enhanced mobile broadband (eMBB) 204, 252, 259
entity-relationship diagrams (ERDs) 69–70
environment 213–14
erbium doped fiber amplifiers (EDFAs) 180, 185
ethical considerations 53
event-based trigger mechanism 57
event-defined uRLLC (EDuRLLC) 204
evolutionary algorithms (EA) 258
expected upper bound (EUB) 224
experiential networked intelligence (ENI) framework 208
explainable AI (XAI) models 73
extended reality (XR) 8–9, 133, 156
extensible RAN (xRAN) 139
Extreme Gradient Boost (XGBoost) algorithm 107–8

FABRIC 192–3
fast Fourier transform (FFT) 142
Federated Continuous Learning framework with a Stacked Broad Learning System (FCL-SBLS) 105
federated learning (FL) 75, 83
5G Alliance for Connected Industries and Automation (5G ACIA) 275
fifth-generation of mobile networks (5G) 204
5G-next generation-radio access network (5G-NG-RAN) 70
finite blocklength (FBL) 29
FL with Secure Bi-Level Optimization (FCL-SBLS) 112
forward error correction (FEC) coding limit 174
four wave mixing (FWM) 181
Free-space optical (FSO) 5
frequency division duplex (FDD) 14

gaussian noise simulation in python (GNPy) 194
generalized extreme value (GEV) distribution 186
generalized signal to noise ratio (GSNR) 174
generative adversarial networks (GANs) 101, 109
Genetic Algorithm for Service mapping with VIrtual Topology design (GASVIT) 208
genetic algorithms 107
global environment for networking innovation (GENI) 192
GnPy model 173
Google Cloud 72
Google Coral Dev Board 60
graph-based approaches 83
graph CNN 113
graph databases 70–1

graphics processing units (GPUs) 190
graph neural networks (GNNs) 73, 162, 258
GraphQL 69
GraphSAGE 95
gRPC network management interface (gNMI) 63

healthcare 292–3
heuristic model 225
hierarchical FL (HFL) 112
hierarchical multi-agent RL 110
higher layer split (HLS) 139
High Performance Machine Type Communications (HMTC) 252
homomorphic encryption 76
human-machine interface (HMI) 167

index modulation (IM) 6
industrial automation 134
industrial Internet of Things (IIoT) 62, 89, 155
Industry 4.0 275
Industry 5.0
 6G enablers and their applicability to E2E digital twins systems 284
 artificial intelligence 288–90
 context awareness 284–5
 digital twin as pillar of 270–1
 from digital twins in Industry 4.0 to Industrial E2E digital twin systems in 275–6
 digital twins in real industrial environments 274–5
 distributed ledger technology federation 287–8
 enablers for industrial digital twins 278
  cloud-to-robot continuum for digital twins 278–9
  computation offloading for digital twin 279
  digital twin as a service 279–80

 resource and service federation 281–2
 robot operating system framework 280–1
 fundamentals and challenges of digital twins for robotic systems 274
 healthcare 292–3
 ICT technologies and adaptation for 271–3
 industrial E2E digital twin systems in collaborative robotic applications 290
 infrastructure behind industrial digital twins 276
  computing and storage 277
  connectivity 277–8
 joint communication and sensing 285–6
 localization and material inspection 290–1
 material handling and logistics 291–2
 open challenges to achieve Industrial E2E digital twin systems 282–4
 safety takeover 292
 semantic orchestration 286–7
 shift from Industry 4.0 to 269–70
 unified role of Industrial E2E digital twin systems in 273–4
InfluxDB 70
information and communications technology (ICT) 9, 269
Information Object Class (IOC) 249
in-line amplifiers (ILAs) 179
input data 54
integer linear programming (ILP) 207
integrated sensing and communication (ISAC) 4, 7
intelligent reflective surfaces (IRSs) 7, 28
intelligent transportation system (ITS) 99, 134, 160

intelligent wireless systems 155
intent-based service provisioning 187
International Telecommunication Union Radiocommunication Sector (ITU-R) 133
internet of everything (IoE) 2, 93
Internet of Things (IoT) 17–18, 27, 133, 160, 239, 269, 278
internet protocol flow information export (IPFIX) 63
interoperability 53
interpretability 113
inverse FFT (iFFT) 142
IP network domain 250–1

JavaScript Object Notation (JSON) 243
joint communication and sensing 285–6, 294
JSON for linked data (JSON-LD) 69

Karush Kuhn Tucker conditions 262
key performance indicators (KPIs) 100, 143, 156, 286
  for 5G and 6G compensation 2
key-value databases 72
k-nearest neighbor (kNN) 70
Knowledge-as-a-Service (KaaS) 102
knowledge graph (KG) 95

light detection and ranging (LiDAR) 5, 283
light emitting diodes (LEDs) 12
light fidelity (LiFi) 5
light-of-sight (LoS) 30, 165
local processing 33
long-range wide-area network (LoRaWAN) 61
long short-term memory (LSTM) 93, 108–9
low Earth orbit (LEO) satellites 7, 12
lower PHY 139
low latency DT 18
  low latency communications 18–19

LowMEP 29
low power wide area (LPWA) networks 277
Lyapunov approach 92
Lyapunov optimization method 97

machine learning (ML) 4, 7–8, 83, 147, 159, 172, 181, 208, 238, 262, 270
Markov Decision Processes (MDPs) 111
Massive Internet of Things (MIoT) 252
massive machine-type communications (mMTC) 204, 259
matched filter and successive interference cancellation (MF-SIC) technique 32
maximum-ratio combining (MRC) 32
media access control (MAC) protocol 61, 137, 142, 155, 249
message passing (MP) 6
message queuing telemetry transport (MQTT) 61
meta-heuristics 207
Microsoft Azure 72
micro-VNFs (mVNF) 205
Mininet 194
Mininet Optical model 173
mixed-integer linear program (MILP) 207
mixed reality (MR) 8, 204
mobile devices 166
mobile edge computing (MEC) 27, 29, 92, 155
modeling layer 179
  analytical *versus* machine learning 180–2
  components 182
    component-level models 182–4
    integration with end-to-end models 184–6

modulation and coding scheme (MCS) 160
MongoDB 72
Monte Carlo approach 258
Moore's law 184
multi-agent deep deterministic policy gradients (MA-DDPGs) 97, 111
multi-agent PPO (MA-PPO) 111
multiagent proximal policy optimization (MAPPO) 98
multi-agent reinforcement learning (MARL) 262
multi-input multi-output (MIMO) 98
multilayer perceptron (MLP) 96
multi-objective optimization algorithms 73
multiple-input and multiple-output (MIMO) 163

Neo4j 70
NETCONF protocol 194
NetFlow v10 63
Netrail Topology 226–9
network configuration protocol (NETCONF) 62
network digital twins (NDTs) 238
network fault detection 149
network function virtualization (NFV) 16, 134, 161, 204, 272
  BtEurope topology 229–30
  deep RL solution for microservice decomposition 211
    action space 215
    decomposition identifier 217–18
    environment 213–14
    granularity criteria 218–19
    overview of proposed model 220–2
    re-architecture of VNF-FG 219–20
    reinforcement learning 211–13
    reward function 215–17
    state space 214–15

digital twin 206–7
DNN architecture 222–3
heuristic model 225
microservices decomposed NFVs 205–6
Netrail Topology 226–9
nodal capacity 230–1
problem statement 209
  constraints 209
  objective 209
time complexity 225–6
networking system layer 145
network model 248
  IP network domain 250–1
  network slicing 251–3
  radio access network 248–50
  transport and computing infrastructure model 253–5
network optimization 240
network resource model (NRM) 248
networks digital twins (NDTs) 83
network slice consumer (NSC) 252
network slice provider (NSP) 252
network slicing 16–17, 161, 251–3, 261
network traffic forecasting and mobility management 147–8
neural network (NN) 14, 100, 107
Next-Generation Mobile Networks Alliance (NGMN) 276
NextG wireless networks (NGWNs) 271–2
nodal capacity 230–1
Nomadic nodes (NNs) 163
non-linear Schrodinger equation (NLSE) 181
non-orthogonal multiple access (NOMA) 6
non-terrestrial networks (NTN) 12
  satellite communications 13
  UAV communications and networks 13–14
NVIDIA Jetson family 60

ontology *versus* data model 67–8
open data 190–5

open emulation tools 190–5
Open Ireland 193–4
open network operating system (ONOS) 191
open radio access network (O-RAN) 16, 133–4
  background on 136
    concept of 139–41
    definition of 141–3
    as enabler for 6G deployment 143–4
    RAN functionalities, building blocks, and disaggregation 136–9
  challenges and future directions 149
    compliance 151
    data annotation 151
    data flow security and privacy 150–1
    real-time synchronization 150
  digital twin 144
    as accelerator toward digitalization 145–6
    definition of 144
    general architecture of 144–5
    interfaces description 143
    motivation and contribution 135–6
open testbeds 190–5
operational and maintenance (O&M) 16
operational problems 261
  capacity sharing for RAN slicing 261–3
  dynamic functional split selection 263–5
optical camera communications (OCC) 5
optical channel monitors (OCMs) 196
optical cross connect (OXC) nodes 189
optical multiplex section (OMS) 190
optical networks 88, 171
  current issues in 173
    issue of limited automation 176–7
    issue of suboptimal network operation 174–6
  DT development for 177
    application layer 187–9
    data collection layer 178–9
    data fusion layer 179
    modeling layer 179–86
    recent work and case studies on optical DTs 189–90
    simulation and virtualization 186
  open testbeds, open emulation tools, and open data 190
    large-scale testbeds 191–4
    open software and open data 194–5
    open source emulation tools 194
optical sensors 283
optical signal to noise ratio (OSNR) 175
optical spectrum as a service (OSaaS) 187
optical testbeds 173, 192
  COSMOS 193
  FABRIC 192–3
  GENI 192
  Open Ireland 193–4
optical wireless communication (OWC) 4–5
OPTIMAIX project 239
Optimization-as-a-Service (OaaS) platform 238
  functional architecture 240
    centralized OaaS platform 241–3
    distributed OaaS platform 243
  network model 248
    IP network domain 250–1
    network slicing 251–3
    radio access network 248–50

transport and computing
  infrastructure model 253–5
OaaS system APIs 243
  design handling 243
  functionality management 246
  OaaS platforms 243–5
  repository management 245–6
use cases 255
  dimensioning problems
    255–60
  operational problems 261–5
  workflow example 246–8
orthogonal frequency division
  multiplexing (OFDM) 6
orthogonal time–frequency space
  (OTFS) 6
output data 54–5
over-estimation problem 212

packet data convergence protocol
  (PDCP) 137, 139,
  142, 249
partial differential equation 181
particle swarm optimization (PSO)
  algorithm 96
physical (PHY) layers 155
physical random access channel
  (PRACH) extraction 142
Physical Resource Blocks (PRBs)
  261
physical twin (PT) 144
points of presences (PoPs) 240
polarization mode dispersion (PMD)
  180
polymorphic learning (PL)
  framework 94
programmable optical transceiver
  (POT) 188
proximal policy optimization (PPO)
  92, 111
PyTorch module 224

Q-learning (QL) 208
quality-of-decision (QoD) 93
quality of experience (QoE) 93, 146,
  148–9

quality of service (QoS) 14, 143,
  148, 160, 207
quality of transmission (QoT) 172

radio access networks (RANs) 16,
  20, 134, 155, 240, 248–50,
  280
radio environment map
  (REM) 163
radio frequency (RF) 136
radio link control (RLC) 137, 142,
  249
radio network information service
  (RNIS) 284
radio planning 157
radio resource control (RRC) 137,
  139, 142, 249
radio resource management (RRM)
  160, 252
radio units (RU) 138
Raspberry Pi 4 59
rate splitting (RS) 6
real-time series database 70
re-configurable intelligent surfaces
  (RIS) 165
re-configurable optical add-drop
  multiplexers (ROADMs) 179,
  182
recurrent neural networks (RNNs)
  73, 162
redundant mVNFs 205
reference signal receive power
  (RSRP) 163
reflective intelligent surfaces (RISs)
  15
reinforcement learning (RL) 28, 83,
  160, 208, 211–13, 262
remote radio unit (RRU) 136
representational state transfer
  configuration protocol
  (RESTCONF) 63
rescue communications 9
resource allocation 96–8
responsible AI 114
reward function 215–17
risk mitigation 186

robotics 269
robot operating system (ROS) framework 280–1
robustness 113
role-based permissions 75
rRMPolicyDedicatedRatio 252
rRMPolicyMaxRatio 252
rRMPolicyMinRatio 252

satellite and aeronautic networks 88–9
satellite communications 13
scalability 113
secure multi-party computation (SMPC) 76
self phase modulation (SPM) 181
semantic communication (SC) 4, 7
semantic orchestration 286–7, 294
semi-supervised learning 183
sensing as a service (SaaS) 287
sensor-based data collection 59–61
service-based architecture (SBA) 272
service data adaptation layer (SDAP) 249
service function chaining (SFC) 204
service level agreements (SLAs) 16, 56, 92, 156, 204, 240, 281
service level objectives (SLOs) 255
service management and orchestration (SMO) entity 142
Shannon's formula 264
signal-to-interference-plus-noise ratio (SINR) 32, 160
signal-to-noise ratio (SNR) 14, 165
simple network management protocol (SNMP) 61
single-board computers (SBCs) 59
single network slice selection assistance information (S-NSSAI) 251
sixth-generation (6G) communication systems 155
  building platforms to train AI models for 159–60
  efficient network slice management and orchestration 161–2
  enabling 6G-based IIoT and industrial 6G use cases 167
  enabling 6G RAN optimization and effective traffic management 162–3
  operation of mobile edge clouds in 166
  optimized planning, service testing, and rapid development of 156–7
  optimizing 6G radio resource management 163–4
  simplifying and accelerating site deployment configuration 157–9
  tackling security and resiliency issues in 160–1
  terahertz wave analysis in support of reconfigurable intelligent surfaces for enhanced 6G performance 164–6
  testing the impact of configuration and function changes 159
sixth generation of mobile networks (6G) 204
6G networks 2, 134
  applications 8
    autonomous transportation 8
    DT 9
    evaluate 6G KPIs and applications 9–10
    extended reality 8–9
    rescue communications 9
    tele-medicine and remote robotic surgery 8
  KPIs 3–4
  technologies 4
    artificial intelligence and machine learning 7–8
    design of fresh waveform, modulation, coding, and information theory methods 5–6

integrated sensing and
communication 7
intelligent reflective surfaces 7
non-orthogonal multiple
access 6
optical wireless
communication 5
semantic communication 7
terrestrial and non-terrestrial
networks collaboration 7
THz communication 4–5
ultra-massive MIMO 6
sliceable bandwidth variable
transceiver (S-BVT) 189
slice-enabled communication
networks (SeCNs) 16–17
slice/service type (SST) 251
software defined networking (SDN)
16, 95, 134, 161, 186, 272
space–air–ground integrated
networks (SAGIN) 14
spectrally efficient frequency domain
multiplexing (SEFDM) 6
stack hosting data adaptation
protocol (SDAP) 142
state space 214–15
stimulated Raman scattering (SRS)
effect 194
successive interference cancellation
(SIC) 6
Successive Shortest Path algorithm
(SSP) 260
superposition coding (SC) 6
supervised learning 160

task offloading optimisation 38
telco-level services 240
Telecom Infra Project (TIP) 140
tele-medicine and remote robotic
surgery 8
temporal convolution network (TCN)
13, 70
terrestrial and non-terrestrial
networks collaboration 7
The Things Network (TTN) services
61

Third Generation Partnership Project
(3GPP) 138
3GPP TR 38.801 functionality split
139
THz communication 4–5, 11–12
time-based trigger mechanism 57
bandwidth and storage constraints
57
granularity and precision 57
real-time monitoring 57
time complexity 225–6
time-sensitive networking (TSN)
scheduling technique 92
TinyML 65
transfer learning (TL) 104, 107, 183,
185
transmission model 30
channel model 30–2
URLLC-based uplink transmission
rate 32
transmit power and computation
resource optimisation 36–8
transmit time intervals
(TTIs) 164

ultra-massive MIMO 6
ultra-reliable and low latency
communication (URLLC) 2,
27–8, 92, 164, 204, 252, 259
motivations and main
contributions 29–30
numerical simulations 38
convergence of proposed
algorithm 39
impact of ESs' processing rate
41–2
impact of required computation
resource 39–40
impact of UEs' processing rate
41
impact of UE transmit power
budget 40–1
simulations setup 38–9
proposed solutions 35
proposed algorithm 38
task offloading optimisation 38

transmit power and computation
  resource optimisation 36–8
system model and problem
  formulation 30
  DT empowered task offloading
    model 32–3
  DT-empowering URLLC-based
    edge networks model 30
  energy and power consumption
    model 34
  problem formulation 34–5
  transmission model 30–2
  UAV deployment 34
unified modeling language (UML)
  69–70
unmanned aerial vehicles (UAVs) 28
  communications and networks
    13–14
  deployment 34
upper confidence bound (UCB) 92
useful data 53
user equipment (UE) measurements
  12
user layer 144

vehicle-to-everything (V2x)
  communications 5, 252
virtualized network functions (VNFs)
  279

virtual network functions (VNFs)
  204
virtual network (VN) model 89, 254
virtual reality (VR) 5, 133, 204
visible light communications (VLC)
  5, 12, 99
VNF-FG 214

Wasserstein distance GAN (WaGAN)
  105, 109
wavelength division multiplexing
  (WDM) systems 184
wavelength selective switches
  (WSSs) 182
web ontology language (OWL) 69
wireless fidelity (WiFi) 5, 61
wireless networks 87
WLAN network information service
  (WNIS) 284–5
Wolf Colony Algorithm (WCA) 104

XGBoost algorithm 102

yet another next generation (YANG)
  61–2
ZigBee 61

Milton Keynes UK
Ingram Content Group UK Ltd.
UKHW051814170624
443981UK00001BA/2